Lecture Notes in Computer Sc

Commenced Publication in 1973
Founding and Former Series Editors:
Gerhard Goos, Juris Hartmanis, and Jan van Leeuwen

Editorial Board

Christos Kaklamanis   Martin Skutella (Eds.)

Approximation and Online Algorithms

5th International Workshop, WAOA 2007
Eilat, Israel, October 11-12, 2007
Revised Papers

 Springer

Volume Editors

Christos Kaklamanis
University of Patras
Department of Computer Engineering and Informatics
GR 26500 Rio, Greece
E-mail: kakl@ceid.upatras.gr

Martin Skutella
TU Berlin
Fak. II - Mathematik und Naturwissenschaften
10623 Berlin, Germany
E-mail: skutella@math.TU-Berlin.DE

Library of Congress Control Number: 2007943462

CR Subject Classification (1998): F.2.2, G.2.1-2, G.1.2, G.1.6, I.3.5, E.1

LNCS Sublibrary: SL 1 – Theoretical Computer Science and General Issues

ISSN   0302-9743
ISBN-10   3-540-77917-5 Springer Berlin Heidelberg New York
ISBN-13   978-3-540-77917-9 Springer Berlin Heidelberg New York

Springer is a part of Springer Science+Business Media

springer.com

© Springer-Verlag Berlin Heidelberg 2008
Printed in Germany

Typesetting: Camera-ready by author, data conversion by Scientific Publishing Services, Chennai, India
Printed on acid-free paper   SPIN: 12224210   06/3180   5 4 3 2 1 0

Preface

The Fifth Workshop on Approximation and Online Algorithms (WAOA 2007) focused on the design and analysis of algorithms for online and computationally hard problems. Both kinds of problems have a large number of applications from a variety of fields. WAOA 2007 took place in Eilat, Israel, during October 11–12, 2007. The workshop was part of the ALGO 2007 event that also hosted ESA 2007, and PEGG 2007. The previous WAOA workshops were held in Budapest (2003), Rome (2004), Palma de Mallorca (2005) and Zurich (2006). The proceedings of these previous WAOA workshops have appeared as LNCS volumes 2909, 3351, 3879 and 4368, respectively.

Topics of interest for WAOA 2007 were: algorithmic game theory, approximation classes, coloring and partitioning, competitive analysis, computational finance, cuts and connectivity, geometric problems, inapproximability results, mechanism design, network design, packing and covering, paradigms for design and analysis of approximation and online algorithms, randomization techniques, real-world applications, and scheduling problems. In response to the call for papers, we received 56 submissions. Each submission was reviewed by at least three referees, and the vast majority by at least four referees. The submissions were mainly judged on originality, technical quality, and relevance to the topics of the conference. Based on the reviews, the Program Committee selected 22 papers.

We are grateful to Andrei Voronkov for providing the EasyChair conference system which was used to manage the electronic submissions, the review process, and the electronic PC meeting. It made our task much easier.

We would also like to thank all the authors who submitted papers to WAOA 2007 as well as the local organizers of ALGO 2007.

November 2007

Christos Kaklamanis
Martin Skutella

Organization

Program Co-chairs

Christos Kaklamanis   University of Patras and RA CTI
Martin Skutella   TU Berlin

Program Committee

Evripidis Bampis   University of Evry
Luca Becchetti   University of Rome "La Sapienza"
Thomas Erlebach   University of Leicester
Naveen Garg   IIT Delhi
Klaus Jansen   University of Kiel
Christos Kaklamanis   University of Patras and RA CTI
Samir Khuller   University of Maryland
Danny Krizanc   Wesleyan University
David Peleg   Weizmann Institute
Giuseppe Persiano   University of Salerno
Kirk Pruhs   University of Pittsburgh
Adi Rosén   CNRS and University of Paris 11
Guido Schäfer   TU Berlin
Martin Skutella   TU Berlin
Roberto Solis-Oba   University of Western Ontario
Leen Stougie   TU Eindhoven and CWI Amsterdam

Additional Referees

Vincenzo Auletta
Yossi Azar
Maria-Florina Balcan
Nikhil Bansal
Andre Berger
Vincenzo Bonifaci
Paul Bonsma
Jit Bose
Janina Brenner
Alberto Caprara
Ioannis Caragiannis
Marek Chrobak
Christine Chung

Fabrizio Grandoni
Tobias Harks
Rafi Hassin
Han Hoogeveen
Panagiotis Kanellopoulos
Haim Kaplan
Samir Khuller
Yoo-Ah Kim
Tracy Kimbrel
Ekkehard Köhler
Felix König
Guy Kortsarz
Elias Koutsoupias

Alantha Newman
Evi Papaioannou
Marco Pellegrini
Paolo Penna
Rajmohan Rajaraman
Rajeev Raman
Rajiv Raman
Ulrich Michael Schwarz
Jiri Sgall
Gennady Shmonin
Rene Sitters
Sebastian Stiller
Tami Tamir

Aparna Das
Amit Deshpande
Florian Diedrich
Leah Epstein
Michal Feldman
Michele Flammini
Stefan Funke
Martin Furer
Rajiv Gandhi
Mohammad Ghodsi

Amit Kumar
Tak-Wah Lam
Stefano Leonardi
Mohammad Mahdian
Azarakhsh Malekian
Euripides Markou
Marios Mavronicolas
Nicole Megow
Pat Morin
David Mount

Orestis Telelis
Ralf Thoele
Marc Uetz
Kasturi Varadarajan
Santosh Vempala
Carmine Ventre
Tjark Vredeveld
Raphael Yuster
Lisa Zhang
An Zhu

Table of Contents

Pricing Commodities, or How to Sell When Buyers Have Restricted Valuations

Robert Krauthgamer[1], Aranyak Mehta[2,*], and Atri Rudra[3,*]

[1] Weizmann Institute, Rehovot, Israel and IBM Almaden, San Jose, CA
robert.krauthgamer@weizmann.ac.il
[2] Google, Inc., Mountain View, CA
aranyak@google.com
[3] University at Buffalo, The State University of New York, NY
atri@cse.buffalo.edu

Abstract. How should a seller price his goods in a market where each buyer prefers a single good among his desired goods, and will buy the cheapest such good, as long as it is within his budget? We provide efficient algorithms that compute near-optimal prices for this problem, focusing on a commodity market, where the range of buyer budgets is small. We also show that our technique (which is based on LP-rounding) easily extends to a different scenario, in which the buyers want to buy all the desired goods, as long as they are within budget.

1 Introduction

Pricing goods to maximize revenue is a critical yet difficult task in almost any market. We study the case of a monopolistic seller (only one seller in the market), a restricted scenario that is already quite challenging. One difficulty is to estimate the demand curves (amount of demand at different prices), but even complete knowledge of the demand curves is sufficient only in rather simple cases, e.g. if the monopolist sells only a single type of good, or if the various goods he sells cater to different sub-markets. In such cases, the revenue-maximizing prices can be determined for each good separately, directly from that good's demand curve.

But what if goods of different types are sold all in the same market? Now, *the seller's own goods could be competing against each other* for the attention of the same buyer. This is generally true of a seller who wants to tap into multiple market segments. For example, Dell sells many models of laptops with varying features catering to varying needs of its consumers, but then it must price the different models carefully so that they do not eat into each other's revenue. As an example on a smaller scale, consider the pricing of movie shows. Different shows are priced differently (for example, matinee vs. evening shows) to attract different audience sections. Again, pricing is critical– a very cheap matinee show might eat into the evening show revenue and decrease the overall revenue.

On the other hand, multiple goods might lead to higher prices by complementing each other. A very visible example is the marketing of Apple's iPod

---

* Work done in part while at the IBM Almaden Research Center.

C. Kaklamanis and M. Skutella (Eds.): WAOA 2007, LNCS 4927, pp. 1–14, 2008.
© Springer-Verlag Berlin Heidelberg 2008

and various accessories. The strategy there is not to sell the iPod in isolation but to offer various accessories. These accessories vary from items that are expensive (for example, a charger) to items that are inexpensive (for example, songs from iTunes). Pricing for revenue maximization becomes computationally complex precisely because of this interaction between different goods. Indeed, Aggarwal, Feder, Motwani and Zhu [1] and also Guruswami, Hartline, Karlin, Kempe, Kenyon and McSherry [2] studied the computational aspects of these pricing problems, showing that in various such settings, computing the optimal prices is NP-hard.

One setting, referred to as *unit-demand consumers* in [2], is where each buyer wants to buy one good out of his desired set, as follows: There are m buyers, each of whom has an arbitrary set of desirable goods and a spending budget. The (single) seller knows the buyers' types (i.e. desired set and budget) and needs to set a price for each of the n goods. Once prices are set, every buyer buys the (single) cheapest good in his set, provided it is within his budget (breaking ties arbitrarily). Another setting, referred to as *single-minded consumers* in [2], differs from the first setting in that now each buyer only wants to buy the desired set as a bundle. That is, once prices are set, every buyer buys the entire set of his desired goods, provided its total cost is within his budget (if not, he buys nothing).

Throughout, we shall assume that the desired set of every buyer has size at most k. As we shall soon see, even the case of small k is nontrivial and interesting. In addition, we shall assume that the goods are available in *unlimited supply*, that is, the seller can sell any number of copies of the item without paying any marginal cost of production.

Several results are known about computing prices that maximize revenue in these two settings. In [1], it is shown that the problem of maximizing revenue in the unit-demand case is not only NP-hard, but APX-hard.[1] For this problem, they also give an $O(\log m)$ approximation algorithm (which uses the best single price). In [2], similar results are shown independently, and it is shown in addition that for the single-minded setting, maximizing revenue is APX-hard and that there is $\log(nm)$ approximation algorithm (which again uses the best single price). Demaine, Feige, Hajiaghayi, and Salavatipour [3] show that the above results are more or less optimal in the general single minded bidder problem – under some complexity assumptions, there is a fixed $\delta > 0$ such that the problem cannot be approximated to within a factor of $\log^\delta n$. Balcan and Blum [4] present a 4-approximation algorithm for single-minded bidders and $k = 2$. Their algorithm extends to larger k, with $O(k)$ approximation. As was observed in [5], their arguments apply to the unit-demand case as well.

These results depict a rather grim landscape (at least computationally) for the problem of pricing to maximize revenue. However, many real-life instances are more specialized, and thus, a more practice-oriented approach is to identify restrictions, under which one can beat the aforementioned $O(\log n)$ factor, or

---

[1] An optimization problem is APX-hard if there exists a constant $\rho > 1$ such that it is NP-hard to approximate the optimum within factor ρ.

better yet, obtain a very small constant-factor approximation. In particular, every percent of improvement counts in practice, requiring us to improve one (small) constant approximation factor to another.

We thus pay special attention to *commoditized markets*,[2] where the range of buyers budgets is restricted to a "small" set $\mathcal{B}$. In one such restriction, $\mathcal{B} = \{1, C\}$ is a doubleton, representing a bimodal market in which buyers are divided into poor and rich. For example, buyers coming from different referring websites such as `lastminutedeals.com` and `hotels.com` might have significantly different budgets for booking a hotel room. As another example, a tourist might be willing to pay for a Broadway show a significantly higher amount than a local. Another motivation for studying such markets could be the low descriptive complexity for the different buyers' budget types, or equivalently a low communication complexity to identify a buyer's budget. In yet another restriction, $\mathcal{B} = [1, C]$ is a small interval, representing a market with little variation, say within 50%, in the valuation of different buyers, and clearly there are numerous examples for such markets. Note that in both cases, the buyers can be completely idiosyncratic regarding their desired goods, as only the budgets are restricted.

1.1 Results and Techniques

Unit-demand setting. In Section 2, we consider inputs with $\mathcal{B} = \{1, C\}$ (i.e., bimodal markets) and $k = 2$ (i.e. a desired set is a pair of goods). On the one hand, the APX-hardness results [1,2] mentioned above are actually shown for such restricted instances (in fact, for $C = 2$). On the other hand, obtaining $2 - \frac{1}{C}$ approximation is rather easy — simply choose the best single price (same for all goods) among $\{1, C\}$ — obviously a very naive solution, but already better than the (more general) 4 approximation that can be derived from [4]. The challenge in this regime is to improve the approximation below 2, and indeed we present an algorithm achieving factor $\frac{3}{2} - \frac{1}{2C}$. Observe that even when C is not too large, this is a significant improvement (e.g. for $C = 2$, from 1.5 to 1.25). This approach easily extends to larger k, in which case the approximation we achieve is $2 - \frac{1}{k} - \frac{k-1}{kC}$.

Our algorithms are based on randomized rounding of a linear programming (LP) relaxation, a powerful paradigm that is often useful for discrete optimization (for example, see the survey of Srinivasan [6]). We "round" the prices suggested by the LP to prices in the discrete ("integral") set $\{1, C\}$. The rationale behind this rounding is that an optimal pricing may always choose prices from the set $\{1, C\}$. However, it is interesting to note that the pricing problem does not require the prices to be "discrete", and thus, the real reason behind our rounding procedure is the following: In contrast with a "standard" randomized rounding algorithm, where the probability (with which we round a variable upwards) depends linearly on the corresponding LP variable, we use a probability that is polynomial in the LP variable. The only other non-linear randomized

---

[2] A commoditized (also commodified) market is one characterized by price-competition with little or no differentiation by brand.

LP rounding algorithms that we are aware of are the approximation algorithm of Goemans and Williamson [7] for MAX SAT, and that for finding the densest k-subgraph problem that is attributed to Goemans [8]. The crux is that at every basic feasible solution of our LP relaxation, all the prices are half-integral [9, Chap. 14] (modulo a normalization factor), and this fact greatly simplifies the choice of the polynomial–in fact, our rounding procedure raises the variables to a power. Interestingly, the value of the power is a function of C.

We further show that our algorithm can be derandomized, and that its approximation matches the LP's integrality gap, and thus it is optimal with respect to this LP. In addition, we observe that in the case where budgets come from an interval $\mathcal{B} = [1, C]$, a simple algorithm achieves $1 + \ln C$ approximation by computing the best single price, and that this factor matches the integrality gap of an LP relaxation that extends the LP mentioned above for the case $\{1, C\}$.

Single-Minded setting. Recently, Khandekar, Könemann and Markakis (Private Communication) have studied the case of single-minded bidders with desired sets of size at most 2, and the same budget for all the buyers, and gave a $4/3$ approximation algorithm. Subsequently (but using independently derived techniques), we found out that our LP rounding approach mentioned above is easy to adapt to this setting as well, achieving $\frac{6+\sqrt{2}}{5+\sqrt{2}} \approx 1.15$ approximation. In Section 3 we briefly present this algorithm, and show a matching integrality gap. Again, this problem is known to be APX-hard because the results of [2] are actually shown for such restricted instances. Further, our algorithm obtains much better approximation than a $3/2$ approximation achievable by choosing the best single price in the set $\{1/2, 1\}$, which was already better than the (more general) 4 approximation of [4].

Online pricing. Finally, we consider in Section 4 inputs with $k = 2$ (and no restriction on the budget). Using a variation of the algorithm designed by Balcan and Blum [4], we design an algorithm that works even in an *online setting*, where goods arrive sequentially (together with the bids of all the buyers interested in that good), and the seller has to determine the price of a good immediately as it arrives. This model may correspond for instance to Comcast cable TV selling video on demand, where new offerings are announced (with prices) on a regular basis. Our algorithm achieves 4 approximation, compared to the best (offline) prices. We note that [4] also give an online pricing algorithm, but in their setting buyers arrive online, and the prices (of a fixed set of goods) need to be updated.

Truthful Mechanisms. We assume throughout the paper that the seller knows the budget of each bidder. We may also be interested in settings where the seller does not know such information about the market. Balcan, Blum, Hartline and Mansour [10] show that every approximation algorithm for revenue maximization can be converted into a truth-revealing mechanism. They design a general technique that loses only an additional factor of $1 + \epsilon$ in the approximation, if certain technical conditions (like sufficiently many bidders) are satisfied. Similarly to [4], we note that this technique is applicable in our setting, and thus converts our algorithms to truthful mechanisms, provided that the number of bidders is at least (roughly) Cn/ϵ^2.

1.2   Related Work

The notion of revenue-maximizing pricing of goods in unlimited supply was introduced by Goldberg, Hartline, Karlin, Saks and Wright [11]. In their setting, the goods were "independent" and hence the optimization problem was trivial, and they focused on designing *truthful* mechanisms to maximize revenue. There have been numerous followup work, and we only mention here results that are directly related to our work.

Guruswami, Hartline, Karlin, Kempe, Kenyon and McSherry [2] considered the problem of revenue maximization in a variety of settings, including both unit-demand and single-minded bidders, and also envy-free pricing of goods in limited supply. As mentioned earlier, they showed logarithmic upper bounds and APX-hardness for both types of bidders. The results for the unit-demand case were also obtained independently by Aggarwal, Feder, Motwani and Zhu [1]. For single-minded bidders, a polylogarithmic hardness result, which complements the result above, was obtained by Demaine, Feige, Hajiaghayi, and Salavatipour [3]. The problem of the single-minded bidder case, where the size of the demand sets was upper bounded by k, was considered by Briest and Krysta [12] who gave an $O(k^2)$ approximation for the problem, and was improved by Blum and Balcan [4] to $O(k)$. For the special case of $k = 2$, they obtain a 4 approximation algorithm.

Another paper that is less directly related but was also a starting point for our work is the work of Bansal, Cheng, Cherniavsky, Rudra, Scheiber, Sviridenko [13], which studies a problem of pricing over time, that was proposed in [2]. A special case of their problem gives another interpretation for the unit-demand setting: The seller is selling just one type of good (in unlimited supply), and does so over a period of n days, and can set a different price on each day. Each of the m buyers has a subset of size k of the n days, which represent the days on which she can purchase the item, and will choose to buy a copy of the good at the cheapest price she sees over the k days. The seller's aim is to maximize revenue.

1.3   Problem Definitions

Our pricing problems involve one seller and m buyers. The seller has a collection V of n goods (also called items). Each $j \in V$ is a digital good, i.e., the seller has 0 marginal cost of production, or equivalently, the number of copies of j is at least the number of buyers m. Once the seller sets the prices of the goods, each buyer will buy a collection of goods, based on his own utility function. The seller's problem is to determine a price p_j of each good $j \in V$ so as to maximize revenue. Depending on the utility functions of the buyers, we have the following variations of the pricing problem. The first variation is our main focus, but we will also show how the techniques we develop also work for the second variation.

1. Unit-demand bidders: We let $UD_k(\mathcal{B})$ denote the problem of item pricing for unit-demand bidders with sets of size at most k, and bids from the set $\mathcal{B}$, as follows. Buyer i has a budget of $u_i \in \mathcal{B}$ and a subset S_i of desirable goods, with $|S_i| \leq k$. He is interested in buying *exactly one* good from S_i, and given prices on the goods, he will buy the cheapest good in S_i, provided that its price is at most u_i. For a price vector $\mathbf{p} = (p_1, ..., p_m)$, let $\pi_i(\mathbf{p})$ be the revenue that the seller obtains from buyer i if the prices are set to $\mathbf{p}$. Thus

$$\pi_i(\mathbf{p}) = \begin{cases} \min\{p_j : j \in S_i\} & \text{if } \min\{p_j : j \in S_i\} \leq u_i \\ 0 & \text{otherwise} \end{cases}$$

Thus the seller's problem is: Find $\mathbf{p}$ so as to maximize $\sum_{i=1}^{n} \pi_i(\mathbf{p})$. We are interested in the following special cases of this problem (defined by different values of k and $\mathcal{B}$): (1) $UD_k([1, C])$ for $C > 1$ and (2) $UD_k(\{1, C\})$ for $C > 1$.

2. Single-minded bidders: We let $SM_k(\mathcal{B})$ denote the problem of item pricing for single minded bidders, who have sets of size k and bids from the set $\mathcal{B}$, as follows. Buyer i has a budget of $u_i \in \mathcal{B}$ and a subset S_i of desirable goods with $|S_i| \leq k$. He is interested in buying *all* the goods in the set S_i. For a price vector $\mathbf{p}$, let π_i be the revenue that the seller obtains from buyer i, if the prices are set to $\mathbf{p}$. Thus

$$\pi_i(\mathbf{p}) = \begin{cases} \sum_{j \in S_i} p_j & \text{if } \sum_{j \in S_i} p_j \leq u_i \\ 0 & \text{otherwise} \end{cases}$$

Again, the seller's problem is: Find $\mathbf{p}$ so as to maximize $\sum_{i=1}^{n} \pi_i(\mathbf{p})$. We will show how our techniques for $UD_2(\{1, C\})$ extend to give an algorithm for the case $SM_2(\{1\})$.

The case of $k = 2$: Pricing on a graph

Following [4], for $k = 2$, $UD_2(\mathcal{B})$ becomes a problem of pricing the vertices of a graph, with the buyers' desired sets corresponding to the edges of the graph. This will be our main focus in our techniques and analysis. We study two settings of budget ranges: $\mathcal{B} = \{1, C\}$ and $\mathcal{B} = [1, C]$, for $C > 1$.

Given a graph $G = (V, E)$ (possibly with self loops and multiple edges), along with edge weights $c_{ij} \in \mathcal{B}$ for every edge $(i, j) \in E$, the goal is to set prices p_i on every vertex i so as to maximize the total revenue. The revenue from an edge $(i, j) \in E$ becomes:

$$\pi_{ij} = \begin{cases} \min(p_i, p_j) & \text{if } \min(p_i, p_j) \leq c_{ij} \\ 0 & \text{otherwise.} \end{cases}$$

The case of $SM_2(\mathcal{B})$, studied in [4], is defined as a pricing problem on a graph analogously.

2   Unit-Demand Buyers in Commoditized Markets

In this section, we look at pricing for unit demand bidders with restricted valuations. We start with valuations restricted to the set $\{1, C\}$ for some $C > 1$. In other words, we are interested in pricing schemes for the $UD_k(\{1, C\})$ model. Our main result is a pricing scheme that generates a revenue within a factor $\frac{(2k-1)C-k+1}{kC}$ of the optimal revenue (Theorem 2). For ease of exposition, we present the proofs for the $k = 2$ case, and mention the result for the general case in Section 2.5.

Our pricing scheme rounds an LP relaxation for the problem. Theorem 1 shows that our rounding algorithm (for the case $k = 2$) has an approximation factor of $\frac{3C-1}{2C}$. Our rounding procedure is tight (optimal) as we show in Section 2.4 that the integrality gap of our LP relaxation is at least $\frac{3C-1}{2C}$.

It is not difficult to verify that the LP in Figure 1 is a relaxation for our pricing problem $UD_2(\{1, C\})$.

$$\max \sum_{(i,j)\in E} \pi_{ij} \quad \text{subject to:}$$

$\forall (i,j) \in E, c_{ij} = C$	$\pi_{ij} \leq 1 + p_i$	(1)
$\forall (i,j) \in E, c_{ij} = C$	$\pi_{ij} \leq 1 + p_j$	(2)
$\forall (i,j) \in E, c_{ij} = 1$	$\pi_{ij} \leq 1$	(3)
$\forall (i,j) \in E, i \neq j, c_{ij} = 1$	$\pi_{ij} \leq 2 - \dfrac{p_i + p_j}{C - 1}$	(4)
$\forall (i,i) \in E, c_{ii} = 1$	$\pi_{ii} \leq 1 - \dfrac{p_i}{C - 1}$	(5)
$\forall i \in V$	$0 \leq p_i \leq C - 1$	(6)
$\forall (i,j) \in E$	$\pi_{ij} \geq 0$	(7)

Fig. 1. LP relaxation for the unit-demand setting

2.1   On the Optimal LP Solutions

We first observe that an optimal basic feasible solution to the LP relaxation is half integral, in the sense that all the p_i variables are from the set $\{0, \frac{C-1}{2}, C-1\}$. Note that the actual "price" set for vertex i is $p_i + 1$.

Proposition 1. *Every optimal basic feasible solution to the LP in Figure 1 is half integral. That is, every extremal optimal assignment to the variables $\{p_i^*\}_{i \in V}$ satisfies the following: $p_i^* \in \{0, \frac{C-1}{2}, C - 1\}$.*

Proof. Let $\{p_i^*\}_{i \in V}$ be the assignments to the $p_{(.)}$ variables in some optimal assignment such that the values are not half integral. We will show that such an assignment is not an extremal optimal solution. In particular, we will exhibit two optimal assignments $p_{(\cdot)}^-$ and $p_{(\cdot)}^+$ such that for all $i \in V$, $p_i^* = \frac{1}{2}(p_i^- + p_i^+)$.

Define the following two subsets of vertices: $V^+ = \{i | \frac{C-1}{2} < p_i^* < C - 1\}$ and $V^- = \{i | 0 < p_i < \frac{C-1}{2}\}$. Note that by the assumption on p^*, $V^- \cup V^+ \neq \emptyset$. Let $\epsilon > 0$ be a small enough number (to be defined later). We define the two related "price" assignments.

$$p_i^+ = \begin{cases} p_i^* + \epsilon & \text{if } i \in V^+ \\ p_i^* - \epsilon & \text{if } i \in V^- \\ p_i^* & \text{otherwise} \end{cases} \qquad p_i^- = \begin{cases} p_i^* - \epsilon & \text{if } i \in V^+ \\ p_i^* + \epsilon & \text{if } i \in V^- \\ p_i^* & \text{otherwise} \end{cases}$$

Obviously, for all $i \in V$, $p_i^* = \frac{1}{2}(p_i^+ + p_i^-)$. To complete the proof, we will show that both $p_{(\cdot)}^+$ and $p_{(\cdot)}^-$ are optimal assignments. Let the "revenue" variables corresponding to $p_{(\cdot)}^*$, $p_{(\cdot)}^+$ and $p_{(\cdot)}^-$ be denoted by $\pi_{(\cdot)}^*$, $\pi_{(\cdot)}^+$ and $\pi_{(\cdot)}^-$. For the sake of contradiction assume w.l.o.g. that $\sum_{(i,j) \in E} \pi_{ij}^- < \sum_{(i,j) \in E} \pi_{ij}^*$. We aim to show that $\sum_{(i,j) \in E} \pi_{ij}^+ > \sum_{(i,j) \in E} \pi_{ij}^*$, which will contradict the optimality of $\pi_{(\cdot)}^*$.

We first set $\epsilon = \frac{1}{4} \min\{\epsilon_1, \epsilon_2, \epsilon_3\}$ where:

$$\epsilon_1 = \min\{|p_i^* - p_j^*| : p_i^* \neq p_j^*, (i,j) \in E \text{ and } c_{ij} = C\},$$

$$\epsilon_2 = \min\{C - 1 - p_i^* : (i,j) \in E \text{ and } c_{ij} = C\},$$

$$\epsilon_3 = \min\left\{|1 - \frac{p_i^* + p_j^*}{C - 1}| : p_i^* + p_j^* \neq C - 1, (i,j) \in E,\right.$$
$$\left. \text{and } c_{ij} = 1\right\}.$$

Note that for $(i,j) \in E$ such that $i, j \notin V^+ \cup V^-$, $\pi_{ij}^* = \pi_{ij}^+ = \pi_{ij}^-$. Let us now consider an edge $(i,j) \in E$ with at least one end point in $V^+ \cup V^-$. To finish the proof, we will show that

$$\pi_{ij}^+ - \pi_{ij}^* = -\left(\pi_{ij}^- - \pi_{ij}^*\right). \tag{8}$$

First assume that $i \neq j$ and $c_{ij} = C$. In this case, $\pi_{ij}^* = 1 + \min(p_i^*, p_j^*)$, $\pi_{ij}^+ = 1 + \min(p_i^+, p_j^+)$ and $\pi_{ij}^- = 1 + \min(p_i^-, p_j^-)$. If $p_i^* = p_j^*$, then by the definitions of ϵ, $p_{(\cdot)}^+$ and $p_{(\cdot)}^-$, (8) is satisfied. If $p_i^* \neq p_j^*$, then by the definition of ϵ if p_i^* (p_j^*) is the minimum price for (i,j), then so are p_i^+ (p_j^+) and p_i^- (p_j^-). Again by the definitions of $p_{(\cdot)}^+$ and $p_{(\cdot)}^-$, (8) is satisfied.

Now let us consider the case when $i \neq j$ and $c_{ij} = 1$. We now consider two subcases. First if $p_i^* + p_j^* \neq C - 1$, then by the definitions of ϵ, $p_{(\cdot)}^+$ and $p_{(\cdot)}^-$, (8) is satisfied. Now if $p_i^* + p_j^* = C - 1$ then either $p_i^* \in V^+$ and $p_j^* \in V^-$ or $p_i^* \in V^-$ and $p_j^* \in V^+$. In all these cases, $\pi_{ij}^* = \pi_{ij}^+ = \pi_{ij}^-$, which in particular implies that (8) is satisfied.

Similarly, one can show that for self loops, (8) is also satisfied.

2.2 A Rounding Algorithm

Consider the following randomized algorithm, where $\tau > 0$ is a parameter (to be chosen later).

Algorithm **Algo**(τ):

1. Solve the LP in Figure 1 and obtain an optimal basic feasible solution with prices variables $\{p_i\}_{i \in V}$.
2. For every $i \in V$, independently assign a price of C with probability $\left(\frac{p_i}{C-1}\right)^{\tau}$ and a price of 1 with probability $1 - \left(\frac{p_i}{C-1}\right)^{\tau}$.

We now analyze the performance of the rounding algorithm above.

Theorem 1. *For every $C > 1$, there is $\tau > 0$ such that **Algo**(τ) is a $(3C - 1)/(2C)$ approximation for the pricing problem with unit-demand bidders, $k = 2$, and budgets from $\mathcal{B} = \{1, C\}$. That is, the expected revenue of **Algo**(τ) is at least $\frac{2C}{3C-1}$ fraction of the optimum for $UD_2(\{1, C\})$.*

Proof. Set $\tau = \frac{1}{2} \log \left(\frac{3C-1}{C-1}\right)$. For notational convenience, we will denote **Algo**(τ) by **Algo**. Let the optimal (extremal) solution of the LP assign prices p_i^* to every vertex i and obtains a revenue of π_{ij}^* from every edge (i, j). We will show that for every edge $(i, j) \in E$, the expected revenue of **Algo** from that edge is at least $\frac{2C}{3C-1} \cdot \pi_{ij}^*$. The result follows from the linearity of expectation. For the rest of the proof, it will be convenient to define, for every $i \in V$, $q_i = \frac{p_i^*}{C-1}$. By Proposition 1, we have $q_i \in \{0, \frac{1}{2}, 1\}$.

Let us first consider the case when $i \neq j$. We have two subcases.

Case 1a: $c_{ij} = 1$. In this case $\pi_{ij}^* \leq \min(1, 2 - q_i - q_j)$, while **Algo** obtains an expected revenue of $0 \cdot (q_i^{\tau} q_j^{\tau}) + 1 \cdot (1 - q_i^{\tau} q_j^{\tau}) = 1 - (q_i q_j)^{\tau}$. When $q_i = q_j = 1$ then both the LP and **Algo** obtain a revenue of 0. When $q_i + q_j = \frac{3}{2}$ then the ratio of the revenue obtained by **Algo** to π_{ij}^* (which is $1/2$) is $2(1 - \frac{1}{2^{\tau}}) > 1 - \frac{1}{2^{2\tau}}$.[3] Finally, when $q_i + q_j \leq 1$, then $\pi_{ij}^* = 1$, while **Algo** obtains the least revenue when $q_i = q_j = \frac{1}{2}$, which implies a ratio of at least $1 - \frac{1}{2^{2\tau}} = \frac{2C}{3C-1}$ in all the possibilities.

Case 1b: $c_{ij} = C$. In this case $\pi_{ij}^* \leq 1 + (C-1)\min(q_i, q_j)$, while **Algo** obtains an expected revenue of $C \cdot (q_i^{\tau} q_j^{\tau}) + 1 \cdot (1 - q_i^{\tau} q_j^{\tau}) = 1 + (C-1)(q_i q_j)^{\tau}$. W.l.o.g. assume that $q_j \geq q_i$. Thus, the ratio of the revenue obtained by **Algo** and π_{ij}^* is at least:

$$\min_{q_i, q_j \in \{0, \frac{1}{2}, 1\}, q_j \geq q_i} \frac{1 + (C-1)(q_i q_j)^{\tau}}{1 + (C-1)\min(q_i, q_j)} \geq \min_{q_i \in \{0, \frac{1}{2}, 1\}} \frac{1 + (C-1)q_i^{2\tau}}{1 + (C-1)q_i}$$

$$= \frac{1 + \frac{C-1}{2^{2\tau}}}{1 + \frac{C-1}{2}} = \frac{2C}{3C-1}. \tag{9}$$

---

[3] To see why this is true set $a = 2^{-\tau}$ and note that we have to show that $2 - 2a > 1 - a^2$, which is true for $a > 1$. The latter inequality is true as $\tau > 0$.

We now consider the case $i = j$. Again we have two sub cases.

Case 2a: $c_{ii} = 1$. In this case $\pi_{ii}^* \leq \min(1, 1 - q_i) = 1 - q_i$, while **Algo** gets a revenue of $0 \cdot q_i^\tau + 1 \cdot (1 - q_i^\tau) = 1 - q_i^\tau$. Thus, the ratio of the revenue of **Algo** to π_{ii}^* is at least

$$\min_{q_i \in \{0, \frac{1}{2}, 1\}} \frac{1 - q_i^\tau}{1 - q_i} = \min(1, 2 - 2^{1-\tau}) \geq 1 - \frac{1}{2^{2\tau}} = \frac{2C}{3C - 1}.$$

Case 2b: $c_{ii} = C$. In this case $\pi_{ii} \leq 1 + (C - 1)q_i$. The expected revenue for **Algo** is $1 \cdot (1 - q_i^\tau) + C \cdot q_i^\tau = 1 + (C - 1)q_i^\tau \geq 1 + (C - 1)q_i^{2\tau}$. Thus, from (9), the ratio is at least $\frac{2C}{3C-1}$.

Thus, in all cases for every edge $(i, j) \in E$, **Algo** obtains an expected revenue of at least $\frac{2C}{3C-1} \cdot \pi_{ij}^*$, as desired.

2.3   Derandomization

Algorithm **Algo**(τ) can be derandomized in a straightforward way using standard techniques. In particular, observe that the analysis of the randomized rounding step only required pairwise independence among the random choices. One can use a small family of pairwise independent random variables (see the survey [14] for such constructions) and exhaustively try all the possibilities in this space.

Alternatively, one can employ the method of conditional expectation [15,6], since the expected revenue after randomized rounding is an easy formula to calculate (given the probabilities).

2.4   A Tight Integrality Gap

Next, we show that Theorem 1 is the best one can hope from any algorithm that rounds the LP. Formally, we prove the following.

Proposition 2. *There exist an instance of* $UD_2(\{1, C\})$ *for which the the integrality gap of the LP in Figure 1 is at least* $\frac{3C-1}{2C}$.

Proof. Consider the graph with two vertices and C parallel edges– one of which has a cost of C and the rest have a cost of 1. (This assumes that C is integral; if however C is not integral, we need to choose an appropriate number of cost 1 edges and cost C edges such that their ratio is C.) The optimal revenue is C. However, the LP can set a price of $\frac{C+1}{2}$ on both the vertices to get a revenue of $\frac{C+1}{2}$ from the cost C edge and a revenue of 1 from each of the cost 1 edges. Thus, the integrality gap is at least

$$\frac{1 \cdot (\frac{C+1}{2}) + (C - 1) \cdot 1}{C} = \frac{3C - 1}{2C}.$$

2.5   The General Case

The results presented for $k = 2$ in the previous sections can be suitably modified to work for the general case. The LP relaxation for general k is the natural one. For example, the constraint (4), the sum $p_i + p_j$ will be replaced by $\sum_{j=1}^{k} p_{i_j}$ for the hyperedge $(p_{i_1}, p_{i_2}, \ldots, p_{i_k})$. The "half-integrality" gap result will now say that the prices are in the set $\{0, (1 - 1/k)(C - 1), C - 1\}$. Finally, we can prove the following counterparts of Theorem 1 and Proposition 2 by straightforward generalizations of their proofs.

Theorem 2. *For every $C > 1$, there an algorithm that is a $\frac{(2k-1)C-k+1}{kC}$ approximation for the pricing problem with unit-demand bidders with demand size at most k and budgets from $\mathcal{B} = \{1, C\}$.*

Proposition 3. *There exist an instance of $UD_k(\{1, C\})$ for which the the integrality gap of the LP used above is at least $\frac{(2k-1)C-k+1}{kC}$.*

2.6   Budget Range $[1, C]$

Another interesting restriction on the range of buyer's budgets is to an interval $\mathcal{B} = [1, C]$, which clearly generalizes the previous doubleton case $\{1, C\}$. For this case, denoted $UD_2([1, C])$, we obtain the following approximation. The proof proceeds by considering the best single price (same price for all the goods) in the range $[1, C]$, and is deferred to the full version.

Proposition 4. *For every $C > 1$, there is a polynomial time $1 + \ln C$ approximation algorithm for the unit-demand pricing problem with $k = 2$ and budgets from $\mathcal{B} = [1, C]$.*

One can try a natural extension of our LP-relaxation technique for $\{1, C\}$ to this more general case $[1, C]$. However, it turns out that the resulting LP has integrality gap $1 + \ln C$, and thus cannot offer improved approximation.

3   Single-Minded Buyers in Commoditized Markets

We now consider the pricing problem for single minded bidders when all the bidders have the same budget, which can be assumed w.l.o.g. to be 1. That is, we are interested in pricing schemes for the $SM_2(\{1\})$ model. We extend our techniques from Section 2 to get a pricing algorithm with an approximation factor of $\frac{6+\sqrt{2}}{5+\sqrt{2}} \approx 1.156$ (Theorem 3). As in the case of single-minded bidders, our rounding procedure is tight (optimal), as we show that this LP relaxation has a matching integrality gap.

It is not difficult to verify the LP in Figure 2 is a relaxation for our problem $SM_2(\{1\})$.

As in the $UD_2(\{1, C\})$, we first observe that an optimal basic feasible solution to the LP relaxation is half integral.

$$\max \sum_{(i,j)\in E} \pi_{ij} \quad \text{subject to:}$$

$\forall (i,i) \in E$	$\pi_{ii} \leq p_i$	(10)
$\forall (i,j) \in E, i \neq j$	$\pi_{ij} \leq p_i + p_j$	(11)
$\forall (i,j) \in E, i \neq j$	$\pi_{ij} \leq 2 - p_i - p_j$	(12)
$\forall i \in V$	$0 \leq p_i \leq 1$	(13)
$\forall (i,j) \in E$	$\pi_{ij} \geq 0$	(14)

Fig. 2. LP relaxation for the single-minded bidders setting

Proposition 5. *Every optimal basic feasible solution to the LP in Figure 2 is half integral. That is, every extremal optimal assignment to the variables $\{p_i^*\}_{i\in V}$ satisfies the following: $p_i^* \in \{0, \frac{1}{2}, 1\}$.*

The proof is very similar to that of Proposition 1 and is omitted (the only possible "tight" edge can be for an edge $(i,j) \in E$, such that $i \neq j$ and $p_i^* + p_j^* = 1$).

We next analyze the following randomized algorithm.

Algorithm **Algo$_{SM}$**:

1. Solve the LP in Figure 2 to obtain an optimal basic feasible solution with price variables $\{p_i\}_{i\in V}$.
2. Fix prices according to the three schemes below and pick the one that generates the maximum revenue.
   (a) Assign a price p_i to vertex i.
   (b) If $p_i \neq 1$, assign a price of p_i to vertex i, else assign a price of $1/2$.
   (c) If $p_i \neq 1/2$, assign a price of p_i to vertex i, else assign a price of 0 with probability $1/\sqrt{2}$ and a price of 1 with probability $1-1/\sqrt{2}$.

Theorem 3. *Algo$_{SM}$ achieves $\frac{6+\sqrt{2}}{5+\sqrt{2}}$ approximation for the pricing problem with single-minded bidders, desired sets of size at most 2, and unit budgets. That is, expected revenue of **Algo$_{SM}$** is at least $\frac{5+\sqrt{2}}{6+\sqrt{2}}$ fraction of the optimum for $SM_2(\{1\})$.*

The proof is deferred to the full version. The rounding procedure above is tight (optimal), as the following proposition shows (proof in the full version).

Proposition 6. *There exist an instance of $SM_2(\{1\})$ for which the the integrality gap of the LP in Figure 2 is at least $\frac{6+\sqrt{2}}{5+\sqrt{2}}$.*

4 An Online 4-Approximation

In this section, we consider the following online version of the $UD_2(\cdot)$ and $SM_2(\cdot)$ problems. Buyers are assumed to be "in the system" at the beginning and the goods arrive in an online fashion. When a good arrives, any buyer who is interested submits a bid and the seller has to price this good before the next good arrives. We assume that every buyer is interested in at most two items and the seller knows the identity of each buyer. Further, the seller knows about the exact set of elements the buyer is interested in only after the buyer had bid for *both* the items he is interested in. The price that a buyer pays follows the same rules as in $UD_2(\cdot)$ and $SM_2(\cdot)$ models respectively. In the graph abstraction of the $UD_2(\cdot)$ and $SM_2(\cdot)$, the online model has the following interpretation. At every step, a vertex in the underlying graph arrives. Once a vertex appears, all the edges incident on it (along with the edge weights) are revealed to the seller. The only way the seller knows about the other end point of an edge is if that vertex had arrived earlier. Under these constraints, the seller has to price every vertex as it arrives, so as to make as much revenue as possible. For the rest of the section, we will only talk about the $UD_2(\cdot)$ model. The discussion holds equally well for the $SM_2(\cdot)$ model (just replace the prices of ∞ by 0).

The algorithm in [4] works in this model *if* when an vertex arrives, the seller has the full information about the edge. That is, if the other end point is in the "future" then the seller also knows about this other end point. We now restate the algorithm in [4] that works in this scenario. Initially with probability $1/2$ decide on "left" or "right". For the ease of exposition, assume that the algorithm chose left. When a vertex (say i) arrives, with probability $1/2$ tag it as a left vertex or a right vertex (unless it is already assigned a tag). If i is a right vertex then assign it a price ∞. Otherwise look at the set of neighbors of i (recall that the seller knows everything about the edge incident on i). If some neighbor j has not arrived yet then assign j one of the tags with equal probability. Let $N'(i)$ denote the set of neighbors of i that are tagged right. Now consider all edges between i and $N'(i)$ and set the price of i to be the best fixed price given that the vertices in $N'(i)$ are priced at infinity By the analysis in [4], this algorithm is 4-competitive.

We now consider the more general model, where the seller has no information about the vertices that are yet to arrive. For this model, we consider the following refinement of the algorithm in [4]. For any vertex i, let p_i^* denote the best fixed price for vertex i, given that all of its neighbors are priced at ∞. Recall that in our online model, once a vertex arrives, the seller knows the weights of all the incident edges. Thus, the seller can calculate the price p_i^*. Given this, the online algorithm is very simple.

> Algorithm: When each vertex i arrives,
>
> - Compute its best fixed price p_i^*.
> - With probability $1/2$ set its price $p_i = \infty$ and with probability $1/2$ set $p_i = p_i^*$.

We have the following performance guarantee (proof in the full version).

Theorem 4. *For the online $UD_2(\cdot)$ model, the algorithm above is 4-competitive in the expected sense.*

References

1. Aggarwal, G., Feder, T., Motwani, R., Zhu, A.: Algorithms for multi-product pricing. In: Díaz, J., Karhumäki, J., Lepistö, A., Sannella, D. (eds.) ICALP 2004. LNCS, vol. 3142, pp. 72–83. Springer, Heidelberg (2004)
2. Guruswami, V., Hartline, J.D., Karlin, A.R., Kempe, D., Kenyon, C., McSherry, F.: On profit-maximizing envy-free pricing. In: Proceedings of the 16th annual ACM-SIAM symposium on Discrete algorithms, Society for Industrial and Applied Mathematics, pp. 1164–1173 (2005)
3. Demaine, E.D., Feige, U., Hajiaghayi, M.T., Salavatipour, M.R.: Combination can be hard: approximability of the unique coverage problem. In: Proceedings of the Seventeenth Annual ACM-SIAM Symposium on Discrete Algorithms (SODA), pp. 162–171 (2006)
4. Balcan, M.F., Blum, A.: Approximation algorithms and online mechanisms for item pricing. In: Proceedings of the 7th ACM conference on Electronic commerce, pp. 29–35. ACM Press, New York (2006)
5. Briest, P., Krysta, P.: Buying cheap is expensive: Hardness of non-parametric multi-product pricing. In: Proceedings of the Eightteenth Annual ACM-SIAM Symposium on Discrete Algorithms (SODA) (2007)
6. Srinivasan, A.: Approximation algorithms via randomized rounding: a survey. In: Karonski, M., Promel, H.J. (eds.) Lectures on Approximation and Randomized Algorithms. Series in Advanced Topics in Mathematics, pp. 9–71. Polish Scientific Publishers PWN (1999)
7. Goemans, M.X., Williamson, D.P.: New 3/4-approximation algorithms for the maximum satisfiability problem. SIAM Journal on Discrete Mathematics 7, 656–666 (1994)
8. Goemans, M.X.: Mathematical programming and approximation algorithms. In: Lecture at Udine School, Undine, Italy (1996)
9. Vazirani, V.V.: Approximation Algorithms. Springer, New York (2001)
10. Balcan, M.F., Blum, A., Hartline, J.D., Mansour, Y.: Mechanism design via machine learning. In: 46th Annual IEEE Symposium on Foundations of Computer Science, IEEE Computer Society, pp. 605–614. IEEE Computer Society, Los Alamitos (2005)
11. Goldberg, A., Hartline, J., Karlin, A., Saks, M., Wright, A.: Competitive Auctions. Games and Economic Behavior 55, 242–269 (2006)
12. Briest, P., Krysta, P.: Single-minded unlimited supply pricing on sparse instances. In: Proceedings of the Seventeenth Annual ACM-SIAM Symposium on Discrete Algorithms (SODA), pp. 1093–1102 (2006)
13. Bansal, N., Cheng, N., Cherniavsky, N., Rudra, A., Scheiber, B., Sviridenko, M.: Pricing to impatient bidders. In: Proceedings of the Seventeenth Annual ACM-SIAM Symposium on Discrete Algorithms (SODA) (to appear 2007)
14. Luby, M., Wigderson, A.: Pairwise independence and derandomization. Found. Trends Theor. Comput. Sci. 1, 237–301 (2006)
15. Alon, N., Spencer, J.H.: The probabilistic method. John Wiley & Sons, Inc, New York (1992)

Improved Lower Bounds for Non-utilitarian Truthfulness

Iftah Gamzu*

School of Computer Science, Tel-Aviv University, Tel-Aviv 69978, Israel
iftgam@post.tau.ac.il

Abstract. One of the most fundamental results in the field of mechanism design states that every utilitarian social choice function admits a mechanism that truthfully implements it. In stark contrast with this finding, when one considers a non-utilitarian social choice function, it turns out that no guarantees can be made, i.e. there are non-utilitarian functions, which cannot be truthfully implemented. In light of this state of affairs, one of the most natural and intriguing objectives of research is to understand the inherent limitations in the infrastructure of truthful mechanisms for non-utilitarian social choice functions.

In this paper, we focus our attention on studying the boundaries imposed by truthfulness for two non-utilitarian multi-parameter optimization problems. The first is the *workload minimization in inter-domain routing* problem, and the other is the *unrelated machines scheduling* problem. Our main findings can be briefly summarized as follows:

1. We prove that any truthful deterministic mechanism, and any universal truthful randomized mechanism for the workload minimization in inter-domain routing problem cannot achieve an approximation guarantee that is better than 2. These results improve the current lower bounds of $(1+\sqrt{5})/2 \approx 1.618$ and $(3+\sqrt{5})/4 \approx 1.309$, which are due to Mu'alem and Schapira [SODA '07].
2. We establish a lower bound of $1 + \sqrt{2} \approx 2.414$ on the achievable approximation ratio of any truthful deterministic mechanism for the unrelated machines scheduling problem when the number of machines is at least 3. This lower bound is comparable to a recent result by Christodoulou, Koutsoupias and Vidali [SODA '07]. Nevertheless, our approach is considerably simpler, and thus may shed some new light on the core of this problem.

1 Introduction

The problems. We study the *workload minimization in inter-domain routing* problem. As input to this problem, we are given a directed graph $G = (V, E)$, such that $n = |V|$, every edge $e \in E$ has a *cost* $c_e \in \mathbb{R}_+$, and there is a designated *target* vertex $t \in V$. An additional ingredient of the input is a set $\mathcal{R}$ of connection

---

* Supported by the Binational Science Foundation, and by the Israel Science Foundation.

requests in which every request $r \in \mathcal{R}$ is characterized by a pair (s_r, d_r) such that s_r is the *source* vertex of the request, and $d_r \in \mathbb{R}_+$ is the *demand* or traffic intensity associated with the request. The objective is to assign a path from s_r to t, for every request r, on which the request's demand will be sent, such that all the paths constitute a *confluent routing tree*. A confluent routing tree is a tree in which all the traffic arriving at any vertex leaves along a single edge. In particular, the goal is to determine a routing tree in which the workload imposed on the "busiest" vertex is minimized, that is a tree T that minimizes $\max_{u \in V} c_u^T \sum_{r \in R_u^T} d_r$, where R_u^T and c_u^T denote the set of requests that route their demand using a path that goes through u in T and the cost of the single edge that leaves vertex u in T, respectively. Remark that the original formulation of the problem [11] is slightly different, e.g. the problem is defined with respect to an undirected graph in which every vertex has a cost function on its outgoing edges. Nevertheless, it is not hard to validate that both formulations are essentially equivalent.

We also consider the *unrelated machines scheduling* problem. An instance of this problem consists of n machines, and m tasks such that the execution time of task j on machine i is determined by the t_{ij} entry of an $n \times m$ matrix t. The objective is to generate an allocation of the tasks to the machines that minimizes the *makespan*, i.e. the maximum completion time of the machines. This goal is equivalent to generating an $n \times m$ allocation matrix x in which every x_{ij} entry is an $\{0, 1\}$-indicator such that $x_{ij} = 1$ if and only if task j is allocated to machine i, every task is assigned to exactly one machine, i.e. $\sum_{i \in [n]} x_{ij} = 1$ for every $j \in [m]$, and $\max_{i \in [n]} \sum_{j \in [m]} x_{ij} t_{ij}$ is minimized.

The setting. In the present paper, we study the aforementioned problems from an *algorithmic mechanism design* [12] point of view. Algorithmic mechanism design studies the design of protocols or mechanisms for algorithmic problems in scenarios where the input is presented by *strategic agents*. Strategic agent might declare a fallacious input in order to manipulate the protocol in a way that will maximize its own utility. A primary interest of algorithmic mechanism design is in the development of *incentive compatible* or *truthful* protocols, which are robust against manipulation by agents, i.e. every agent is rationally motivated to truthfully report its input. Particulary, in this paper, we concentrate on lower bounding the achievable approximation guarantee of any truthful protocol for the problems under consideration. Note that in the workload minimization in inter-domain routing problem, every vertex is assumed to be controlled by a strategic agent, which may be dishonest about the costs of the vertex's outgoing edges, and in the unrelated machines scheduling problem, every machine is assumed to be controlled by a strategic agent, which may be untruthful about the execution times of the tasks on the corresponding machine.

1.1   Our Results

Workload minimization in inter-domain routing. We establish that any truthful deterministic mechanism for the workload minimization in inter-domain routing problem cannot achieve an approximation guarantee that is better than 2. Additionally, we reinforce this inapproximability result by demonstrating that no randomized mechanism, which is truthful in the universal sense, can obtain an approximation ratio better than 2. These results improve upon the lower bounds presented by Mu'alem and Schapira [11], which are $(1 + \sqrt{5})/2 \approx 1.618$ and $(3+\sqrt{5})/4 \approx 1.309$, respectively. The specifics of these findings are presented in Section 3.

Unrelated machines scheduling. We prove that any truthful deterministic mechanism for the unrelated machines scheduling problem cannot yield an approximation guarantee that is better than $1 + \sqrt{2} \approx 2.414$ for input instances that have at least 3 machines. This result is equal to the lower bound exhibited recently by Christodoulou, Koutsoupias and Vidali [4]. Notwithstanding, our approach is significantly simpler. In particular, we demonstrate how to bypass the so-called *geometrical structure of mechanisms*, which seems to be imperative in their proof. This result appears in Section 4.

1.2   Related Work

The workload minimization in inter-domain routing problem models one of the most fundamental problems in the design of routing protocols. Specifically, it captures the problem of establishing a routing tree for a network (e.g. the Internet), in which no single autonomous system (AS) is excessively congested. This problem was introduced by Mu'alem and Schapira [11], who posed it as a natural extension to the *inter-domain routing* problem, which was formulated by Feigenbaum, Papadimitriou, Sami, and Shenker [7]. Mu'alem and Schapira proved that this problem cannot be approximated to within factors of $(1+\sqrt{5})/2$ and $(3 + \sqrt{5})/4$ by any truthful deterministic mechanism and any randomized mechanism that is truthful in the universal sense, respectively. In addition, they designed a simple truthful deterministic mechanism that obtains an approximation ratio of n. They also considered the single-parameter variant of the problem, in which all the outgoing edges of a vertex have the same cost, and demonstrated that this variant can be solved in an optimal way by a truthful deterministic exponential-time mechanism.

The problem of unrelated machines scheduling is one of the most classical and general variants in the field of scheduling and as such, it has been given extensive attention in past years, both from an algorithmic point of view and from a game-theoretic one. From a pure algorithmic point of view, the problem is known to admit 2-approximation algorithms (see e.g. [10,14,1]), and is known to be $\frac{3}{2}$-hard to approximate in polynomial time, unless P = NP [10]. The mechanism design version of the problem originates in the pioneering work of Nisan and Ronen [12]. They proposed a polynomial-time truthful deterministic mechanism, which is a member of the VCG family [15,5,8], that achieves an approximation guarantee of n, and also proved that no deterministic mechanism can obtain an approximation ratio that is better than 2. They also conjectured that this gap will be resolved by showing that no deterministic mechanism can attain approximation ratio better than n. Finally, they demonstrated that randomization may help to obtain better outcome by presenting a randomized truthful mechanism for a two machines scenario, which has an approximation ratio of $\frac{7}{4}$. Recently, Mu'alem and Schapira [11] extended the last result, and devised a randomized truthful mechanism for any n machines, which achieves an approximation ratio of $\frac{7n}{8}$. In the same work, they also established a lower bound of $2 - \frac{1}{n}$ on the achievable approximation guarantee of any randomized mechanism. Correspondingly, Christodoulou, Koutsoupias and Vidali [4] proved that any truthful deterministic mechanism cannot yield an approximation guarantee that is better than $1 + \sqrt{2}$ for 3 machines or more. A concurrent line of work, initiated independently by Christodoulou, Koutsoupias and Kovács [3] and Lavi and Swamy [9], studied special variants of the unrelated machines scheduling problem. Christodoulou, Koutsoupias and Kovács considered the fractional variant of the problem, and devised a deterministic polynomial-time truthful mechanism that attains an approximation ratio of $\frac{n+1}{2}$, while proving that this fractional variant cannot be truthfully approximated within a factor better than $2 - \frac{1}{n}$. Lavi and Swamy researched the "low-high" variant of the problem, in which the execution time of every tasks is either "low" or "high", and designed a 3-approximation truthful-in-expectation mechanism. They also presented a truthful deterministic 2-approximation mechanism for the case that all the tasks share the same "low" and "high" values, and demonstrated that in this case no truthful deterministic mechanism can achieve an approximation ratio better than 1.14.

2   Preliminaries

In this section, we present a brief introduction to the field of algorithmic mechanism design, and then turn to describe a key property, which every truthful deterministic mechanism must satisfy. This property is fundamental to our approach as our lower bound proofs are built upon it. Remark that this section strives to provide a succinct description of the relevant definitions and results of Bikhchandani et al. [2], and hence the keen reader is encouraged to refer to the aforesaid paper for a more comprehensive presentation of the underlying concepts.

We begin by outlining the nature of questions that algorithmic mechanism design studies. In an algorithmic mechanism design problem setup, there is a set of n *strategic agents* and a finite set of *outcomes* A. Every agent i has a private *type* represented by a *valuation function* $v_i : A \to \mathbb{R}_+$, where $v_i \in V_i$, and V_i denotes the domain of all valid types of agent i. Note that each agent is only interested to maximize its own gain, and thus may be dishonest when reporting its type. The main interest of algorithmic mechanism design is to generate a *social choice mechanism* that is *truthful*. Essentially, the goal is to design an *allocation algorithm* $f : V_1 \times \ldots \times V_n \to A$, and a *payment scheme* $p : V_1 \times \ldots \times V_n \to \mathbb{R}^n$ such that each agent's dominant strategy is to truthfully report its type to the mechanism $M = (f, p)$.

We now turn to describe a property that every truthful deterministic mechanism must satisfy. Notice that this reduces the goal of establishing a lower bound on the achievable approximation ratio of truthful deterministic mechanisms to that of proving a lower bound for a restricted class of allocation algorithms, which constitute truthful mechanisms. Note that we will henceforth refer to such allocation algorithms as *truthful allocation algorithms*.

Definition 1. Let $v = (v_1, \ldots, v_n)$ be a tuple of valuations, and $v' = (v_i', v_{-i})$ be the tuple of valuations obtained by replacing the valuation function of agent i in v from v_i to v_i'. In addition, let f be an allocation algorithm such that $a = f(v)$ and $b = f(v')$. The allocation algorithm f is said to be *weakly monotone* if

$$v_i(a) + v_i'(b) \leq v_i'(a) + v_i(b) ,$$

for every $i \in [n]$, and every valid valuations tuples v and v'.[1]

Theorem 2 ([2]). *If $M = (f, p)$ is a truthful mechanism then f must be weakly monotone.*

Remark that Saks and Yu [13] proved that the weak monotonicity property is not only necessary for truthfulness, but for convex domains is also sufficient. Nonetheless, this fact will not be utilized in the context of this paper.

3   Workload Minimization in Inter-domain Routing

In this section, we study the workload minimization in inter-domain routing problem, and establish a lower bound of 2 on the achievable approximation guarantee of any truthful deterministic mechanism, and any universal truthful randomized mechanism. Prior to describing the finer details of our approach, we provide an interpretation of the weak monotonicity theorem, i.e. Theorem 2, to the problem under consideration. Bear in mind that the valuation function of the agent that controls vertex u satisfies $v_u(a) = d \cdot c_e$, where d is the total traffic that goes through u in the routing tree defined by the outcome a, e is the single edge that leaves vertex u in that routing tree, and c_e is its cost.

Corollary 3. *Suppose we are given two input instances for the workload minimization in inter-domain routing problem, which only differ in the cost functions on the edges. Specifically, suppose that the cost functions c and c' only disagree on vertex u's outgoing edges costs. Every truthful allocation algorithm must satisfy that if*

– e and e' are two outgoing edges of u,

---

[1] Note that this definition applies for cases in which the valuation function of every agent represents cost induced on that agent, and the interest of every agent is to minimize its cost, e.g. it applies for the problems under consideration. For cases in which the valuation function of every agent corresponds to profit, the inequality is in the opposite direction, i.e. $\geq$ instead of $\leq$.

– *d units of traffic are routed through e in the routing tree generated w.r.t. c,*
– *d' units of traffic are routed through e' in the routing tree generated w.r.t. c',*

then

$$d(c_e - c'_e) + d'(c'_{e'} - c_{e'}) \leq 0 \ .$$

We now turn to argue about the deterministic lower bound. In particular, we prove that any truthful allocation algorithm has a "poor" input instance, for which it generates a routing tree whose workload value is bounded away from the optimal workload value by a factor of at least 2.

Theorem 4. *The approximation ratio of any truthful deterministic mechanism for the workload minimization in inter-domain routing problem cannot be better than 2.*

Proof. Consider a truthful allocation algorithm for the problem under consideration, and suppose that its input is the directed graph schematically described in Figure 1(a), and the set of requests is $\mathcal{R} = \{r_1, r_2\} = \{(s_1, 1), (s_2, 1)\}$.

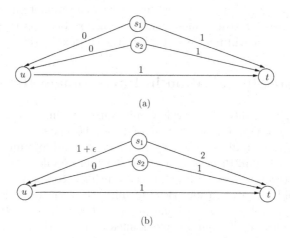

(a)

(b)

Fig. 1. The deterministic lower bound instances

Notice that if the algorithm routes both requests using the vertex u, the obtained routing tree has a workload value of 2. Also notice that an optimal routing tree for this instance has a workload value of 1, e.g. the routing tree, which consists of the edge set $\{(s_1, t), (s_2, t)\}$. Therefore, in such case, we infer that the algorithm cannot have an approximation ratio better than 2. Consequently, we will assume, throughout the remainder of this proof and without loss of generality, that the algorithm routes the request r_1 through the edge (s_1, t). Now, suppose the algorithm is given as input the directed graph schematically described in Figure 1(b), and the same set of requests as before. Remark that the only difference between this input instance, and the aforementioned input instance is the costs of the edges that leave s_1. Specifically, $c_{(s_1,t)} = 2$, and $c_{(s_1,u)} = 1 + \epsilon$, where $0 < \epsilon < 1$ is constant. We claim the algorithm, given this input instance, must also route the request r_1 through the edge $e = (s_1, t)$. This follows from the observation that if the request r_1 is routed through the edge $e' = (s_1, u)$, we yield a contradiction to Corollary 3, since $d = 1$, $d' = 1$, $c_e = 1$, $c'_e = 2$, $c'_{e'} = 1 + \epsilon$ and $c_{e'} = 0$. Consequently, the workload value of the routing tree generated by the algorithm is at least 2, whereas the optimal routing tree has a workload value of $1+\epsilon$, e.g. the routing tree that comprises the edge set $\{(s_1, u), (u, t), (s_2, t)\}$. This establishes that the algorithm cannot have an approximation guarantee better than $\frac{2}{1+\epsilon}$. Since one can select any positive constant $\epsilon \to 0$, the theorem follows. ∎

In the following, we reinforce the last theorem by establishing a lower bound of 2 for universally truthful randomized mechanisms. Note that such mechanisms are defined as a probability distribution over truthful deterministic mechanisms [12,6]. Our approach is based on Yao's minimax principle [16]. In the context of our setting, this principle states that the approximation ratio of the best universal truthful randomized mechanism is equal to the approximation ratio of the best deterministic truthful mechanism under a worst-case input distribution. Accordingly, we exhibit a probability distribution over input instances for which any deterministic truthful mechanism cannot attain an approximation guarantee better than 2.

Theorem 5. *The approximation ratio of any universal truthful randomized mechanism for the workload minimization in inter-domain routing problem cannot be better than 2.*

Proof. Let I denote the input instance, which consists of the directed graph schematically described in Figure 2, and the set of requests $\mathcal{R} = \{r_1, r_2, \ldots, r_k\} = \{(s_1, 1), (s_2, 1), \ldots, (s_k, 1)\}$. Additionally, let I_j be the input instance that is nearly I, but has different costs to the edges that leave s_j. Specifically, the costs of the corresponding edges in I_j are $c_{(s_j,t)} = 2$, and $c_{(s_j,u)} = 1 + \epsilon$, where $0 < \epsilon < 1$ is constant. Finally, let P be a probability distribution over the set of instances $\{I, I_1, \ldots, I_k\}$ such that every instance is picked with probability $\frac{1}{k+1}$.

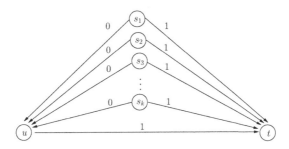

Fig. 2. The randomized lower bound instance

Consider a truthful allocation algorithm for the problem under consideration, and let us analyze its performance on the set of input instances $\{I, I_1, \ldots, I_k\}$ with probability distribution P. We now consider two cases, depending on the structure of the routing tree generated by the algorithm for the instance I.

Case I: The algorithm does not route any of the requests through u. Notice that the algorithm generates a routing tree $T = \bigcup_{i=1}^{k}(s_i, t)$ whose workload value is optimal for I. However, one can apply arguments similar to those used in Theorem 4, and yield that the algorithm cannot obtain an approximation guarantee better than $\frac{2}{1+\epsilon}$ for any of the instances $\{I_1, \ldots, I_k\}$. In particular, one can easily verify that given the input instance I_j, the algorithm must route the request r_j using the edge (s_j, t). Consequently, the expected approximation ratio of the algorithm on the input distribution P is at least $1 \cdot \frac{1}{k+1} + \frac{2}{1+\epsilon} \cdot \frac{k}{k+1} > 2\frac{k}{(k+1)(1+\epsilon)}$. Since one can utilize graph instances for which $k \to \infty$, and may select any constant $\epsilon \to 0$, the theorem follows for this case.

Case II: The algorithm routes $1 \leq q \leq k$ requests through u. Let $Q \subseteq \mathcal{R}$ be the set of requests that the algorithm routes through u. Notice that the workload value of the routing tree generated by the algorithm is q, and accordingly the approximation ratio of the algorithm is q. Additionally, one can apply arguments similar to those used in Theorem 4, and yield that the algorithm cannot attain a better than $\frac{2}{1+\epsilon}$-approximation for any of the instances $\{I_j : r_j \in \mathcal{R} \backslash Q\}$. Hence, the expected approximation ratio of the algorithm on the input distribution P is at least $q \cdot \frac{1}{k+1} + \frac{2}{1+\epsilon} \cdot \frac{k-q}{k+1} + 1 \cdot \frac{q}{k+1} > 2\frac{k}{(k+1)(1+\epsilon)}$, and thus the theorem follows also for this case. ∎

4   Unrelated Machines Scheduling

In this section, we establish a lower bound of $1 + \sqrt{2}$ on the approximation ratio of any truthful deterministic mechanism for the unrelated machines scheduling problem when the number of machines is at least 3. Before we turn to portray the details of our approach, we provide an abstraction of the weak monotonicity theorem, i.e. Theorem 2, to the problem under consideration. Remark that the valuation function of the agent that controls machine i satisfies $v_i(a) = \sum_{j \in [m]} x_{ij} t_{ij}$, where x_{ij} indicates if task j is allocated to machine i in the outcome a, and t_{ij} is the execution time of task j on machine i.

Corollary 6. *Let t and t' be input matrices for the unrelated machines scheduling problem, which differ only in the execution times of machine i. Every truthful allocation algorithm that generates the allocation matrices x and x' w.r.t. t and t' must satisfy*

$$\sum_{j \in [m]} (x_{ij} - x'_{ij})(t_{ij} - t'_{ij}) \leq 0 .$$

In the following, we exploit Corollary 6 to derive two simple lemmas, which will later enable us to demonstrate the desired lower bound. Remark that the first lemma extends Lemma 1 of [4].

Lemma 7. *Let x and x' be allocation matrices generated by a truthful allocation algorithm w.r.t. input matrices t and t', which differ only in the execution times of machine i. If every task realizes one of the following cases*

- Case I: *task j can only be allocated to machine i w.r.t. both t and t'.*[2]
- Case II: *$t'_{ij} > t_{ij}$ and $x_{ij} = 0$.*
- Case III: *$t'_{ij} < t_{ij}$ and $x_{ij} = 1$.*

then x and x' must agree on the allocation of machine i.

Proof. If task j can only be allocated to machine i, then clearly $x_{ij} = x'_{ij} = 1$. Also notice that this implies that the corresponding weak monotonicity term, i.e. $(x_{ij} - x'_{ij})(t_{ij} - t'_{ij})$, equals 0, and hence it does not contribute to the left hand side of the requirement in Corollary 6, that is $\sum_{j \in [m]}(x_{ij} - x'_{ij})(t_{ij} - t'_{ij})$.

Focusing on the other two cases, one can easily validate that every corresponding weak monotonicity term can only be nonnegative. For example, if task j satisfies $t'_{ij} > t_{ij}$ and $x_{ij} = 0$ then the corresponding weak monotonicity term reduces to $(0 - x'_{ij})(t_{ij} - t'_{ij})$. This term is nonnegative as $(t_{ij} - t'_{ij}) < 0$, and $x_{ij} \in \{0, 1\}$. Consequently, in order to satisfy the weak monotonicity theorem, it must follow that $x'_{ij} = x_{ij}$ for all the corresponding tasks. ∎

Lemma 8. *Let x and x' be allocation matrices generated by a truthful allocation algorithm w.r.t. input matrices t and t', and let $\{k, \ell\}$ be a set of two tasks. If $t_{ik} = t_{i\ell} = a$ for some $a > 1$, $t'_{ik} = t'_{i\ell} = 1$, all other execution times of t and t' are identical, and x assigns exactly one of the tasks in $\{k, \ell\}$ to machine i, then x' must assigns at least one of the tasks in $\{k, \ell\}$ to machine i.*

Proof. Notice that in order to satisfy Corollary 6 in the aforesaid settings, x'_{ik} and $x'_{i\ell}$ must fulfill $(1 - x'_{ik} - x'_{i\ell})(a - 1) \leq 0$. Accordingly, x' must assign at least one of the tasks in $\{k, \ell\}$ to machine i. ∎

We are now ready to prove the aforementioned lower bound. Essentially, we prove that any truthful allocation algorithm has a "bad" input matrix, for which it generates an allocation whose makespan value is bounded away from the optimal makespan value by a factor of at least $1 + \sqrt{2}$. Remark that for ease of presentation, we may apply Lemma 7 to input matrices t and t', in which some execution times of machine i are the same. Nevertheless, the understanding between us is that there is a tiny change in these execution times, which we neglect in order to keep the expressions simple, that satisfies the restrictions imposed by the lemma.

Theorem 9. *The approximation ratio of any truthful deterministic mechanism for the unrelated machines scheduling problem with at least 3 machines cannot be better than $1 + \sqrt{2}$.*

---

[2] We say that task j can *only* be allocated to machine i w.r.t. t if $t_{ij} \neq \infty$, and $t_{\ell j} = \infty$ for every $\ell \in [n] \setminus \{i\}$.

Proof. Consider a truthful allocation algorithm for the unrelated machines scheduling problem, and suppose that its input is the following 3-machines 5-tasks matrix

$$t = \begin{pmatrix} 0 & \infty & \infty & \sqrt{2} & \sqrt{2} \\ \infty & 0 & \infty & \sqrt{2} & \sqrt{2} \\ \infty & \infty & 0 & \sqrt{2} & \sqrt{2} \end{pmatrix} .$$

Note that this input matrix admits two distinct task allocations up to symmetries, i.e. name changes of the machines. These two possible tasks allocations are

$$x = \begin{pmatrix} 0^* & \infty & \infty & \sqrt{2}^* & \sqrt{2}^* \\ \infty & 0^* & \infty & \sqrt{2} & \sqrt{2} \\ \infty & \infty & 0^* & \sqrt{2} & \sqrt{2} \end{pmatrix} , \text{ and } y = \begin{pmatrix} 0^* & \infty & \infty & \sqrt{2}^* & \sqrt{2} \\ \infty & 0^* & \infty & \sqrt{2} & \sqrt{2}^* \\ \infty & \infty & 0^* & \sqrt{2} & \sqrt{2} \end{pmatrix} ,$$

where every superscript $*$ denotes an assignment of the column corresponding task to the row corresponding machine. We now consider two cases, depending on which allocation is generated by the algorithm.

Case I: x is generated by the algorithm. Lets consider the matrix t', which is identical to t with a single exception that is $t'_{11} = \sqrt{2}$. By Lemma 7, we know that the allocation generated by the algorithm for the first machine when t' is the input matrix cannot change. Namely, the tasks allocation is

$$\begin{pmatrix} \sqrt{2}^* & \infty & \infty & \sqrt{2}^* & \sqrt{2}^* \\ \infty & 0^* & \infty & \sqrt{2} & \sqrt{2} \\ \infty & \infty & 0^* & \sqrt{2} & \sqrt{2} \end{pmatrix} .$$

This allocation has a value of $3\sqrt{2}$, while it is easy to verify that the optimal allocation has a value of $\sqrt{2}$. Consequently, this proves that the algorithm cannot have an approximation ratio better than $3 > 1 + \sqrt{2}$.

Case II: y is generated by the algorithm. Lets consider the matrix t' that has the same execution times as t with two exceptions, which are $t'_{14} = t'_{15} = 1$. By Lemma 8, we know that the allocation generated by the algorithm when t' is the input matrix must assign at least one of the tasks $\{4, 5\}$ to the first machine. Accordingly, the tasks allocation, up to symmetry, is either

$$x' = \begin{pmatrix} 0^* & \infty & \infty & 1^* & 1^* \\ \infty & 0^* & \infty & \sqrt{2} & \sqrt{2} \\ \infty & \infty & 0^* & \sqrt{2} & \sqrt{2} \end{pmatrix} , \text{ or } y' = \begin{pmatrix} 0^* & \infty & \infty & 1^* & 1 \\ \infty & 0^* & \infty & \sqrt{2} & \sqrt{2}^* \\ \infty & \infty & 0^* & \sqrt{2} & \sqrt{2} \end{pmatrix} .$$

Again, we regard two cases, depending on which allocation is generated.

Case IIa: x' is generated by the algorithm. Lets consider the matrix t'', which is identical to t' with a single difference that is $t''_{11} = \sqrt{2}$. By Lemma 7, we know that the allocation generated by the algorithm for the first machine when t'' is the input matrix cannot change. Consequently, the tasks allocation is

$$\begin{pmatrix} \sqrt{2}^{*} & \infty & \infty & 1^{*} & 1^{*} \\ \infty & 0^{*} & \infty & \sqrt{2} & \sqrt{2} \\ \infty & \infty & 0^{*} & \sqrt{2} & \sqrt{2} \end{pmatrix} .$$

Notice that this allocation has a value of $\sqrt{2}+2$, whereas the optimal allocation has a value of $\sqrt{2}$. Therefore, this establishes that the algorithm cannot have an approximation guarantee better than $\frac{\sqrt{2}+2}{\sqrt{2}} = 1 + \sqrt{2}$.

Case IIb: y' is generated by the algorithm. Lets consider a two-step transition. First, consider the input matrix t'' that is alike t' with a single exception, which is $t''_{14} = 0$. By Lemma 7, we know that the allocation generated by the algorithm for the first machine cannot change. Hence, the tasks allocation, up to symmetry, is

$$\begin{pmatrix} 0^{*} & \infty & \infty & 0^{*} & 1 \\ \infty & 0^{*} & \infty & \sqrt{2} & \sqrt{2}^{*} \\ \infty & \infty & 0^{*} & \sqrt{2} & \sqrt{2} \end{pmatrix} .$$

Second, consider the input matrix t''', which is identical to the matrix t'' with a single change that is $t'''_{22} = 1$. Again, by Lemma 7, we know that the allocation generated by the algorithm for the second machine cannot change. Hence, the tasks allocation for the second machine is

$$\begin{pmatrix} 0 & \infty & \infty & 0 & 1 \\ \infty & 1^{*} & \infty & \sqrt{2} & \sqrt{2}^{*} \\ \infty & \infty & 0 & \sqrt{2} & \sqrt{2} \end{pmatrix} .$$

This allocation has a value of $1+\sqrt{2}$, while it is easy to validate that the optimal allocation has a value of 1. Thus, this proves that the algorithm cannot have an approximation guarantee better than $1 + \sqrt{2}$.   ∎

Acknowledgments

I would like to thank Yossi Azar for his valuable comments on earlier drafts of this paper. Furthermore, I would like to thank the anonymous referees for their constructive suggestions.

References

1. Azar, Y., Epstein, A.: Convex programming for scheduling unrelated parallel machines. In: Proceedings 37th ACM Symposium on Theory of Computing, pp. 331–337 (2005)
2. Bikhchandani, S., Chatterji, S., Lavi, R., Mu'alem, A., Nisan, N., Sen, A.: Weak monotonicity characterizes deterministic dominant-strategy implementation. Econometrica 74(4), 1109–1132 (2006)

3. Christodoulou, G., Koutsoupias, E., Kovács, A.: Mechanism design for fractional scheduling on unrelated machines. In: Proceedings 34th International Colloquium on Automata, Languages and Programming (2007)
4. Christodoulou, G., Koutsoupias, E., Vidali, A.: A lower bound for scheduling mechanisms. In: Proceedings 18th annual ACM-SIAM Symposium on Discrete Algorithms, pp. 1163–1170 (2007)
5. Clarke, E.H.: Multipart pricing of public goods. Public Choice 8, 17–33 (1971)
6. Dobzinski, S., Nisan, N., Schapira, M.: Truthful randomized mechanisms for combinatorial auctions. In: Proceedings 38th ACM Symposium on Theory of Computing, pp. 644–652 (2006)
7. Feigenbaum, J., Papadimitriou, C.H., Sami, R., Shenker, S.: A bgp-based mechanism for lowest-cost routing. Distributed Computing 18(1), 61–72 (2005)
8. Groves, T.: Incentives in teams. Econemetrica 41(4), 617–631 (1973)
9. Lavi, R., Swamy, C.: Truthful mechanism design for multi-dimensional scheduling via cycle monotonicity. In: Proceedings 8th ACM Conference on Electronic Commerce, pp. 252–261 (2007)
10. Lenstra, J.K., Shmoys, D.B., Tardos, É.: Approximation algorithms for scheduling unrelated parallel machines. Mathematical Programming 46(3), 259–271 (1990)
11. Mu'alem, A., Schapira, M.: Setting lower bounds on truthfulness. In: Proceedings 18th annual ACM-SIAM Symposium on Discrete Algorithms, pp. 1143–1152 (2007)
12. Nisan, N., Ronen, A.: Algorithmic mechanism design. Games and Economic Behavior 35, 166–196 (2001)
13. Saks, M.E., Yu, L.: Weak monotonicity suffices for truthfulness on convex domains. In: Proceedings 6th ACM Conference on Electronic Commerce, pp. 286–293 (2005)
14. Shmoys, D.B., Tardos, É.: An approximation algorithm for the generalized assignment problem. Mathematical Programming 62(3), 461–474 (1993)
15. Vickery, W.: Counterspeculation, auctions and competitive sealed tender. Journal of Finance 16, 8–37 (1961)
16. Yao, A.: Probabilistic computations: Toward a unified measure of complexity. In: Proceedings 18th Annual IEEE Symposium on Foundations of Computer Science, pp. 222–227 (1977)

Buyer-Supplier Games: Optimization over the Core

Nedialko B. Dimitrov* and C. Greg Plaxton**

University of Texas at Austin
1 University Station C0500
Austin, Texas 78712–0233
{ned,plaxton}@cs.utexas.edu

Abstract. In a buyer-supplier game, a special type of assignment game, a distinguished player, called the buyer, wishes to purchase some combinatorial structure. A set of players, called suppliers, offer various components of the structure for sale. Any combinatorial minimization problem can be transformed into a buyer-supplier game. While most previous work has been concerned with characterizing the core of buyer-supplier games, in this paper we study optimization over the set of core vectors. We give a polynomial time algorithm for optimizing over the core of any buyer-supplier game for which the underlying minimization problem is solvable in polynomial time. In addition, we show that it is hard to determine whether a given vector belongs to the core if the base minimization problem is not solvable in polynomial time. Finally, we introduce and study the concept of focus point price, which answers the question: If we are constrained to play in equilibrium, how much can we lose by playing the wrong equilibrium?

1 Introduction

In this paper, we study the core of a large set of games, a subset of assignment games, which we term buyer-supplier games [3,22] [23, Chapter 6]. We are primarily concerned with efficient computations over the set of vectors belonging to the core of buyer-supplier games. Before diving into an overview of buyer-supplier games, we present some connections between our work and the existing literature.

1.1 Related Work

Though suggested by Edgeworth as early as 1881 [8], the notion of the core was formalized by Gillies and Shapley [11,21], extending von Neumann and Morgenstern's work on coalitional game theory [24]. Recently, Goemans and Skutella studied the core of a cost sharing facility location game [12]. In their paper, Goemans and Skutella are primarily interested in using core vectors as a cost sharing

---

* Supported by an MCD Fellowship from the University of Texas at Austin.
** Supported by NSF Grants CCR–0310970 and ANI–0326001.

C. Kaklamanis and M. Skutella (Eds.): WAOA 2007, LNCS 4927, pp. 27–40, 2008.

indicator, to decide how much each customer should pay for opening the facility used by the customer. Goemans and Skutella show that, in general, the core of the cost sharing facility location game they study is empty. In contrast, for the buyer-supplier games we study, the core is always nonempty. Additionally, in our work we do not view vectors in the core as an indication of cost shares but rather as rational outcomes of negotiation amongst the players in the buyer-supplier game. Pál and Tardos extend the work of Goemans and Skutella by developing a mechanism for the cost sharing facility location game which uses the concept of an approximate core [15].

There has been great interest in comparing the game's best outcome to the best equilibrium outcome, where the term best is based on some objective function. For example, one may wish to compare the outcome maximizing the net utility for all players in the game against the best possible Nash equilibrium, with respect to net utility. Papadimitriou termed one such comparative measure as the price of anarchy [16]. Roughgarden and Tardos have studied the price of anarchy in the context of routing [18,19,20].

In this paper, we introduce a quantity with a similar motivation to that of the price of anarchy. Solution concepts often yield multiple predictions, or equilibria. In actual game play, however, only one of the equilibria can be chosen by the game's players. Experiments show that conditions outside the game, such as societal pressures or undue attention to a specific player, focus the players' attention on the point of a single equilibrium, which then becomes the outcome of the game. This is a common notion in game theory called the focus point. A player may receive different payoffs in different equilibria. How much is the player willing to pay for a good focus point? We define the *focus point price* with respect to a given player as the difference between the maximum and minimum equilibrium payoffs to the player. Stated succinctly, focus point price answers the question: If we are constrained to play in equilibrium, how much can we lose by playing the wrong equilibrium?

Recently, Garg et al. studied transferable utility games they call coalitional games on graphs [10]. Coalitional games on graphs are a proper subset of buyer-supplier games, which can be derived by setting the buyer's internal cost, Bcost, to zero (see Section 1.3 and Lemma 1). For some buyer-supplier games, for example the buyer-supplier facility location game, it does not appear that the game can be described with Bcost fixed to zero.

Garg et al. study the concepts of "frugality" and "agents are substitutes." They show that suppliers are substitutes if and only if the core of the game forms a lattice. In buyer-supplier games, suppliers are not always substitutes. In Lemma 4, we show that if suppliers are substitutes, we can optimize over the core by solving a polynomially sized linear program. Garg et al. and, more recently, Karlin et al. study and characterize the frugality certain auction mechanisms; the focus point price concept introduced in this paper is quite different from frugality [13].

A third difference between Garg et al. and this work comes from the fact that, similarly to the economics literature, Garg et al. are mainly concerned with the characterization of the core: When does the core form a lattice? How do core vectors relate to auctions? We, on the other hand, are mainly concerned with characterizing optimization over the core. Our main results are in the flavor of Deng and Papadimitriou, in that we are interested with the complexity of computing using game theoretic characterizations [6].

Faigle and Kern study optimization over the core for submodular cost partition games [9]. Faigle and Kern exhibit a generic greedy-type algorithm for optimization of any linear function over the core of partition games whose value function is both submodular and *weakly increasing*, a property they define.

The greedy framework of Faigle and Kern captures certain buyer-supplier games, such as the buyer-supplier minimum spanning tree game. However, even some buyer-supplier games derived from problems that admit greedy solutions, such as the buyer-supplier shortest path game, are not amenable to the approach of Faigle and Kern. In this paper, we do not restrict ourselves to greedy algorithms. By making use of the ellipsoid method, we are able to give polynomial time algorithms for optimization over the core of any buyer-supplier game for which the underlying minimization problem is solvable in polynomial time.

To provide the reader with a simple, concrete example of optimization over the core of a buyer-supplier game, towards the end of this paper, we focus our attention on the buyer-supplier minimum spanning tree game. We give a simple greedy algorithm for this problem, which is a minor extension of Kruskal's minimum spanning tree algorithm. A greedy algorithm is provided by the work of Faigle and Kern, but their exposition involves a good deal of machinery. Our exposition is completely elementary.

Several methods, apart from buyer-supplier games, are known for transforming a combinatorial optimization problem into a game. The cores of these transformations have also been extensively studied. For example, Deng et al. show results on core non-emptyness, distinguishability of core vectors, and finding core vectors for one such transformation [5]. Caprara et al. continue the work of Deng et al. by considering a certain optimization over the set of core vectors for this alternate transformation [4].

1.2   Main Contributions

There has been increased interest from the theoretical computer science community in game theory. While problem-specific solutions may give us insight, to leverage the full power of decades of study in both research areas, we must find generic computational solutions to game theoretic problems. Indeed, others have already realized this need [1,17]. In this paper, we continue this line of work by deriving generic results for computing with core solutions in a large class of games.

The core of buyer-supplier games in the transferable utility setting is characterized by Shapley and Shubik [22]. As a minor contribution, we extend their result by showing that the core in the non-transferable utility setting is the same as the core with transferable utilities. Our primary contributions are as follows:

1. While previous work in the economics literature has concentrated on characterizing the core of buyer-supplier games and relating core vectors to

auctions, our main interest is in optimizing over the set of core vectors [3]. We provide a generally applicable algorithm, based on the ellipsoid method, for optimizing over the core. If the original minimization problem is solvable in polynomial time, we show that it is possible to optimize linear functions of core vectors in polynomial time.

2. We fully characterize optimization over the core of buyer-supplier games by using a polynomial time reduction to show that if the original minimization problem is not solvable in polynomial time, it is impossible, in polynomial time, to test if an arbitrary vector is in the core of the buyer-supplier game.

3. We introduce the concept of focus point price. Our positive computational results give a polynomial time algorithm for computing the buyer's focus point price in buyer-supplier games when the underlying minimization problem is solvable in polynomial time. When the underlying minimization problem is not solvable in polynomial time, we show that it is impossible to approximate the buyer's focus point price to within *any* multiplicative factor.

1.3 Overview of Buyer-Supplier Games

The definition of a buyer-supplier game, given in Section 2.1, is self-contained and does not require an argument. However, it is also possible to transform a combinatorial minimization problem into a buyer-supplier game. Consider a combinatorial minimization problem of the following form. We have some finite set of elements $\mathcal{C}$. We designate some subsets of $\mathcal{C}$ as feasible. To capture feasibility, we use a predicate $P : 2^{\mathcal{C}} \rightarrow \{0, 1\}$, where the predicate is one on all feasible subsets of $\mathcal{C}$. With each feasible set $\mathcal{A} \subseteq \mathcal{C}$, we associate a nonnegative cost $f(\mathcal{A})$. The combinatorial minimization problem can then be captured by the function $\mathrm{MinProb} : 2^{\mathcal{C}} \rightarrow \Re_{+}$ defined by

$$\mathrm{MinProb}(\mathcal{B}) = \min_{\substack{\mathcal{A} \subseteq \mathcal{B} \\ P(\mathcal{A}) = 1}} f(\mathcal{A})$$

where $\Re_{+}$ denotes the nonnegative real numbers.

To transform the above minimization problem into a buyer-supplier game, we associate a player with each element of $\mathcal{C}$; we call such players suppliers. We also add another player whom we call the buyer. In the game, the buyer wishes to purchase a feasible subset of $\mathcal{C}$. The suppliers, on the other hand, are offering their membership to the buyer's set at a price.

To fully specify the game's model of a realistic interaction, we let M designate the maximum investment the buyer is willing to spend on a feasible set. We decompose f such that $f(\mathcal{A}) = \mathrm{Bcost}(\mathcal{A}) + \sum_{a \in \mathcal{A}} \tau(a)$, where $\tau(a)$ is the internal cost for supplier a to be present in the buyer's set and $\mathrm{Bcost}(\mathcal{A})$ is the internal cost to the buyer for purchasing this specific feasible set. In general, many such decompositions are possible, and they produce different games. However, when specifically applying the core solution concept, Lemma 1 shows that all such decompositions are equivalent. Though it is not necessary, to remove special cases in our statements, it is convenient to let $\mathrm{Bcost}(\mathcal{A}) = M$ when $\mathcal{A} = \emptyset$ or $\mathcal{A}$ is not feasible.

Now that we have determined the internal costs for the buyer and the suppliers, we can specify the game. The buyer-supplier game is specified by the tuple $(\mathcal{C}, \tau, \mathrm{Bcost})$. The strategy set for the buyer is the power set of $\mathcal{C}$. By playing $\mathcal{A} \subseteq \mathcal{C}$, the buyer chooses to purchase the membership of the suppliers in $\mathcal{A}$. The strategy set for every supplier $a \in \mathcal{C}$ is the nonnegative real numbers, indicating a bid or payment required from the buyer for the supplier's membership.

For any supplier $a \in \mathcal{C}$, we let $\beta(a)$ denote the associated bid. Let $\mathcal{A}$ be the set of suppliers chosen by the buyer. The payoff for the buyer is $M - \mathrm{Bcost}(\mathcal{A}) - \sum_{a \in \mathcal{A}} \beta(a)$. The payoff for a supplier not in $\mathcal{A}$ is 0. The payoff for a supplier a in $\mathcal{A}$ is $\beta(a) - \tau(a)$.

Since we are applying the solution concept of the core, one may think of the game play as follows. All the players in the game sit down around a negotiating table. All the players talk amongst themselves until they reach an agreement which cannot be unilaterally and selfishly improved upon by any subset of the players. Once such an agreement is reached, game play is concluded. Since no subset of the players can unilaterally and selfishly improve upon the agreement, rationality binds the players to follow the agreement.

The fully formal definition of a buyer-supplier game is given in Section 2.1. The transformation process described above can be used to create buyer-supplier games from most combinatorial minimization problems. For example, minimum spanning tree, Steiner tree, shortest path, minimum set cover, minimum cut, single- and multi-commodity flow can all be used to instantiate a buyer-supplier game. As a concrete example and interpretation of a buyer-supplier game, consider the buyer-supplier minimum spanning tree game. In this game, a company owns factories on every node of a graph. The company wishes to connect the factories by purchasing edges in the graph. Each edge is owned by a unique supplier player. Each supplier has an internal cost associated with the company's usage of the edge. The company has a maximum amount of money it is willing to spend on purchasing edges. Depending on the transportation conditions of a particular edge, the company may have some internal cost associated with choosing that particular edge. The buyer-supplier game paradigm yields similarly natural games when applied to other minimization problems.

In this paper we will be concerned with efficient computation over the set of core vectors. For the rest of the paper, when we say polynomial time, we mean time polynomial in the size of the parameter $\mathcal{C}$, which is also polynomial in the number of players of the buyer-supplier game.

1.4   Organization of the Paper

In Section 2 we define buyer-supplier games and the core of a game. In Section 3 we characterize the core of buyer-supplier games. In Section 4 we give positive computational results, namely the generic algorithm for optimizing over the set of core vectors. In Section 5 we give negative computational results by showing polynomial time equivalence between several related problems. In Section 6 we give the problem-specific combinatorial algorithm for the buyer-supplier game arising from the minimum spanning tree problem.

2   Definitions

2.1   Buyer-Supplier Games

Let $\mathcal{C}$ be a finite set and M be a nonnegative real number. Let τ be a function from $\mathcal{C}$ to $\Re_+$. Let Bcost be a function from $2^{\mathcal{C}}$ to $\Re_+$ such that $\mathrm{Bcost}(\emptyset) = M$. The simplifying condition that $\mathrm{Bcost}(\emptyset) = M$ is not required. We explain the condition's purpose later in this section. For $\mathcal{A} \subseteq \mathcal{C}$, let $\mathrm{Eval}(\tau, \mathrm{Bcost}, \mathcal{A})$ denote $\mathrm{Bcost}(\mathcal{A}) + \sum_{a \in \mathcal{A}} \tau(a)$. For $\mathcal{A} \subseteq \mathcal{C}$, let $\mathrm{MinEval}(\tau, \mathrm{Bcost}, \mathcal{A})$ denote $\min_{\mathcal{B} \subseteq \mathcal{A}} \mathrm{Eval}(\tau, \mathrm{Bcost}, \mathcal{B})$. We will omit the parameters τ and Bcost from the functions $\mathrm{Eval}(\tau, \mathrm{Bcost}, \mathcal{A})$ and $\mathrm{MinEval}(\tau, \mathrm{Bcost}, \mathcal{A})$ when the value is clear.

Given a tuple $(\mathcal{C}, \tau, \mathrm{Bcost})$, we proceed to define a buyer-supplier game. Associate a player with each element of $\mathcal{C}$. Call the players in $\mathcal{C}$ suppliers. Let there also be another player, μ, whom we call the buyer. Let $\mathcal{P} = \mathcal{C} \cup \{\mu\}$ be the set of players for the buyer-supplier game.

The strategy for supplier a is a tuple $(\beta(a), p_a)$ with $\beta(a) \in \Re_+$ and $p_a : \mathcal{P} \to \Re_+$. The first element, $\beta(a)$, represents supplier a's bid to the buyer, requiring the buyer to pay $\beta(a)$ for using the supplier's services. The second element, p_a, represents the nonnegative side payments supplier a chooses to make to the game's players. By $p_a(b)$ we denote the side payment a makes to player b.

The strategy for the buyer, μ, is a tuple $(\mathcal{A}, p_\mu)$ where $\mathcal{A} \in 2^{\mathcal{C}}$ and $p_\mu : \mathcal{P} \to \Re_+$. The first element, $\mathcal{A}$, represents the suppliers chosen by the buyer for a purchase. Similarly to a supplier, the second element, p_μ, represents the nonnegative side payments the buyer chooses to make to the game's players.

For each player $a \in \mathcal{P}$ we denote the player's strategy set by $\mathcal{S}_a$. For a set of players $\mathcal{A} \subseteq \mathcal{P}$, we denote the set of strategies $\bigotimes_{a \in \mathcal{A}} \mathcal{S}_a$ by $\mathcal{S}_\mathcal{A}$. We call elements of $\mathcal{S}_\mathcal{A}$ strategy vectors. We index strategy vectors from $\mathcal{S}_\mathcal{A}$ by the elements of $\mathcal{A}$.

We now define the utility function for each player. Suppose strategy $s \in \mathcal{S}_\mathcal{P}$ is played. Specifically, suppose that $(\mathcal{A}, p_\mu) \in \mathcal{S}_\mu$ and $(\beta(a), p_a) \in \mathcal{S}_a$ for each $a \in \mathcal{C}$ are played. The utility function for buyer is $u_\mu(s) = M - [\mathrm{Bcost}(\mathcal{A}) + \sum_{a \in \mathcal{A}} \beta(a)] + [\sum_{b \in \mathcal{P}} p_b(\mu) - \sum_{b \in \mathcal{P}} p_\mu(b)]$. The utility for a supplier a in $\mathcal{A}$ is $u_a(s) = \beta(a) - \tau(a) + [\sum_{b \in \mathcal{P}} p_b(a) - \sum_{b \in \mathcal{P}} p_a(b)]$. The utility for a supplier a not in $\mathcal{A}$ is $u_a(s) = [\sum_{b \in \mathcal{P}} p_b(a) - \sum_{b \in \mathcal{P}} p_a(b)]$.

Interpreting, the buyer begins with a total of M utility and chooses to make a purchase from each supplier in $\mathcal{A}$. The buyer gives $\beta(a)$ to each supplier $a \in \mathcal{A}$ and loses an extra $\mathrm{Bcost}(\mathcal{A})$ from the initial M utility. Each supplier a in $\mathcal{A}$ receives the bid payment from the buyer and loses $\tau(a)$ because the supplier must perform services for the buyer. The distribution of sidepayments completes the utility functions. The requirement that $\mathrm{Bcost}(\emptyset) = M$ lets the strategy $\emptyset$ stand as a "don't play" strategy for the buyer. To remove the requirement, we could introduce a specific "don't play" strategy to the buyer's strategy set, however this creates a special case in most of our proofs.

Let the sidepayment game we have defined be denoted SP. Let NOSP denote the same game with the additional requirement that all sidepayments be fixed to zero. In other words, in NOSP we restrict the strategy set for each $a \in \mathcal{P}$ so that p_a is identically zero.

2.2   Game Theoretic Definitions

All of the definitions in this section closely follow those of Shubik [23, Chapter 6].

We call a vector in $\Re^{|\mathcal{P}|}$, indexed by $a \in \mathcal{P}$, a *payoff vector*.

Let π be a payoff vector and s be a strategy vector in $\mathcal{S}_{\mathcal{A}}$ for $\mathcal{A} \subseteq \mathcal{P}$. Let t be any strategy vector in $\mathcal{S}_{\mathcal{P}}$ such that the projection of t onto the coordinates in $\mathcal{A}$ is equal to s. If for all t and for all $a \in \mathcal{A}$ we have $\pi_a \leq u_a(t)$, we say that the players in $\mathcal{A}$ can *guarantee* themselves payoffs of at least π by playing s.

We use Shubik's alpha theory to define our characteristic sets [23, pp. 134-136]. Thus for a set of players $\mathcal{A} \subseteq \mathcal{P}$, we define the characteristic set, $V(\mathcal{A})$, to be the set of all payoff vectors π such that there is a strategy vector $s \in \mathcal{S}_{\mathcal{A}}$, possibly dependent on π, with which the players in $\mathcal{A}$ can guarantee themselves payoffs of at least π. In the transferable utility setting, SP, the characteristic sets can be replaced with a characteristic function. Given the definitions of the utility functions in Section 2.1, the characteristic function $\tilde{V}(\mathcal{A})$ for a set of players $\mathcal{A}$ is equal to $M - \mathrm{MinEval}(\tau, \mathrm{Bcost}, \mathcal{A} - \{\mu\})$.

We say that a set $\mathcal{A} \subseteq \mathcal{P}$ of *players are substitutes* if $\tilde{V}(\mathcal{P}) - \tilde{V}(\mathcal{P} - \mathcal{B}) \geq \sum_{a \in \mathcal{B}} \tilde{V}(\mathcal{P}) - \tilde{V}(\mathcal{P} - \{a\})$ for all $\mathcal{B} \subseteq \mathcal{A}$.

We say that a payoff vector π dominates a payoff vector ν through a set $\mathcal{A} \subseteq \mathcal{P}$ if $\pi_a > \nu_a$ for all $a \in \mathcal{A}$. In other words, π dominates ν through $\mathcal{A}$ when each player in $\mathcal{A}$ does better in π than in ν.

For a set of players $\mathcal{A} \subseteq \mathcal{P}$, we define $D(\mathcal{A})$ as the set of all payoff vectors which are dominated through $\mathcal{A}$ by a payoff vector in $V(\mathcal{A})$. Interpreting, the players in $\mathcal{A}$ would never settle for a payoff vector $\pi \in D(\mathcal{A})$ since they can guarantee themselves higher payoffs than those offered in π.

The *core* of a game consists of all $\pi \in V(\mathcal{P})$ such that $\pi \notin D(\mathcal{A})$ for all $\mathcal{A} \subseteq \mathcal{P}$.

3   A Characterization of the Core

The characterazation of the core of buyer-supplier games in the transferable utility setting was done by Shapley and Shubik [22]. In this seciton, we show the surprising result that the same characterization holds in the non-transferable utility setting. In general, it is not the case that the core of the transferable utility and non-transferable utility versions of a game are the same. For example, the buyer may be able to use bribes to alter the bidding strategies of some suppliers, and thus reduce the bids of other suppliers. The following theorem characterizes the core of buyer-supplier games.

Theorem 1. *A payoff vector π is in the core of a buyer-supplier game defined by $(\mathcal{C}, \tau, \mathrm{Bcost})$ if and only if it satisfies*

$$\pi_a \geq 0 \qquad\qquad \text{for all } a \in \mathcal{P}, \quad (1)$$

$$\sum_{a \in \mathcal{A}} \pi_a \leq \mathrm{MinEval}(\tau, \mathrm{Bcost}, \mathcal{C} - \mathcal{A}) - \mathrm{MinEval}(\tau, \mathrm{Bcost}, \mathcal{C}) \quad \text{for all } \mathcal{A} \subseteq \mathcal{C}, \quad (2)$$

$$\pi_\mu = M - \mathrm{MinEval}(\tau, \mathrm{Bcost}, \mathcal{C}) - \sum_{a \in \mathcal{C}} \pi_a. \qquad\qquad (3)$$

Because of space considerations, here and in the rest this paper we choose to present the intuition and a proof sketch for most of the stated results. Fully detailed proofs of all results are presented in the companion technical report [7].

We take as a given the result by Shapley and Shubik, which shows that under transferable utilities, the core is characterized by Theorem 1.

The intuition for the equivalence of the transferable utility core and the non-transferable utility core is as follows. Consider a payoff vector π satisfying Equation (1). A set of suppliers can only guarantee zero payoffs for themselves. Thus, for a set of players $\mathcal{A}$ to be able to truly improve upon the payoffs given in π, the buyer must be in $\mathcal{A}$. However, if the buyer is in $\mathcal{A}$, the players in $\mathcal{A}$ can simulate sidepayments amongst themselves by having the suppliers in $\mathcal{A}$ alter their bids to the buyer. Thus, the sidepayments do not add any additional power to the set of players $\mathcal{A}$.

As a corollary to Theorem 1, we have the following lemma, which shows that the core does not change depending on the decomposition chosen in the transformation from a combinatorial minimization problem to a buyer-supplier game.

Lemma 1. Let $\text{Bcost}^*(\mathcal{A}) = \sum_{a \in \mathcal{A}} \tau(a) + \text{Bcost}(\mathcal{A})$. The core of the buyer supplier-games defined by $(\mathcal{C}, \tau, \text{Bcost})$ and $(\mathcal{C}, 0, \text{Bcost}^*)$ is the same.

4   Polynomial Time Optimization over the Core Vectors

We define the separation problem on a set of linear inequalities $\mathcal{A}$ as follows. Given a vector π, if π satisfies all of the inequalities in $\mathcal{A}$, then do nothing; otherwise, output a violated inequality $a \in \mathcal{A}$. It is well known that the separation problem is polynomial time equivalent to linear function optimization over the same set of inequalities [14, p. 161].

Let $(\mathcal{C}, \tau, \text{Bcost})$ define a buyer-supplier game. In this section, to simplify the notation, we will omit the parameter Bcost from Eval and MinEval since it is fixed by the buyer-supplier game.

In this section, we will analyze an algorithm to solve the separation problem for the exponentially sized set of inequalities given in Equations (1), (2), and (3). We now give the algorithm, which we call the separation algorithm. Given the payoff vector π as input,

1 Iterate over Equations (1) and (3) to check that they hold. If some equation does not hold, output that equation and halt.
2 Compute $\mathcal{F} \subseteq \mathcal{C}$ such that $\text{Eval}(\tau, \mathcal{F}) = \text{MinEval}(\tau, \mathcal{C})$. If there is some $a \in \mathcal{C} - \mathcal{F}$ with $\pi_a > 0$, output the inequality from Equation (2) corresponding to $\{a\}$ and halt.
3 Define $\hat{\tau}(a) = \tau(a) + \pi_a$ for $a \in \mathcal{C}$. Now, compute $\hat{\mathcal{F}} \subseteq \mathcal{C}$ such that $\text{Eval}(\hat{\tau}, \hat{\mathcal{F}}) = \text{MinEval}(\hat{\tau}, \mathcal{C})$. If $\text{Eval}(\hat{\tau}, \hat{\mathcal{F}}) < \text{Eval}(\hat{\tau}, \mathcal{F})$, output the inequality from Equation (2) corresponding to $\mathcal{F} - \hat{\mathcal{F}}$. Otherwise, halt.

Theorem 2. If given an input $\hat{\tau} : \mathcal{C} \to \Re_+$ it is possible to compute both $\text{Eval}(\hat{\tau}, \mathcal{A})$ for any $\mathcal{A} \subseteq \mathcal{C}$ and $\mathcal{F} \subseteq \mathcal{C}$ such that $\text{Eval}(\hat{\tau}, \mathcal{F}) = \text{MinEval}(\hat{\tau}, \mathcal{C})$

in polynomial time, then the separation problem for Equations (1), (2), and (3) is solvable in polynomial time. By the equivalence of separation and optimization, optimizing any linear function of π over Equations (1), (2), and (3) is also possible in polynomial time.

Proof. It is clear that given the theorem's assumptions, the separation algorithm runs in polynomial time. The statement follows from Lemmas 2 and 3.

Lemma 2. *If the separation algorithm returns an inequality on input π, then π violates the returned inequality.*

Proof. If the algorithm returns an inequality in step 1, then the inequality is violated since the algorithm performed a direct check.

If the algorithm returns an inequality in step 2, then the inequality is violated since $\pi_a > 0$, but $\text{MinEval}(\tau, \mathcal{C} - a) = \text{MinEval}(\tau, \mathcal{C}) = \text{Eval}(\tau, \mathcal{F})$.

Suppose the algorithm returns an inequality in step 3. Thus, $\text{Eval}(\hat{\tau}, \hat{\mathcal{F}}) < \text{Eval}(\hat{\tau}, \mathcal{F})$. By applying the definitions of Eval and $\hat{\tau}$, we have $\sum_{a \in \hat{\mathcal{F}}} \pi_a + \text{Eval}(\tau, \hat{\mathcal{F}}) < \sum_{a \in \mathcal{F}} \pi_a + \text{Eval}(\tau, \mathcal{F})$.

Since the algorithm reaches step 3, we know that $\pi_a = 0$ for all $a \in \mathcal{C} - \mathcal{F}$. Thus, we have $\sum_{a \in \hat{\mathcal{F}} \cap \mathcal{F}} \pi_a + \text{Eval}(\tau, \hat{\mathcal{F}}) < \sum_{a \in \mathcal{F}} \pi_a + \text{Eval}(\tau, \mathcal{F})$, which in turn gives $\text{Eval}(\tau, \hat{\mathcal{F}}) - \text{Eval}(\tau, \mathcal{F}) < \sum_{a \in \mathcal{F} - \hat{\mathcal{F}}} \pi_a$.

Let $\mathcal{A} = \mathcal{F} - \hat{\mathcal{F}}$. From the algorithm, we know that the set $\mathcal{F}$ satisfies $\text{Eval}(\tau, \mathcal{F}) = \text{MinEval}(\tau, \mathcal{C})$. Since $\hat{\mathcal{F}} \subseteq \mathcal{C} - \mathcal{A}$, the definition of MinEval implies that $\text{MinEval}(\tau, \mathcal{C} - \mathcal{A}) \leq \text{Eval}(\tau, \hat{\mathcal{F}})$. Thus, we have $\text{MinEval}(\tau, \mathcal{C} - \mathcal{A}) - \text{MinEval}(\tau, \mathcal{C}) \leq \text{Eval}(\tau, \hat{\mathcal{F}}) - \text{Eval}(\tau, \mathcal{F}) < \sum_{a \in \mathcal{A}} \pi_a$, which shows that the inequality output by the algorithm is violated.

Lemma 3. *If π violates some inequality in Equations (1), (2), and (3), then the separation algorithm run on input π returns an inequality.*

Proof. If the violation is in Equations (1) or (3), the violated inequality will be output by the direct check in step 1. If some inequality is output by step 2, we are done. Otherwise, since steps 1 and 2 output no inequality, we know that $\pi_a = 0$ for all $a \in \mathcal{C} - \mathcal{F}$, where $\mathcal{F}$ is as computed in the algorithm.

Now, suppose the inequality from Equation (2) for set $\mathcal{A} \subseteq \mathcal{C}$ is violated. In other words, we have, $\sum_{a \in \mathcal{A}} \pi_a > \text{MinEval}(\tau, \mathcal{C} - \mathcal{A}) - \text{MinEval}(\tau, \mathcal{C})$. Let $\mathcal{B}$ be such that $\text{Eval}(\tau, \mathcal{B}) = \text{MinEval}(\tau, \mathcal{C} - \mathcal{A})$.

Thus, we have $\sum_{a \in \mathcal{A}} \pi_a > \text{MinEval}(\tau, \mathcal{C} - \mathcal{A}) - \text{MinEval}(\tau, \mathcal{C}) = \text{Eval}(\tau, \mathcal{B}) - \text{Eval}(\tau, \mathcal{F})$.

Since $\pi_a = 0$ for all $a \in \mathcal{C} - \mathcal{F}$, we have $\text{Eval}(\tau, \mathcal{F}) + \sum_{a \in \mathcal{F} \cap \mathcal{A}} \pi_a > \text{Eval}(\tau, \mathcal{B})$.

Adding $\sum_{a \in \mathcal{F} - \mathcal{A}} \pi_a$ to both sides of the above inequality and substituting the definition of Eval, we have $\text{Bcost}(\mathcal{F}) + \sum_{a \in \mathcal{F}} \tau(a) + \sum_{a \in \mathcal{F}} \pi_a > \text{Bcost}(\mathcal{B}) + \sum_{a \in \mathcal{B}} \tau(a) + \sum_{a \in \mathcal{F} - \mathcal{A}} \pi_a$.

Since $\pi_a = 0$ for all $a \in \mathcal{C} - \mathcal{F}$ and $\mathcal{B} \subseteq \mathcal{C} - \mathcal{A}$, we can alter the right hand side of the above inequality to get $\text{Bcost}(\mathcal{F}) + \sum_{a \in \mathcal{F}} \tau(a) + \sum_{a \in \mathcal{F}} \pi_a > \text{Bcost}(\mathcal{B}) + \sum_{a \in \mathcal{B}} \tau(a) + \sum_{a \in \mathcal{B}} \pi_a + \sum_{a \in \mathcal{F} - \mathcal{A} - \mathcal{B}} \pi_a$.

By applying the definition of $\hat{\tau}$ and Eval, we have $\text{Eval}(\hat{\tau}, \mathcal{F}) > \text{Eval}(\hat{\tau}, \mathcal{B}) + \sum_{a \in \mathcal{F} - \mathcal{A} - \mathcal{B}} \pi_a$. We know that $\pi_a \geq 0$ for all $a \in \mathcal{P}$ since the algorithm does not output anything in step 1. Thus, $\text{Eval}(\hat{\tau}, \mathcal{F}) > \text{Eval}(\hat{\tau}, \mathcal{B}) \geq \text{MinEval}(\hat{\tau}, \mathcal{C}) = \text{Eval}(\hat{\tau}, \hat{\mathcal{F}})$, where $\hat{\mathcal{F}}$ is as computed in the algorithm. So, step 3 outputs an inequality.

The following lemma illustrates a key difference between Garg et al. and this work.

Lemma 4. *If suppliers are substitutes, then all but the $|\mathcal{C}|$ singleton equations of Equation (2) are not constraining. Thus, if suppliers are substitutes, optimization over the core of the buyer-supplier game is reduced to solving a polynomially sized linear program.*

Proof. Suppose that the suppliers are substitutes. By the definition of suppliers are substitutes, we have that $\tilde{V}(\mathcal{P}) - \tilde{V}(\mathcal{P} - \mathcal{A}) \geq \sum_{a \in \mathcal{A}} [\tilde{V}(\mathcal{P}) - \tilde{V}(\mathcal{P} - \{a\})]$ for all $\mathcal{A} \subseteq \mathcal{C}$. By the definition of $\tilde{V}$, we have $\text{MinEval}(\tau, \text{Bcost}, \mathcal{C} - \mathcal{A}) - \text{MinEval}(\tau, \text{Bcost}, \mathcal{C}) \geq \sum_{a \in \mathcal{A}} [\text{MinEval}(\tau, \text{Bcost}, \mathcal{C} - \{a\}) - \text{MinEval}(\tau, \text{Bcost}, \mathcal{C})]$ for all $\mathcal{A} \subseteq \mathcal{C}$. This implies that if the singleton equations in Equation (2) are satisfied, then so are all equations in Equation (2). Thus, if suppliers are substitutes, we may drop all non-singleton equations from Equation (2) and reduce the number of inequalities to a polynomial in the number of players. $\qquad\blacksquare$

5   Inapproximability of Optimization over Core Solutions

Consider a buyer-supplier game defined by $(\mathcal{C}, \tau, \text{Bcost})$. We introduced the concept of the focus point price in the introduction. The concept leads us to ask the natural question: What is the difference between the best and worst core outcome for the buyer? In other words, the value of interest is the solution to the linear program: maximize $\sum_{a \in \mathcal{C}} \pi_a$ subject to Equations (1), (2), and (3). This natural question leads us to define the focus point price (FFP) problem as follows: on input $(\mathcal{C}, \tau, \text{Bcost})$, output the optimal value of the afore mentioned linear program.

Define the Necessary Element (NEL) problem as follows. Given parameters $(\mathcal{C}, \tau, \text{Bcost})$ return TRUE if there exist an element $a \in \mathcal{C}$ such that for all $\mathcal{F} \subseteq \mathcal{C}$ satisfying $\text{Eval}(\tau, \text{Bcost}, \mathcal{F}) = \text{MinEval}(\tau, \text{Bcost}, \mathcal{C})$ we have $a \in \mathcal{F}$. Otherwise, return FALSE.

Define the OPT-SET problem as follows. Given parameters $(\mathcal{C}, \tau, \text{Bcost})$, return $\mathcal{F}$ such that $\text{Eval}(\tau, \text{Bcost}, \mathcal{F}) = \text{MinEval}(\tau, \text{Bcost}, \mathcal{C})$.

In this section, we will show that the FPP problem, the OPT-SET problem and the NEL problem are polynomial time equivalent. Again, because of space considerations we choose to present some intuition and a proof sketch. For the fully detailed proofs, see the companion technical report [7].

For a fixed tuple $(\mathcal{C}, \tau, \text{Bcost})$ we say we extend the tuple to contain a *shadow element* for an element $a \subseteq \mathcal{C}$ by creating the extended tuple $(\hat{\mathcal{C}}, \hat{\tau}, \text{Bcost}^*)$, where $\hat{\mathcal{C}} = \mathcal{C} \cup b$ with $b \notin \mathcal{C}$; $\hat{\tau}$ is the same as τ with the addition that $\hat{\tau}(b) = \tau(a)$;

and for $\mathcal{A} \subseteq \hat{\mathcal{C}}$, if $b \notin \mathcal{A}$, then $\mathrm{Bcost}^*(\mathcal{A}) = \mathrm{Bcost}(\mathcal{A})$, otherwise $\mathrm{Bcost}^*(\mathcal{A}) = \mathrm{Bcost}((\mathcal{A} - \{b\}) \cup \{a\})$. We call b the *shadow element* corresponding to a.

The *full shadow extension* of $(\mathcal{C}, \tau, \mathrm{Bcost})$ is the tuple $(\hat{\mathcal{C}}, \hat{\tau}, \mathrm{Bcost}^*)$ resulting from extending $(\mathcal{C}, \tau, \mathrm{Bcost})$ to contain a shadow element for each element in $\mathcal{C}$.

First, we reduce OPT-SET to NEL. To show the result, we analyze the following algorithm, which we call the shadow algorithm.

On input $(\mathcal{C}, \tau, \mathrm{Bcost})$,

1 Let $(\hat{\mathcal{C}}, \hat{\tau}, \mathrm{Bcost}^*)$ be the full shadow extension of $(\mathcal{C}, \tau, \mathrm{Bcost})$.
2 For each $a \in \mathcal{C}$
   - Remove a's corresponding shadow element from $\hat{\mathcal{C}}$.
   - Run NEL on $(\hat{\mathcal{C}}, \hat{\tau}, \mathrm{Bcost}^*)$.
   - If the return value is TRUE, then add the shadow element back to $\hat{\mathcal{C}}$.
   - If the return value is FALSE, then remove a from $\hat{\mathcal{C}}$.
3 Return $\hat{\mathcal{C}} \cap \mathcal{C}$. In other words, we return all elements from $\mathcal{C}$ remaining in $\hat{\mathcal{C}}$, disregarding any shadow elements.

Lemma 5. *Let $(\mathcal{C}, \tau, \mathrm{Bcost})$ be the input to the shadow algorithm. Also, let $(\hat{\mathcal{C}}, \hat{\tau}, \mathrm{Bcost}^*)$ be the full shadow extension of $(\mathcal{C}, \tau, \mathrm{Bcost})$. If for all $\mathcal{A} \subseteq \hat{\mathcal{C}}$ the NEL problem on input $(\mathcal{A}, \hat{\tau}, \mathrm{Bcost}^*)$ is solvable in polynomial time, then the OPT-SET problem on input $(\mathcal{C}, \tau, \mathrm{Bcost})$ is solvable in polynomial time.*

Given the lemma assumptions, a simple analysis shows that the shadow algorithm runs in polynomial time. The rest of the proof comes in two steps. First, the shadow algorithm maintains the invariant $\mathrm{MinEval}(\tau, \mathrm{Bcost}, \mathcal{C}) = \mathrm{MinEval}(\hat{\tau}, \mathrm{Bcost}^*, \hat{\mathcal{C}})$. This is true because we only remove an element from $\hat{\mathcal{C}}$ if there is an optimal set that does not contain the element. Second, if a remains in $\hat{\mathcal{C}}$ at the end of the iteration associated with a, then it can be shown that a is contained in all subsets of $\hat{\mathcal{C}} \cap \mathcal{C}$ that are solutions to the OPT-SET problem on input $(\mathcal{C}, \tau, \mathrm{Bcost})$.

The following lemma captures the relationship between the FPP problem and the NEL problem.

Lemma 6. *The solution to the FPP problem on input $(\mathcal{C}, \tau, \mathrm{Bcost})$ is 0 if and only if the solution to the NEL problem on input $(\mathcal{C}, \tau, \mathrm{Bcost})$ is FALSE. Thus, if it is possible to approximate the the FPP problem on input $(\mathcal{C}, \tau, \mathrm{Bcost})$ within any multiplicative factor in polynomial time, then the NEL problem on input $(\mathcal{C}, \tau, \mathrm{Bcost})$ is solvable in polynomial time.*

The intuition behind this lemma is that if the solution to NEL is TRUE, then there is some element a that is in all OPT-SET solutions on input $(\mathcal{C}, \tau, \mathrm{Bcost})$. In this case, the solution to the FPP problem is at least the difference between the value of an OPT-SET solution on input $(\mathcal{C}, \tau, \mathrm{Bcost})$ and the value of an OPT-SET solution on input $(\mathcal{C} - \{a\}, \tau, \mathrm{Bcost})$. On the other hand, if the solution to NEL is FALSE, then the right hand sides of all singleton equations from Equation (2) are zero, and thus the FPP problem solution is also zero.

A set of $(\mathcal{C}, \tau, \text{Bcost})$ instances is *proper* if the following conditions hold:

- Given that $(\mathcal{C}, \tau, \text{Bcost})$ is in the set, then so is $(\mathcal{C}, \hat{\tau}, \text{Bcost})$, where $\hat{\tau}(a) = \tau(a) + \pi_a$ for a vector $\pi \in \Re_+^{|\mathcal{C}|}$.
- Given that $(\mathcal{C}, \tau, \text{Bcost})$ is in the set, then so is $(\mathcal{A}, \hat{\tau}, \text{Bcost}^*)$, where $\mathcal{A}$ is a subset of $\hat{\mathcal{C}}$ and $(\hat{\mathcal{C}}, \hat{\tau}, \text{Bcost}^*)$ is the full shadow extension of $(\mathcal{C}, \tau, \text{Bcost})$.

The definition of proper instances has a natural interpretation when applied to the transformations of combinatorial minimization problems to buyer-supplier games. For example, for the shortest path problem, the first condition implies that the set of instances is closed with respect to lengthening the edges of the graph. On the other hand, the second condition implies that the set of instances is closed with respect to adding parallel edges or removing a subset of the edges.

The results of Section 4 and the relationships we have given in this section lead us to the following theorem.

Theorem 3. *On a proper set of instances, the separation problem over Equations (1), (2), and (3), the NEL problem and the OPT-SET problem are polynomial time equivalent.*

Lemma 6 in combination with Theorem 3 gives us the following inapproximability result.

Lemma 7. *On a proper set of instances, if it is not possible to solve the OPT-SET problem in polynomial time, it is not possible to approximate the solution to the FPP problem to within any multiplicative factor in polynomial time.*

6   A Complementary Combinatorial Algorithm

In this section, we present an efficient combinatorial algorithm for solving the FPP problem for the buyer-supplier minimum spanning tree (MST) game.

Let a graph $G = (\mathcal{V}, \mathcal{E})$ and edge weights $w : \mathcal{E} \to \Re_+$ be given. Let MSTVal : $2^{\mathcal{E}} \to \Re_+$ be a function that takes as input a set of the edges $\mathcal{A} \subseteq \mathcal{E}$ and returns the weight of the minimum spanning tree of the graph induced by the edges of $\mathcal{A}$. If no spanning tree exists, MSTVal returns ∞.

By the transformation in Section 1 and Lemma 1 in the buyer-supplier minimum spanning tree game, we have $\mathcal{C} = \mathcal{E}$, $\tau(a) = w(a)$, and $\text{Bcost}(\mathcal{A}) = M$ if $\mathcal{A}$ does not connect all nodes in $\mathcal{V}$, or 0 otherwise. We omit the parameters τ and Bcost from MinEval, since they are fixed by the game.

Call the linear program from the FPP problem for the given game LP1, and let its optimal value be O_1. Consider the linear program: maximize $\sum_{b \in \mathcal{C}} \pi_b$ subject to $\sum_{b \in \mathcal{A}} \pi_b \leq \text{MinProb}(\mathcal{C} - \mathcal{A}) - \text{MinProb}(\mathcal{C})$ for all $\mathcal{A} \subseteq \mathcal{C}$ and $\pi_b \geq 0$ for all $b \in \mathcal{C}$. Call the linear program from the previous sentence LP2, and let its optimal value be O_2.

We are able to prove the following relationship between LP1 and LP2. If $\text{MinProb}(\mathcal{C}) \geq M$, then $O_1 = 0$. If $\text{MinProb}(\mathcal{C}) < M$ and $O_2 \leq M - \text{MinProb}(\mathcal{C})$, then $O_1 = O_2$. If $\text{MinProb}(\mathcal{C}) < M$ and $O_2 > M - \text{MinProb}(\mathcal{C})$, then

$O_1 = M - \mathrm{MinProb}(\mathcal{C})$. When considering the FPP problem arising from the buyer-supplier game for a specific minimization problem, it may often be helpful to consider LP2 instead of LP1. In fact, the combinatorial algorithm we present finds the optimal value for LP2.

The key insight behind the combinatorial algorithm for the FPP problem for the buyer-supplier MST game is the following. Let T be an MST of G. Suppose edges e_1 and e_2 are edges in T. Suppose the removal of the individual edge e_1 (e_2) increases the MST cost by λ_1 (λ_2). Then, the removal of both edges increases the MST cost by at least $\lambda_1 + \lambda_2$. This insight leads Bikhchandani et al. to show that for the buyer-supplier MST game, suppliers are substitutes [2]. Their result along with Lemma 4 shows that the singleton inequalities of LP2 are an optimal basis. Thus, all our combinatorial algorithm must calculate is the increase in the MST cost associated with the removal of each edge in T.

We give a modified Kruskal Algorithm which can be used to compute the optimal value of LP2. The modifications are as follows. Throughout the algorithm's execution we will keep an auxiliary set of edges, $\mathcal{A}$, which is initially empty. When edge e is added to the minimum spanning forest, also add e to the set $\mathcal{A}$. Suppose edge e is rejected from addition to the minimum spanning forest because it creates a cycle. Let the cycle created be $H = (\mathcal{V}', \mathcal{E}')$. For each edge $a \in \mathcal{E}' - \{e\}$, if $a \in \mathcal{A}$, label a with $w(e) - w(a)$ and remove a from $\mathcal{A}$. The labels computed by the algorithm are the required increases in the MST cost.

References

1. Archer, A., Tardos, E.: Truthful mechanisms for one-parameter agents. In: Proceedings of the 42nd Annual IEEE Symposium on Foundations of Computer Science, pp. 482–491 (October 2001)
2. Bikhchandani, S., de Vries, S., Schummer, J., Vohra, R.: Linear programming and vickrey auctions. In: Mathematics of the Internet: E-Auction and Markets, pp. 75–115. Springer, New York (2002)
3. Bikhchandani, S., Ostroy, J.: The package assignment model. Journal of Economic Theory 107, 377–406 (2002)
4. Caprara, A., Letchford, A.N.: Computing good allocations for combinatorial optimization games (2006), Available from `http://www.lancs.ac.uk/staff/letchfoa/pubs.htm`
5. Deng, X., Ibaraki, T., Nagamochi, H.: Algorithmic aspects of the core of combinatorial optimization games. Mathematics of Operations Research 24, 751–766 (1999)
6. Deng, X., Papadimitriou, C., Safra, S.: On the complexity of equilibria. In: Proceedings of the 34th annual ACM symposium on Theory of Computing, Montreal, Quebec, pp. 67–71 (2002)
7. Dimitrov, N.B., Plaxton, C.G.: Buyer-supplier games: Core characterization and computation. Technical Report TR–06–19, Department of Computer Science, University of Texas at Austin (April 2006)
8. Edgeworth, F.Y.: Mathematical psychics, an essay on the application of mathematics to the moral sciences. A. M. Kelley, New York (1961)
9. Faigle, U., Kern, W.: On the core of ordered submodular cost games. Mathematical Programming 87, 483–499 (2000)

10. Garg, R., Kumar, V., Rudra, A., Verma, A.: Coalitional games on graphs: core structure, substitutes and frugality. In: Proceedings of the 4th ACM conference on Electronic Commerce, San Diego, CA, pp. 248–249 (2003)
11. Gillies, D.B.: Some Theorems on n-Person Games. PhD thesis, Princeton University (1953)
12. Goemans, M.X., Skutella, M.: Cooperative facility location games. Journal of Algorithms 50, 194–214 (2004)
13. Karlin, A.R., Kempe, D., Tamir, T.: Beyond VCG: frugality of truthful mechanisms. In: Proceedings of the 46th Annual IEEE Symposium on Foundations of Computer Science, pp. 615–624 (October 2005)
14. Nemhauser, G.L., Wolsey, L.A.: Integer and Combinatorial Optimization. John Wiley and Sons, New York (1988)
15. Pál, M., Tardos, E.: Group strategy proof mechanisms via primal-dual algorithms. In: Proceedings of the 44th Annual IEEE Symposium on Foundations of Computer Science, pp. 584–593 (October 2003)
16. Papadimitriou, C.H.: Algorithms, games, and the internet. In: Proceedings of the 33rd Annual ACM Symposium on Theory of Computing, pp. 749–753 (July 2001)
17. Papadimitriou, C.H., Roughgarden, T.: Computing equilibria in multi-player games. In: Proceedings of the 16th Annual ACM-SIAM symposium on Discrete algorithms, pp. 82–91(January 2005)
18. Roughgarden, T.: Selfish Routing and the Price of Anarchy. MIT Press, Cambridge (2005)
19. Roughgarden, T., Tardos, E.: How bad is selfish routing? J. ACM 49, 236–259 (2002)
20. Roughgarden, T., Tardos, E.: Bounding the inefficiency of equilibria in nonatomic congestion games. Games and Economic Behaviour 47, 389–403 (2004)
21. Shapley, L.S.: Notes on the n-person game III: Some variants of the von Neumann-Morgenstern definition of solution. In: Research memorandum, RAND Corporation, Santa Monica, CA (1952)
22. Shapley, L.S., Shubik, M.: The assignment game i: The core. International Journal of Game Theory 1, 111–130 (1972)
23. Shubik, M.: Game Theory In The Social Sciences. MIT Press, Cambridge (1984)
24. von Neumann, J., Morgenstern, O.: Theory of Games and Economic Behavior. Princeton University Press, Princeton (1953)

Very Large-Scale Neighborhoods with Performance Guarantees for Minimizing Makespan on Parallel Machines*

Tobias Brueggemann[1], Johann L. Hurink[1], Tjark Vredeveld[2],
and Gerhard J. Woeginger[3]

[1] Dept. of Applied Mathematics, University of Twente, P.O. Box 217, 7500 AE
Enschede, The Netherlands
j.l.hurink@utwente.nl
[2] Dept. of Quantitative Economics, Maastricht University, P.O. Box 616, 6200 MD
Maastricht, The Netherlands
t.vredeveld@ke.unimaas.nl
[3] Dept. of Mathematics and Computer Science, Eindhoven University of Technology,
P.O. Box 513, 5600 MB Eindhoven, The Netherlands
gwoegi@win.tue.nl

Abstract. We study the problem of minimizing the makespan on m parallel machines. We introduce a very large-scale neighborhood of exponential size (in the number of machines) that is based on a matching in a complete graph. The idea is to partition for every machine the set of assigned jobs into two sets by some fixed rule and then to reassign these $2m$ parts such that every machine gets exactly two parts. The split neighborhood consists of all possible reassignments of the parts and a best neighbor can be calculated in $\mathcal{O}(m \log m)$ by determining a perfect matching with minimum maximal edge weight.

We examine local optima in the split neighborhood and in combined neighborhoods consisting of the split and other known neighborhoods and derive performance guarantees for these local optima.

1   Introduction

In this paper, we consider the following multiprocessor scheduling problem. Given are n jobs each of which has to be scheduled on one of m identical parallel machines. The time it takes for a job j to be fully processed is denoted by p_j. A machine can process at most one job at a time, and a job may not be preempted. The goal is to schedule the jobs in such a way that the *makespan* is minimized, i.e., we want the last job to complete as early as possible. In the standard notation of [9], this problem is denoted as $P||C_{\max}$.

This problem is known to be strongly NP-hard for m being part of the input [8]. Therefore, we search for approximate solutions. If an algorithm is guaranteed

---

* Supported by the Netherlands Organization for Scientific Research (NWO) grant 613.000.225 (Local Search with Exponential Neighborhoods) and by BSIK grant 03018 (BRICKS: Basic Research in Informatics for Creating the Knowledge Society).

C. Kaklamanis and M. Skutella (Eds.): WAOA 2007, LNCS 4927, pp. 41–54, 2008.

to deliver a solution that has value at most ρ times the optimal solution value, we call it a *ρ-approximation* algorithm; the value ρ is called the *(worst-case) performance guarantee*. A well known approximation algorithm for the problem under consideration is the *LPT-algorithm* due to Graham [10]: starting from an empty schedule, we select the job with longest processing time among the unscheduled jobs and schedule this job on the machine with currently minimal workload. This LPT-algorithm has a performance guarantee of $4/3 - 1/3m$.

Another way to find approximate solutions is through *local search*, see e.g. [1]. These methods iteratively search through the set of feasible solutions. Starting from an initial solution, a local search procedure moves from one feasible solution to a neighboring one until some stopping criteria are met. The choice of a suitable neighborhood function has an important influence on the performance of local search. The simplest form of local search is *iterative improvement*, also called local improvement or, in the case of minimization problems, descent algorithms. This method iteratively chooses a better solution in the neighborhood of the current one, and it stops when no better solution is found. The final solution is called a *local optimum*.

Recently, there has been an increasing interest in the quality of local optima and the time needed to obtain these local optima through iterative improvement. For the parallel machine scheduling problem under consideration, Finn and Horowitz [7] showed that a so-called move-optimal solution is guaranteed to deliver a solution with value no more than $2 - 2/(m + 1)$ times the optimal makespan. Moreover, this bound is tight [12]. Brucker et al. [3] showed that the iterative improvement procedures needs $\mathcal{O}(n^2)$ moves to come to a local optimal solution, and this bound is tight [11]. For performance guarantees of local search methods regarding makespan minimization, we refer to [12,11]. For the objective of minimizing total weighted completion time, Brueggemann et al. [4] give a performance guarantee of $3/2 - 1/2m$ for move-optimal schedules.

Over the last years, very large-scale neighborhoods have received considerable attention. These neighborhoods mostly contain up to an exponential number of solutions, but allow a polynomial exploration. A survey about very large-scale neighborhood techniques is given by Ahuja et al. [2] and Deĭneko and Woeginger [6] present an overview of very large-scale neighborhoods for the traveling salesman and quadratic assignment problem.

In Section 2, we define a very large-scale neighborhood, the so-called *split-neighborhood*, and in the following sections we investigate its worst-case behavior. In Section 3, we show that a split-optimal solution has the same performance guarantee as a simple move-optimal solution. In Sections 4 and 5, we give performance guarantees on combined move-optimal and split-optimal solutions. If we combine the two neighborhoods in the most straightforward way, we show that the performance guarantee marginally improves but is still essentially 2, whereas a better combination leads to a performance guarantee of $3/2$.

2   Neighborhoods

As mentioned in the introduction, an important part of local search algorithms is the definition of the neighborhood on which the method operates. Before discussing the neighborhoods, we first describe the used representation of a schedule. As the sequence in which the jobs are processed does not influence the makespan of a schedule for a given assignment, we represent a schedule by such an assignment of jobs to machines, $A : J \to \{1, \ldots, m\}$, where $J = \{1, \ldots, n\}$ denotes the set of jobs. Each assignment leads to a partition of the set of jobs into m disjoint subsets $M_1^A, \ldots, M_m^A$, where $M_i^A = \{j \in J : A(j) = i\}$ is the set of jobs scheduled on machine i. Abusing terminology, we use "schedule A" for any schedule complying to the assignment A. If there is no ambiguity, we write M_i for M_i^A. The *workload* of machine i is denoted by

$$L_i^A = \sum_{j \in M_i} p_j,$$

and this workload is equal to the completion time of the last job scheduled on machine i. Again, if there is no ambiguity, we write L_i for L_i^A. Hence, for a given assignment A of jobs to machines, the makespan is equal to the machine with maximum workload:

$$C_{\max}^A = \max_i L_i^A.$$

We call such a machine with maximum workload *a critical machine*.

The move-neighborhood. Probably the most basic neighborhood is the *move-neighborhood*. Given a schedule A, we select a job j, scheduled on machine h, and a machine $i \neq h$. The move neighbor, A', is obtained by moving job j to machine i, i.e., $M_h^{A'} = M^A \setminus \{j\}$, $M_i^{A'} = M^A \cup \{j\}$, and $M_k^{A'} = M^A$ for $k \neq h, i$. The set of all move neighbors of schedule A is denoted by $\mathcal{N}_{\text{move}}(A) = \{A' : A' \text{ is a move neighbor of } A\}$. We call an assignment A *move-optimal* if for all move neighbors A', $C_{\max}^A \leq C_{\max}^{A'}$ and, in case of $C_{\max}^A = C_{\max}^{A'}$, the number of critical machines in A is at most the number of critical machines in A'. Finn and Horowitz [7] gave the following upper bound on the performance guarantee of move-optimal assignments.

Theorem 1 ([7]). *Let A be a move-optimal assignment, and let $C_{\max}^*$ denote the optimal makespan. Moreover, let $n_k = \max\{ |M_i| : L_i = C_{\max}^A \}$ denote the maximum number of jobs on a critical machine in the assignment A. Then*

$$C_{\max}^A \leq \frac{n_k m}{(n_k - 1)m + 1} C_{\max}^*.$$

Moreover, if $n_k = 1$, then $C_{\max}^A = C_{\max}^*$.

The bound in Theorem 1 attains its maximum for $n_k = 2$, yielding a performance guarantee of $2 - 2/(m + 1)$. This bound has been proven tight by Schuurman and Vredeveld [12], see Figure 1 for the assignments attaining this bound.

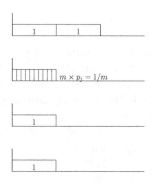

Fig. 1. Worst-case move-optimal schedule

The split-neighborhood. The *split-neighborhood* is of exponential size in the number of machines. The basis of this neighborhood is a *split-operator* that partitions the set of jobs assigned to a machine i into two disjoint sets, i.e., $\mathsf{split}(M_i) = (M_{i1}, M_{i2})$, where M_i denotes the set of jobs scheduled on machine i and (M_{i1}, M_{i2}) is a partition of the set M_i into two disjoint subsets. We assume w.l.o.g. that $L_{i1} = \sum_{j \in M_{i1}} p_j \geq \sum_{j \in M_{i2}} p_j = L_{i2}$. We refer to the sets M_{i1} and M_{i2} of a machine i (or a set M_i) as the *left* and *right part*, respectively.

If we use the split-operator on all sets M_i given by an assignment A, we obtain $2m$ parts M_{i_1} and M_{i_2} for $i = 1, \ldots, m$. Abusing notation, we denote the set of these $2m$ parts by

$$\mathsf{split}(A) = \big\{ M_{i1}, M_{i2} : \mathsf{split}(M_i) = (M_{i1}, M_{i2}) \text{ for } i = 1, \ldots, m \big\}.$$

We call an assignment A' a *split-neighbor* of A, if A' can be obtained by assigning the jobs of exactly two of the $2m$ parts from *split*(A) to each machine. The neighborhood $\mathcal{N}_{split}$ of an assignment A is denoted by

$$\mathcal{N}_{\mathrm{split}}(A) := \{ A' : A' \text{ is neighbor of } A \}.$$

Although the size of the neighborhood is exponentially large in the number of machines, the following fact tells us that the best neighbor, one with lowest makespan and fewest number of critical machines among all neighbors with lowest makespan, can be found in $\mathcal{O}(m \log m)$ time.

Fact 1. *Given $2m$ numbers $a_1 \geq \ldots \geq a_{2m}$. A perfect matching of these numbers such that the maximum of the sum of two matched numbers is minimized, is obtained by matching a_i to a_{2m+1-i}. Moreover, this matching minimizes the number of matched pairs whose sum equals this maximum.*

In other words, an optimal solution for the bottleneck assignment problem is obtained by ordering the cost-matrix of the assignment problem, so that it fulfills the bottleneck Monge property. Thus, we obtain the best neighbor of the neighborhood $\mathcal{N}_{split}(A)$ by first rearranging the $2m$ parts $T_1, \ldots, T_{2m}$ in $\mathsf{split}(A)$, so that for the sum of processing times of the parts holds $L_{T_1} \geq \ldots \geq L_{T_{2m}}$ and, then, assigning the jobs of the parts T_i and T_{2m+1-i} to machine i for $i = 1, \ldots, 2m$.

We call an assignment *split-optimal* if for all $A' \in \mathcal{N}_{\text{split}}(A)$, $C_{\text{max}}^A \leq C_{\text{max}}^{A'}$ and in case $C_{\text{max}}^A = C_{\text{max}}^{A'}$ the number of critical machines in A is at most the number of critical machines in A'. Of course, the quality of a split-optimal solution depends on the split-operator. For most of the presented results, we only assume that the split-operator produces a move-optimal partition. That is, for any job $j \in M_{i1}$, we have that $L_{i2} + p_j \geq L_{i1}$. Such a split-operator we call a move-optimal split-operator.

Combinations of move and split-neighborhood. As we will see in the following section, a split-optimal assignment needs not to be move-optimal. Hence, we also consider assignments that are both move- and split-optimal. These local optima may however be improved by first applying a move-operator leading to a schedule with the same makespan and, then applying a split-operator leading to a better schedule. Therefore, we define a *lexicographic-move-optimal* assignment. For a given assignment A and $A' \in \mathcal{N}_{\text{move}}(A)$, we reorder the machines in A and A' so that

$$L_1^A \geq \ldots \geq L_m^A \quad \text{and}$$
$$L_1^{A'} \geq \ldots \geq L_m^{A'}.$$

The assignment A' is called *lexicographically better* than A, if there exists a machine k such that

$$\begin{aligned} L_i^{A'} &= L_i^A \quad \text{for } i = 1, \ldots, k - 1, \\ L_k^{A'} &< L_k^A. \end{aligned} \tag{1}$$

We say that A is *lexicographic-move-optimal*, or *lexmove-optimal*, if there exists no move neighbor $A' \in \mathcal{N}_{\text{move}}(A)$ that is lexicographically better than A.

Note that the move-optimal assignment A in Figure 1 is also lexmove-optimal. Therefore, the move-optimal and the lexmove-optimal assignments have the same performance guarantee. As will be seen in the following, the performance guarantee of an assignment that is both, lexmove-optimal and split-optimal, is better than that of a move- and split-optimal assignment.

3   Performance Guarantee on Split-Optimal Assignments

Recall that we assume that a split-operator on a set M_i produces two parts M_{i1} and M_{i2} satisfying $L_{i1} \geq L_{i2}$. Moreover, for a critical machine k in a split-optimal schedule, we have the following property for any machine i:

$$\begin{aligned} &\text{if } L_{i1} \geq L_{k1} \text{ then } L_{i2} \leq L_{k2} \text{ holds}, \\ &\text{if } L_{i1} < L_{k1} \text{ then } L_{i2} \geq L_{k2} \text{ holds}. \end{aligned} \tag{2}$$

The first statement follows from the fact that k is a critical machine, and the second statement holds since A is split-optimal.

The performance guarantee of a split-optimal assignment, using a move-optimal split-operator, does not improve on the bound obtained by move-optimal assignment.

Theorem 2. *Let A be a split-optimal assignment using a move-optimal split-operator. Then the makespan of A is bounded by $C_{\max}^A \leq (2 - \frac{2}{m+1})C_{\max}^*$, where $C_{\max}^*$ denotes the value of the optimal makespan.*

Proof. W.l.o.g. we assume that $C_{\max}^A = 1$. Let k be a critical machine, i.e. $L_k = 1$. If $\sum_j p_j \geq mL_{k1} + L_{k2}$, then the optimal makespan can be bounded from below by $C_{\max}^* \geq \frac{1}{m}\sum_j p_j \geq L_{k1} + L_{k2}/m$. Using the fact that $L_{k1} + L_{k2} = 1$, we have

$$\frac{C_{\max}^A}{C_{\max}^*} \leq \frac{m}{(m-1)L_{k1}+1} \leq \frac{2m}{m+1} = 2 - \frac{2}{m+1},$$

as the above expression is maximized for minimal L_{k1} and by $L_{k1} \geq L_{k2}$ we know that $L_{k1} \geq 1/2$.

On the other hand, if $\sum_j p_j < mL_{k1} + L_{k2}$, then a machine l with minimal load satisfies

$$L_l \leq \sum_{i \neq k} L_i/(m-1) < L_{k1}. \tag{3}$$

Moreover, by (2), we know that (3) implies $L_{l2} \geq L_{k2}$. Hence,

$$L_{k1} > L_l \geq 2L_{l2} \geq 2L_{k2}. \tag{4}$$

From the fact that a move-optimal split-operator is used to obtain the sets M_{k1} and M_{k2}, we know that for all $j \in M_{k1}$, $L_{k2} + p_j \geq L_{k1}$. Therefore, from (4) it follows that $p_j > \frac{1}{2}L_{k1}$ for all $j \in M_{k1}$. Hence, M_{k1} contains only one job and

$$C_{\max}^A = L_k = L_{k1} + L_{k2} < \frac{3}{2}L_{k1} \leq \frac{3}{2}C_{\max}^*,$$

as $C_{\max}^* \geq p_j$ for all $j \in J$ and thus $C_{\max}^* \geq L_{k1}$.

For $m \geq 3$, the theorem is proven, as $\frac{3}{2} \leq 2 - 2/(m+1)$. For $m = 2$, it follows from (4) that

$$C_{\max}^* \geq \frac{1}{2}\sum_j p_j \geq \frac{1}{2}(L_k + 2L_{l2}) \geq \frac{1}{2}(L_{k1} + 3L_{k2}) = \frac{3}{2} - L_{k1},$$

where the last equality follows from the fact that $L_{k2} = 1 - L_{k1}$. Moreover, as $C_{\max}^* \geq L_{k1}$ due to the fact that M_{k1} contains only one job, we have

$$C_{\max}^* \geq \max\{L_{k1}, \frac{3}{2} - L_{k1}\},$$

which is minimal for $L_{k1} = \frac{3}{4}$. Therefore, for $m = 2$, we have

$$C_{\max}^A \leq \frac{4}{3}C_{\max}^*. \qquad \square$$

To show that the analysis is tight, consider the following instance consisting of m jobs with processing time 1 and m jobs with processing time $1/m$. In the split-optimal assignment A, we schedule on every machine one job with processing time 1 and on the first machine all jobs with processing time $1/m$ are scheduled. It is easy to check that this assignment is split-optimal for a move-optimal split-operator and it has makespan $C_{max}^A = 2$. In an optimal assignment A^*, we schedule on every machine one job with processing time 1 and one with processing time $1/m$. The optimal makespan is $C_{max}^* = 1 + 1/m$, and thus $C_{max}^A = 2m/(m+1)C_{max}^*$. See Figure 2 for an illustration.

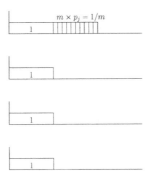

Fig. 2. A split-optimal assignment

4   Split-Optimal and Move-Optimal Assignments

The worst-case instance for split-optimal assignments, showing the tightness of the analysis in the previous section, is obviously not move-optimal. This raises the question whether a combination of the two neighborhoods gives a better performance guarantee, which is answered in Theorem 3, for move-optimal split-operators.

Lemma 1. *Let A be a move-optimal and split-optimal assignment using a move-optimal split-operator. If there exists a critical machine k such that*

$$\sum_j p_j < mL_{k1} + L_{k2},$$

then $C_{max}^A = C_{max}^*$, where C_{max}^* denotes the optimal makespan.

Proof. Let l be a machine with minimal load. Then, we know by (3) that $L_l < L_{k1}$. By move-optimality of the assignment A, we know that for any job $j \in M_k$

$$L_l + p_j \geq L_k = L_{k1} + L_{k2}.$$

Hence, $p_j > L_{k2}$ for $j \in M_k$, and thus M_{k2} contains no job at all, i.e., $L_{k2} = 0$. It follows from the move-optimal split-operator, that whenever M_{k2} is empty, M_{k1} contains only one job, j_1. Hence, $C_{max}^A = L_k = p_{j_1} \leq C_{max}^*$. $\square$

From this lemma, it follows that we only have to consider cases in which the total load on all machines is large enough. Moreover, if L_{k1} is large enough, we can actually prove a bound on the makespan of this local optimal assignment, which is better than the guarantee in Theorem 3.

Lemma 2. *Let A be a move-optimal and split-optimal assignment obtained by using a move-optimal split-operator. If there exists a critical machine k, satisfying $L_{k1} \geq \frac{2}{3}L_k$, then the makespan of A can be bounded by*

$$C^A_{\max} \leq \frac{3m}{2m+1}C^*_{\max},$$

where $C^*_{\max}$ denotes the optimal makespan.

Proof. By Lemma 1, we only have to consider the case that $\sum_j p_j \geq mL_{k1}+L_{k2}$. Hence, the optimal makespan can be bounded from below by $C^*_{\max} \geq \frac{1}{m}\sum_j p_j \geq \frac{(m-1)L_{k1}+L_k}{m}$. As, $C^A_{\max} = L_k$, we thus have

$$\frac{C^A_{\max}}{C^*_{\max}} \leq \frac{m}{(m-1)L_{k1}+L_k} \leq \frac{3m}{2m+1},$$

where the last inequality is due to $L_{k1} \geq 2/3L_k$. $\square$

Let k be a critical machine. Before we prove the performance guarantee on a move- and split-optimal assignment A, we first partition the set of machines into several classes.

$$\begin{aligned}
S_< &= \{\, i : L_{i1} < L_{k1}\,\}, \\
S_\geq &= \{\, i : L_{i1} \geq L_{k1}\,\}, \\
S_{\text{multi}} &= \{\, i \in S_\geq : |M_{i1}| \geq 2\,\}, \\
S_{\text{single}} &= S_\geq \setminus (S_{\text{multi}} \cup \{\,k\,\}).
\end{aligned} \tag{5}$$

That is, $S_<$ is the set of machines that have a left part which is smaller than L_{k1}. This set of remaining machines is again partitioned in one set containing all machines that have at least two jobs in the left part and the remaining machines in $S_\geq \setminus \{k\}$ containing exactly one job in the left part. Note that, $S_\geq \setminus \{k\} = S_{\text{multi}} \cup S_{\text{single}}$.

The load of a machine in each of the above classes can be bounded as follows.

Lemma 3. *Let A be a move-optimal and split-optimal schedule, for a move-optimal split-operator and let k be a critical machine in this assignment. Moreover, let $S_<$ and S_{multi} be as defined in (5). Then:*

$$\begin{aligned}
L_i &\geq 2(C^A_{\max} - L_{k1}) &&\text{for}\quad i \in S_<, \\
L_i &\geq \tfrac{3}{2}L_{k1} &&\text{for}\quad i \in S_{\text{multi}}.
\end{aligned}$$

Proof. Consider a machine $i \in S_<$. Then by property (2), we know that $L_{i1} < L_{k1}$ implies that $L_{i2} \geq L_{k2}$. Moreover, as $L_{i1} \geq L_{i2}$, we have that $L_i \geq 2L_{i2} \geq 2L_{k2} = 2(1 - L_{k1})$.

For $i \in S_{\text{multi}}$ let $j_s \in M_{i1}$ be the smallest job in the left part of machine i. As M_{i1} contains at least two jobs, we know that $p_{j_s} \leq \frac{1}{2}L_{i1}$. Due to the move-optimality of the split-operator, we also know that $L_{i2} \geq L_{i1} - p_{j_s} \geq \frac{1}{2}L_{i1}$. Hence, $L_i \geq \frac{3}{2}L_{i1} \geq \frac{3}{2}L_{k1}$. $\qquad\square$

Lemma 4. *Let A be a move-optimal and split-optimal assignment for a move-optimal split-operator and k be a critical machine in A. Moreover, let $S_<$ be as defined in (5). If $L_{k1} \leq 2/3L_k$ and $|S_<| \geq 1$, then*

$$\frac{C^A_{\max}}{C^*_{\max}} \leq \begin{cases} \frac{2m}{m+2} & \text{for } m \geq 4, \\ \frac{3m}{2m+1} & \text{for } m \leq 3, \end{cases}$$

where $C^*_{\max}$ denotes the optimal makespan.

Proof. Using Lemma 3, we can bound the optimal makespan by

$$mC^*_{\max} \geq \sum_j p_j \geq C^A_{\max} + 2|S_<|(C^A_{\max} - L_{k1}) + (m - 1 - |S_<|)L_{k1}$$

$$\geq C^A_{\max} + (2C^A_{\max} - 3L_{k1})|S_<| + (m-1)L_{k1} \geq 3C^A_{\max} + (m-4)L_{k1}, \tag{6}$$

where the last inequality is due to $L_{k1} \leq 2/3L_k = 2/3C^A_{\max}$. For $m \geq 4$, the expression in (6) is minimized for L_{k1} minimal, whereas for $m \leq 3$, it is minimized for L_{k1} maximal. Using the fact that $1/2C^A_{\max} \leq L_{k1} \leq 2/3C^A_{\max}$, we have

$$\frac{C^A_{\max}}{C^*_{\max}} \leq \begin{cases} \frac{2m}{m+2} & \text{for } m \geq 4, \\ \frac{3m}{2m+1} & \text{for } m \leq 3. \end{cases} \qquad\square$$

Lemma 5. *Let A be a move-optimal and split-optimal assignment for a move-optimal split-operator and k be a critical machine in A. Moreover, let S_{multi} be as defined in (5). If $L_{k1} \leq 2/3L_k$ and $|S_{multi}| \geq 2$, then*

$$\frac{C^A_{\max}}{C^*_{\max}} \leq \frac{2m}{m+2},$$

where $C^*_{\max}$ denotes the optimal makespan.

Proof. Consider a move-optimal and split-optimal assignment A for a move-optimal split-operator and let $S_<$, S_{multi}, and S_{single} be as defined in (5). For $L_{k1} \leq 2/3L_k$, we know from Lemma 3 that for $i \in S_<$, $L_i \geq L_{k1}$. Hence, using Lemma 3, we can bound the optimal makespan by

$$C^*_{\max} \geq \frac{C^A_{\max} + (|S_{multi}|/2 + m - 1)L_{k1}}{m} \geq \frac{C^A_{\max} + mL_{k1}}{m} \geq \frac{2+m}{2m}C^A_{\max},$$

where the second inequality is due to $|S_{multi}| \geq 2$ and the last to $L_{k1} \geq C^A_{\max}/2$. $\qquad\square$

Theorem 3. *A move-optimal and split-optimal assignment, obtained by a move-optimal split-operator, has a performance guarantee of $2 - \frac{4}{m+3}$.*

Proof. Let A be a move-optimal and split-optimal assignment for a move-optimal split-operator and let k be a critical machine in A. Assume that $C^A_{\max} = 1$. Since $\frac{3m}{2m+1} \leq 2 - \frac{4}{m+3}$, by Theorem 1 we only need to consider assignments A in which a critical machine k contains exactly two jobs. Moreover, by Lemma 2, we may restrict ourselves to the case that $L_{k1} \in [\frac{1}{2}, \frac{2}{3}]$, and by Lemma 1, we may assume that $\sum_j p_j \geq mL_{k1} + L_{k2}$.

As $\max\{\frac{2m}{m+2}, \frac{3m}{2m+1}\} \leq 2 - \frac{4}{m+3}$, due to Lemma 4 we can restrict ourselves to the case that there is no machine i with $L_{i1} < L_{k1}$. Due to Lemma 5 we assume that there is at most one machine i with $|M_{i1}| \geq 2$. Note that if no such machine exists, there are m jobs of length at least L_{k1} and one job of length $1 - L_{k1} \leq L_{k1}$. Then, by the pigeonhole principle $C^*_{\max} = C^A_{\max}$. Hence, we assume that there is exactly one machine s with $|M_{s1}| \geq 2$.

Let j_1 be the smallest job in M_{s1}. If $p_{j_1} \geq \frac{m+3}{2m+2} - L_{k2}$, then there are $m - 1$ jobs of length L_{k1}, one job of length $L_{k2} = 1 - L_{k1} \leq L_{k1}$ and at least two jobs of length $\frac{m+3}{2m+2} - L_{k2}$, and by the pigeonhole principle, we know that $C^*_{\max} \geq \frac{m+3}{2m+2}$.

On the other hand, if $p_{j_1} \leq \frac{m+3}{2m+2} - (1 - L_{k1})$, we can bound the load of the right part of machine s by $L_{s2} \geq L_{s1} - p_{j_1}$ Hence, using the fact that $L_{k1} \geq 1/2$, we can bound the total workload by

$$\sum_j p_j \geq L_s + \sum_{i \neq s} L_i \geq (m-1)L_{k1} + 1 - L_{k1} + L_{s1} + L_{s2}$$

$$\geq (m-2)L_{k1} + 1 + 2L_{s1} - p_{j_1} \geq mL_{k1} + 1 - p_{j_1}$$

$$\geq (m-1)L_{k1} + 2 - \frac{m+3}{2m+2} \geq \frac{m-1}{2} + 2 - \frac{m+3}{2m+2}$$

$$= \frac{m^2 + 3m}{2m+2}.$$

This implies that the optimal makespan can be bounded by

$$C^*_{\max} \geq \frac{1}{m} \sum_j p_j \geq \frac{m+3}{2m+2},$$

and thus we obtain

$$\frac{C^A_{\max}}{C^*_{\max}} \leq \frac{2m+2}{m+3}. \qquad \square$$

For instances with an odd number of machines, the analysis of the previous theorem is tight. If we schedule m jobs of length 1 and m jobs of length $2/(m+1)$ as illustrated by the assignment A in Figure 3, we obtain a move-optimal and split-optimal assignment for a move-optimal split-operator, with makespan $C^A_{\max} = 2$. In the optimal schedule, all machines have the same workload and $C^*_{\max} = 1 + \frac{2}{m+1}$. For the split-optimality of this example, we need that the left part of machine M_2 has workload equal to 1. Therefore, this example only works for an odd number of machines. For even number of machines, a lower bound on the performance guarantee is $\frac{2m}{m+2}$. This bound is obtained by an instance with m jobs of size 1 and $m - 1$ jobs of size $2/m$. In the move-optimal and split-optimal assignment, these jobs are scheduled simular as in Figure 3.

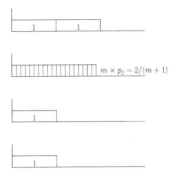

Fig. 3. A split and move-optimal assignment for odd number of machines

5   Split-Optimal and Lexicographic-Move-Optimal Assignments

In the previous section, we have seen that the performance guarantee of a move-optimal and split-optimal assignment marginally improves on the performance guarantee of only a move-optimal or only a split-optimal assignment. Moreover, the example, showing the tightness of the guarantee for an odd number of machines, is not lexicographic-move-optimal. Therefore, in this section we consider the lexmove-optimal and split-optimal assignments.

For lexmove-optimal assignments, we have the following fact.

Fact 2. *Let A be an assignment of the jobs to the machines and let l be a machine with minimal workload. A schedule represented by A is lexmove-optimal if and only if, for all machines i and all jobs $j \in M_i$ it holds*

$$L_l + p_j \geq L_i.$$

In this section, we only consider the LPT-algorithm as the split-operator. Remember that the LPT-algorithm sorts the jobs in non-increasing size and then iteratively assigns a job to the set with minimal workload. In this way, we obtain a partition $\mathsf{LPT}(M_i) = (M_{i1}, M_{i2})$ that is move-optimal. Therefore, we can apply Lemma 1–5.

Lemma 6. *Let A be a lexmove-optimal and split-optimal assignment for a move-optimal split-operator. Let k be a critical machine and l be a machine with minimal load. Moreover, let $C^*_{\max}$ denote the optimal makespan. Then, if $l \in S_< \cup S_{multi}$,*

$$\frac{C^A_{\max}}{C^*_{\max}} \leq \frac{3m}{2m+1}.$$

Proof. Assume w.l.o.g. that $C^A_{\max} = 1$. By Lemmas 1 and 2, we can restrict ourselves to the case that $\sum_j p_j \geq mL_{k1} + L_{k2}$ and $L_{k1} \leq 2/3$.

If $l \in S_<$, we know from Lemma 3 that $L_l \geq 2(1 - L_{k1}) \geq 2/3$ and if $l \in S_{\text{multi}}$, it follows from Lemma 3 that $L_l \geq \frac{3}{2}L_{k1} \geq 3/4 \geq 2/3$. Hence, we have

$$C_{\max}^* \geq \frac{1}{m}\left(1 + \frac{2(m-1)}{3}\right) = \frac{2m+1}{3m}. \qquad \square$$

By this lemma, we know that in order to prove the performance guarantee of $3/2$ in Theorem 4, we can restrict ourselves to local optimal schedules with $l \in S_{\text{single}}$. Moreover, as $L_l \geq 2/3C_{\max}^A$ implies that $C_{\max}^* \geq 2/3C_{\max}^A$, we assume in the remainder of this section that $L_l < 2/3C_{\max}^A$.

In the proof of Theorem 4, we use the concept of *blocking jobs*.

Definition 1. *We call a job j a blocking job, if $p_j + L_{k1} \geq 2/3C_{\max}^A$, where L_{k1} is the load of the left part of a critical machine.*

Note that if, in some schedule, a blocking job is assigned to the same machine as a job of size at least L_{k1}, then the makespan of this schedule will be at least $2/3C_{\max}^A$. The idea of the proof of the following theorem is that the total volume of blocking jobs is large enough so that if no blocking job is assigned on the same machine as a job of size at least L_{k1}, then the makespan of the schedule is also at least $2/3C_{\max}^A$.

Theorem 4. *Let A be a lexmove-optimal and split-optimal assignment using the LPT-algorithm as split-operator. Then,*

$$C_{\max}^A \leq \frac{3}{2}C_{\max}^*,$$

where, $C_{\max}^*$ denotes the optimal makespan.

Proof. Let A be a lexmove-optimal and split-optimal assignment and let k be a critical machine and l a machine with minimal load. By Theorem 1 we may assume w.l.o.g. that $|M_k| = 2$. Moreover, by Lemma 1 and 2, we restrict ourselves to the case that $\sum_j p_j \geq mL_{k1} + L_{k2}$ and $L_{k1} \leq 2/3L_k$. Finally, we define the sets $S_<$, S_{multi}, and S_{single} as in (5). Then, by Lemma 6 we assume that $l \in S_{\text{single}}$ and $L_l < 2/3C_{\max}^A$.

Under these assumptions, we claim that for a machine $i \in S_< \cup S_{\text{multi}}$ the sum of processing times of blocking jobs, scheduled on this machines is at least $2/3C_{\max}^A$. None of the blocking jobs, which A assigns to a machine in $S_< \cup S_{\text{multi}}$, can be scheduled together with a job of size at least L_{k1} in a schedule with makespan smaller than $2/3C_{\max}^A$. Thus, in such a schedule all these jobs need to be distributed over $|S_< \cup S_{\text{multi}}|$ machines, as each machine in $S_{\text{single}} \cup \{k\}$ processes at least one job with processing time at least L_{k1}. From our claim, it now follows that in every feasible schedule the blocking jobs lead on at least one machine to a workload of at least $2/3C_{\max}^A$. Hence, we always have $C_{\max}^* \geq 2/3C_{\max}^A$, and the theorem is proven.

To prove our claim, first consider a machine $i \in S_<$. From Lemma 3 and lexmove-optimality of A, it follows that a job $j \in M_i$ has processing time $p_j \geq$

$L_i - L_l \geq 4/3C_{\max}^A - 2L_{k1}$. Hence, $p_j + L_{k1} \geq 4/3C_{\max}^A - L_{k1} \geq 2/3C_{\max}^A$, as $L_{k1} \leq 2/3L_k = 2/3C_{\max}^A$, and j is a blocking job. As each job $j \in M_i$ is a blocking job, the total load of blocking jobs scheduled on machine $i \in S_<$ is $L_i \geq 2(C_{\max}^A - L_{k1}) \geq 2/3C_{\max}^A$.

Now, consider a machine $i \in S_{\text{multi}}$, with $|M_{i1}| \geq 3$. The smallest job in the left part, say $j_0 \in M_{i1}$ has length at most $p_{j_0} \leq L_{i1}/|M_{i1}|$. By move-optimality of the split-operator, we know that the load of the right part can be bounded by

$$L_{i2} \geq L_{i1} - p_{j_0} \geq \frac{|M_{i1}| - 1}{|M_{i1}|}L_{i1} \geq \frac{2}{3}L_{i1}.$$

Hence, by lexmove-optimality of the assignment, we know that any job $j \in M_i$ has processing time $p_j \geq L_i - L_l \geq \frac{5}{3}L_{i1} - 2/3C_{\max}^A$. Thus $p_j + L_{k1} \geq \frac{8}{3}L_{k1} - 2/3C_{\max}^A \geq 2/3C_{\max}^A$. Hence, each job $j \in \{i \in S_{\text{multi}} : |M_{i1}| \geq 3\}$ is a blocking job, and the total processing times of the blocking jobs assigned to such a machine i is $L_i \geq 2/3C_{\max}^A$.

Finally, consider a machine $i \in S_{\text{multi}}$, with $|M_{i1}| = 2$, say $M_{i1} = \{j_1, j_2\}$ with $p_{j_1} \geq p_{j_2}$. By move-optimality of the split-operator, we know that $L_{i2} \geq p_{j_1}$, and by lexmove-optimality of the assignment A, we also know that $L_l \geq L_i - p_{j_2} = L_{i2} + p_{j_1} \geq 2p_{j_1}$. Hence, $p_{j_1} \leq L_l/2 \leq 1/3C_{\max}^A$.

This implies that $p_{j_2} = L_{i1} - p_{j_1} \geq L_{k1} - 1/3C_{\max}^A \geq 1/6C_{\max}^A$ and p_{j_2} is a blocking job, as $L_{k1} + 1/6C_{\max}^A \geq 2/3C_{\max}^A$. Moreover, due to the fact that the LPT-algorithm is used as a split-operator, we know that there exists at least one job $j \in M_{i2}$ in the right part of machine i with $p_j \geq p_{j_2}$. Hence, M_i contains at least three blocking jobs, j_1, j_2, and j_3, and the total processing time of these three jobs is at least

$$p_{j_1} + p_{j_2} + p_{j_3} \geq L_{k1} + 1/6C_{\max}^A \geq 2/3C_{\max}^A,$$

which completes the proof. $\square$

A lower bound on the performance guarantee is given in Figure 4. Let $\delta = \frac{1}{3m-4}$ and consider the instance consisting of $2m - 2$ jobs with processing time 3δ, one

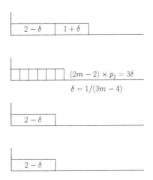

Fig. 4. A lexmove- and split-optimal schedule A

job of size $1 + \delta$ and $m - 1$ jobs of length $2 - \delta$. The assignment A as depicted in Figure 4 is lexmove-optimal and split-optimal and has makespan $C^A_{\max} = 3$, whereas the optimal makespan is $C^*_{\max} = 2 + 2\delta$. This yields a ratio of The ratio between the lexmove- and split-optimal schedule depicted in this figure and the optimal makespan is $\frac{C^A_{\max}}{C^*_{\max}} = \frac{3m-4}{2m-2} = \frac{3}{2} - \frac{1}{2m-2}$.

References

1. Aarts, E.H.L., Lenstra, J.K. (eds.): Local search in combinatorial optimization. John Wiley & Sons, Chichester (1997)
2. Ahuja, R.K., Özlem, E., Orlin, J.B., Punnen, A.P.: A survey of very large-scale neighborhood search techniques. Discrete Applied Mathematics 123, 75–102 (2002)
3. Brucker, P., Hurink, J.L., Werner, F.: Improving local search heuristics for some scheduling problems II. Discrete Applied Mathematics 72, 47–69 (1997)
4. Brueggemann, T., Hurink, J.L., Kern, W.: Quality of move-optimal schedules for minimizing total weighted completion time. Operations Research Letters 34, 583–590 (2006)
5. Brueggemann, T., Hurink, J.L., Vredeveld, T., Woeginger, G.J.: Very large-scale neighborhoods with performance guarantees for minimizing makespan on parallel machines, Tech. Report No. 1801, University of Twente, Dep. of Mathematical Sciences (2006)
6. Deĭneko, V.G. Woeginger, G.J.: A study of exponential neighborhoods for the travelling salesman problem and for the quadratic assignment problem. Mathematical Programming, Series A 87, 519–542 (2000)
7. Finn, G., Horowitz, E.: A linear time approximation algorithm for multiprocessor scheduling. BIT 19, 312–320 (1979)
8. Garey, M.R., Johnson, D.S.: Complexity results for multiprocessor scheduling under resource constraints. SIAM Journal on Computing 4, 397–411 (1975)
9. Graham, R.L., Lawler, E.L., Lenstra, J.K., Rinnooy Kan, A.H.G.: Optimization and approximation in deterministic sequencing and scheduling: a survey. Annals of Discrete Mathematics 5, 287–326 (1979)
10. Graham, R.L.: Bounds on multiprocessing anomalies. SIAM Journal on Applied Mathematics 17, 416–429 (1969)
11. Hurkens, C.A.J., Vredeveld, T.: Local search for multiprocessor scheduling: How many moves does it take to a local optimum? Operations Research Letters 31, 137–141 (2003)
12. Schuurman, P., Vredeveld, T.: Performance guarantees of local search for multiprocessor scheduling. Informs Journal on Computing 19, 52–63 (2007)

A 3/2-Approximation for the Proportionate Two-Machine Flow Shop Scheduling with Minimum Delays

Alexander A. Ageev[*]

Sobolev Institute of Mathematics, pr. Koptyuga 4, Novosibirsk, Russia
ageev@math.nsc.ru

Abstract. We study the two-machine flow shop problem with minimum delays. The problem is known to be strongly NP-hard even in the case of unit processing times and to be approximable within a factor of 2 of the length of an optimal schedule in the general case. The question whether there exists a polynomial-time algorithm with a better approximation ratio has been posed by several researchers but still remains open. In this paper we improve the above bound to $\frac{3}{2}$ for the special case of the problem when both operations of each job have equal processing times (this case of flow shop is known as the proportionate flow shop). Our analysis of the algorithm relies upon a nontrivial generalization of the lower bound established by Yu for the case of unit processing times.

1 Introduction

In the two-machine flow shop problem with minimum delays there are two machines available from time zero onwards for processing n independent jobs. Each machine can process at most one job at a time. Each job j consists of two operations; the second operation can start at least l_j time units after the completion of the first operation. The first (second) operation has to be executed by machine 1 (machine 2) and processing the first (second) operation takes time a_j (b_j). All input numbers a_j, b_j, and l_j are assumed to be nonnegative integers. The objective is to minimize the makespan, or the schedule length, that is the maximum job completion time. As in [13], we denote this problem by $F2 \mid l_j \mid C_{\max}$.

The problems with minimum delays arise, in particular, in manufacturing where there may be a transportation time from one production facility to another, and in computer systems where the output of a task on one processor may require a communication time so as to become the input to a subsequent task on another processor.

Related results. The first result related to $F2 \mid l_j \mid C_{\max}$ is due to Johnson [6] who presents an $O(n \log n)$ algorithm for solving the problem without delays. Kern and Nawijn [8] consider a single-machine problem with two operations

---

[*] Supported by the Russian Foundation for Basic Research (project codes 06-01-00255, 05-01-00960).

C. Kaklamanis and M. Skutella (Eds.): WAOA 2007, LNCS 4927, pp. 55–66, 2008.

per jobs and intermediate minimum delays. Following the extension [13] of the three-field notation scheme introduced by Graham et al. [5] we denote this problem by $1 \mid l_j \mid C_{\max}$. Yu et al. [12,13] show that this problem is equivalent to $F2 \mid l_j \mid C_{\max}$. Kern and Nawijn [8] show that $1 \mid l_j \mid C_{\max}$ is NP-hard in the ordinary sense. This result is strengthened to NP-hardness in the strong sense for $F2 \mid l_j \mid C_{\max}$ by Lenstra [9], for $F2 \mid l_j, a_j = b_j \mid C_{\max}$ by Dell'Amico and Vaessens [4], and for $F2 \mid l_j \in \{0, l\}, a_j = b_j \mid C_{\max}$ by Yu [12]. Yu et al. [13] prove that $F2 \mid l_j, a_i = b_j = 1 \mid C_{\max}$ is NP-hard in the strong sense. Dell'Amico [3] presents several 2-approximation algorithms with running time $O(n \log n)$ where n is the number of jobs. The question whether there exists a polynomial-time algorithm with a better approximation ratio has been posed by several researchers (see, for example, Strusevich [11]) but still remains open. It is also an open question whether $F2 \mid l_j \mid C_{\max}$ is APX-hard.

The two-machine flow shop problem with minimum delays is closely related to the two-machine flow shop problem with exact delays, which differs from the former only by the requirement that the second operation of each job $j \in J$ must start processing exactly l_j time units after the first operation has been completed. In [1,2], we refer to this problem as $F2 \mid \text{exact} l_j \mid C_{\max}$. In [2], in particular, it is proved that $F2 \mid \text{exact } l_j \mid C_{\max}$ admits a 3-approximation while its special case $F2 \mid \text{exact } l_j, a_j = b_j \mid C_{\max}$ can be solved by a 2-approximation algorithm, which in fact provides a 2-approximation for $F2 \mid l_j, a_j = b_j \mid C_{\max}$ as well. In [2], we also show that $F2 \mid \text{exact } l_j, a_j = b_j \mid C_{\max}$ cannot be approximated within a factor of $1.5 - \varepsilon$ provided that P$\neq$NP. In [1], $F2 \mid \text{exact } l_j, a_j = b_j = 1 \mid C_{\max}$ is shown to be approximable within a factor of $\frac{3}{2}$. Since we prove it over the lower bound that is also valid for $F2 \mid l_j, a_j = b_j = 1 \mid C_{\max}$, the algorithm provides a $\frac{3}{2}$-approximation for this problem as well. Unfortunately, this algorithm does not admit a constant-factor extension even to the case when $a_j \equiv b_j \equiv a$.

Our result. It can be shown that $F2 \mid l_j \mid C_{\max}$ admits a simple α-approximation where

$$\alpha = 1 + \frac{\min\{\sum_j a_j, \sum_j b_j\}}{\max\{\sum_j a_j, \sum_j b_j\}}.$$

Note that $\alpha = 2$ if and only if $\sum_j a_j = \sum_j b_j$. So this provides no improvement of Dell-Amico's approximations even for the important case of identical machines, i.e., for problem $F2 \mid l_j, a_j = b_j \mid C_{\max}$, which the subject of this paper. Our main result is that $F2 \mid l_j, a_j = b_j \mid C_{\max}$ can be approximated within a factor of $\frac{3}{2}$, which provides a substantially improved approximation for this problem. The algorithm is simple and can be implemented in $O(n^2)$ time.

Overview of the paper. It is clear that in problem $F2 \mid l_j \mid C_{\max}$ we may restrict ourselves by searching just for the schedules in which both machine continuously execute the jobs (in the paper we refer to them as continuous schedules). If a continuous schedule cannot be shorten by a parallel shifting of the operations, then we call it an early schedule. The early schedules are uniquely defined by the orders in which the machines execute jobs. These orders can be represented by a pair of job permutations $[(i_1, \ldots, i_n), (j_1, \ldots, j_n)]$. In any feasible schedule some job j can be critical, i.e., its second operation starts exactly l_j time units after the completion of the first operation. The importance of critical jobs is that given the starting time of such a job on machine 1 and the pair of job permutations, we are able to compute the length of the schedule.

Our algorithm first orders the jobs in non-decreasing order of $a_j + l_j$. Assuming that the jobs are indexed in this order it then constructs n early schedules σ_k generated by the permutations $[(k+1, \ldots, n, 1, \ldots, k), (1, \ldots, n)]$, $k = 1, \ldots, n$, and finally outputs a shortest schedule among them.

The schedules σ_k have a remarkable property: either job k, or job n is critical in σ_k. This fact leads to an exact formula (Lemma 4) expressing the length of σ_k as a function of k and the instance data. The function has the form $\max\{X_k, Y_k\}$, where (roughly speaking) X_k is the length of the schedule σ_k if job n is critical while Y_k is that when job k is critical. The proof of Theorem 1 is based on the fact that X_k, Y_k, and $X_k - Y_k$ are monotone functions of k, which implies that either $Y_k \geq X_k$ for all k (Case 1) or there exists a threshold $r \in \{1, \ldots, n-1\}$ such that such that $Y_r < X_r$ and $Y_k \geq X_k$ for all $k = r+1, \ldots, n$ (Case 2). When analyzing these cases we heavily rely upon our nontrivial generalization of the lower bound established by Yu for the case of unit processing times (Lemma 2). This lemma is a key ingredient of our approach. Its proof is based on a "relaxation" of problem $F2 \mid l_j, a_j = b_j \mid C_{\max}$ by problem $F2 \mid l_j, a_j = b_j = 1 \mid C_{\max}$ with unit processing times. The final part of the argument of Case 2 makes use of a trick similar to that used in the analysis of the $\frac{3}{2}$-approximation in [1].

2   Definitions, Notation, and Helpful Observations

Before proceeding to the main part we introduce the basic notation and make some observations and assumptions that do not restrict generality. An instance of $F2 \mid l_j \mid C_{\max}$ includes a set of jobs $J = \{1, \ldots, n\}$. Each job $j \in J$ consists of two operations $O_{1,j}$ and $O_{2,j}$ whose processing tomes will be denoted by a_j and b_j, respectively. We assume that a_j and b_j are positive integers for all $j \in J$. For each $j \in J$ operation $O_{2,j}$ is separated from $O_{1,j}$ by a delay of length at least l_j time units. For any $j \in J$, we denote by $\sigma(1, j)$ and $\sigma(2, j)$ the starting times of $O_{1,j}$ and $O_{2,j}$, respectively. Sometimes we will represent jobs $j \in J$ by the triples (a_j, l_j, b_j).

Note that a schedule σ is feasible if and only if

$$\sigma(2, j) \geq \sigma(1, j) + a_j + l_j$$

for all $j \in J$. For a schedule $\sigma = (\sigma(1, 1), \sigma(2, 1), \ldots, \sigma(2, n))$, denote by $C_j(\sigma)$ the completion time of job j in σ; then $C_j(\sigma) = \sigma(2, j) + b_j$. The length of a schedule σ is defined as $C_{\max}(\sigma) = \max_{j \in J} C_j(\sigma)$. Denote by $C^*_{\max}$ the length of a shortest schedule.

We set $A = \sum_{j \in J} a_j$, $B = \sum_{j \in J} b_j$, and $L = \max_{j \in J} l_j$. When considering problem $F2 \mid l_j, a_j = b_j \mid C_{\max}$ we assume that for each job $j \in J$ the processing times of both operations are a_j and so $A = \sum_{j \in J} a_j = \sum_{j \in J} b_j$.

Let σ be a schedule of $F2 \mid l_j \mid C_{\max}$. We say that a job j is *critical* in σ if $\sigma(2, j) = \sigma(1, j) + a_j + l_j$. Observe that if a schedule σ does not contain critical jobs, i.e., $\sigma(2, j) > \sigma(1, j) + a_j + l_j$ for all jobs $j \in J$, then by setting $\sigma'(j) := \sigma(j)$, $\sigma'(2, j) := \sigma(2, j) - \min\{\sigma(2, j) - \sigma(1, j) - a_j - l_j : j \in J\}$ we get another feasible schedule σ' with

$$C_{\max}(\sigma') = C_{\max}(\sigma) - \min\{\sigma(2, j) - \sigma(1, j) - a_j - l_j : j \in J\} < C_{\max}(\sigma).$$

Thus any optimal schedule has a critical job.

Any feasible schedule σ generates a pair of permutations (φ, ψ) of the set of jobs J such that φ specifies the order of operations on machine 1 and ψ specifies that on machine 2. More specifically, $\varphi(k)$ ($\psi(k)$) is the k-th job executed by machine 1 (machine 2). Note that for any $j \in J$, $\varphi^{-1}(j)$ ($\psi^{-1}(j)$) means the order number in which job j is processed on machine 1 (machine 2). We will further represent the permutations φ and ψ by the sequences $(\varphi(1), \ldots, \varphi(n))$ and $(\psi(1), \ldots, \psi(n))$. It is obvious that any feasible schedule σ can be transformed into another feasible schedule $\overline{\sigma}$ with $C_{\max}(\sigma) \geq C_{\max}(\overline{\sigma})$ and such that in $\overline{\sigma}$ the jobs on both machines are processed continuously in the same order as in σ. We shall call such schedules *continuous* schedules. By the definition if σ is a continuous schedule with job permutations (φ, ψ), then for any $k = 2, \ldots, n$,

$$\sigma(1, \varphi(k)) = \sigma(1, \varphi(k - 1)) + a_{\varphi(k-1)},$$
$$\sigma(2, \psi(k)) = \sigma(2, \psi(k - 1)) + b_{\psi(k-1)}.$$

A continuous schedule associated with a pair (φ, ψ) is depicted in Fig 1. A continuous schedule in which machine 1 starts processing at time 0 and at least one job is critical is called *early*. Observe that given a pair of orders (φ, ψ), the

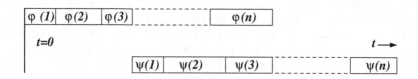

Fig. 1.

early schedule is uniquely defined by the pair and has minimum length among all feasible schedules associated with (φ, ψ). We will denote this schedule by $[\varphi, \psi]$. From the above it follows that the set of early schedules contains an optimal schedule.

Note that given a pair of permutations (φ, ψ), the early schedule $\sigma = [\varphi, \psi]$ can be found in linear time. Indeed, to find σ it suffices to determine the time when machine 2 starts executing, i.e., $x = \sigma(2, \psi(1))$. Since σ is a feasible continuous schedule, for any $k = 1, \ldots, n$ we have

$$\sigma(2, \psi(k)) = x + \sum_{i=1}^{k-1} b_{\psi(i)} \geq \sigma(1, \varphi(k)) + a_{\varphi(k)} + l_{\varphi(k)}.$$

Thus

$$x = \min\{\sigma(1, \varphi(k)) + a_{\varphi(k)} + l_{\varphi(k)} - \sum_{i=1}^{k-1} b_{\psi(i)} : k = 1, \ldots, n\},$$

which can be found in linear time.

3 Lower Bound

Before proceeding to the algorithm we present a lower bound for $C_{\max}^*$ that will play a crucial role in establishing an upper bound on the approximation ratio of our algorithm. We deduce it from a lower bound for $F2 \mid l_j, a_j = b_j = 1 \mid C_{\max}$ due to Yu et al. [12,13] (see Lemma 2 in [13]). For completeness we provide it with a proof in this section.

Lemma 1 (Yu et al.[12]). *For any instance of* $F2 \mid l_j, a_j = b_j = 1 \mid C_{\max}$,

$$C_{\max}^* \geq n + 1 + \left\lceil \frac{\sum_{j \in J} l_j}{n} \right\rceil \tag{1}$$

Proof. Let σ be a feasible continuous schedule with the jobs permutations φ and ψ. Then for any $j \in J$,

$$C_{\max}(\sigma) \geq \sigma(1, j) + a_j + l_j + \sum_{k=\psi^{-1}(j)}^{n} b_{\psi(k)}$$

$$= \sum_{k=1}^{\varphi^{-1}(j)} a_{\varphi(k)} + l_j + \sum_{k=\psi^{-1}(j)}^{n} b_{\psi(k)}.$$

By taking into account that $a_j = b_j \equiv 1$, it follows that

$$C_{\max}(\sigma) \geq \frac{1}{n} \left(\sum_{j \in J} \left(\sum_{k=1}^{\varphi^{-1}(j)} a_{\varphi(k)} + l_j + \sum_{k=\psi^{-1}(j)}^{n} b_{\psi(k)} \right) \right)$$

$$= \frac{1}{n} \left(\sum_{j \in J} \sum_{k=1}^{\varphi^{-1}(j)} a_{\varphi(k)} + \sum_{j \in J} \sum_{k=\psi^{-1}(j)}^{n} b_{\psi(k)} \right) + \frac{\sum_{j \in J} l_j}{n}$$

$$= \frac{1}{n} \left(\sum_{j \in J} \varphi^{-1}(j) + \sum_{j \in J} (n - \psi^{-1}(j) + 1) \right) + \frac{\sum_{j \in J} l_j}{n}$$

$$= \frac{1}{2n} \left(n(n+1) + n(n+1) \right) + \frac{\sum_{j \in J} l_j}{n}$$

$$= n + 1 + \frac{\sum_{j \in J} l_j}{n}.$$

It is easy to see that the above argument does not extend to problem $F2 \mid l_j, a_j = b_j \mid C_{\max}$. Nevertheless, the following result shows that the above lemma generalizes to this problem.

Lemma 2. *For any instance of $F2 \mid l_j, a_j = b_j \mid C_{\max}$ and $J' \subseteq J$,*

$$C_{\max}^* \geq \sum_{j \in J'} a_j + \left\lceil \frac{\sum_{j \in J'} a_j^2}{\sum_{j \in J'} a_j} + \frac{\sum_{j \in J'} a_j l_j}{\sum_{j \in J'} a_j} \right\rceil. \tag{2}$$

Proof. Indeed, since any lower bound for the contracted instance on a subset of jobs $J' \subseteq J$ is a lower bound for the original instance, it suffices to prove the lemma for the case of $J' = J$, i.e., to show that

$$C_{\max}^* \geq A + 1 + \frac{\sum_{j \in J} a_j^2}{A} - 1 + \frac{\sum_{j \in J} a_j l_j}{A}$$

$$= A + 1 + \frac{\sum_{j \in J} a_j(a_j + l_j - 1)}{A}.$$

Let $\mathcal{I}$ be an instance of $F2 \mid l_j, a_j = b_j \mid C_{\max}$. Consider an instance $\mathcal{I}_1$ of $F2 \mid l_j, a_j = b_j = 1 \mid C_{\max}$ that is obtained from $\mathcal{I}$ by the following transformation: replace each job $j \in J$ by the set of a_j identical jobs $\{\sum_{i=1}^{j-1} a_i + 1, \ldots, \sum_{i=1}^{j} a_i\}$ with parameters $(1, a_j + l_j - 1, 1)$. Thus the set of jobs J_1 in $\mathcal{I}_1$ is

$$J_1 = \bigcup_{j \in J} \{\sum_{i=1}^{j-1} a_i + 1, \ldots, \sum_{i=1}^{j} a_i\} = \{1, \ldots, A\}.$$

Let σ be an early schedule of $\mathcal{I}$ with permutations φ and ψ. Construct a schedule τ of $\mathcal{I}_1$ by setting

$$\tau(1, i) = \sigma(1, j) + i - \sum_{k=1}^{j-1} a_k - 1,$$

$$\tau(2, i) = \sigma(2, j) + i - \sum_{k=1}^{j-1} a_k - 1$$

for all $i \in \{\sum_{k=1}^{j-1} a_k + 1, \ldots, \sum_{k=1}^{j} a_k\}$ and $j \in J$. Fig. 2 shows an example of a schedule σ of three jobs $(4, 4, 4)$, $(3, 7, 3)$, and $(2, 0, 2)$ and the schedule τ of 9 jobs with parameters $(1, 7, 1)$, $(1, 9, 1)$, and $(1, 0, 1)$ obtained from σ. By the construction the lengths of σ and τ coincide. It follows that the length of a shortest schedule in $\mathcal{I}$ is at least the length of a shortest schedule in $\mathcal{I}_1$. However, by the construction of $\mathcal{I}_1$ and Lemma 1 the latter is bounded from below by

$$A + 1 + \frac{\sum_{j \in J} a_j(a_j + l_j - 1)}{A},$$

which proves the lemma.

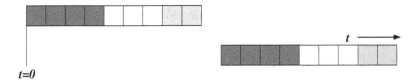

Fig. 2. An example of a schedule σ and the schedule τ obtained from σ

4   Algorithm

In this section we present our algorithm for $F2 \mid l_j, a_j = b_j \mid C_{\max}$. Remind that this case of the problem remains strongly NP-hard even if $l_j \in \{0, l\}$ or $a_j = b_j = 1$ for all $j \in J$ [12,13]. We now proceed to the description of the algorithm.

Algorithm Min_Delay

Step 0. Sort the jobs in nondecreasing order of $a_j + l_j$. For convenience, we further assume that

$$a_1 + l_1 \le a_2 + l_2 \le \ldots \le a_n + l_n. \tag{3}$$

Step k $(k = 1, \ldots, n)$. Construct the schedule $\sigma_k = [\varphi_k, \psi]$ where $\psi = (1, 2, \ldots, n)$ and $\varphi_k = (k + 1, \ldots, n, 1, \ldots, k)$ if $k \le n - 1$ and $\varphi_n = (1, \ldots, n)$. (The schedule σ_k is depicted in Fig. 3.)

Output a schedule σ having the shortest length among those in $\{\sigma_1, \ldots, \sigma_n\}$.

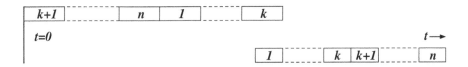

Fig. 3.

Note that Step 0 takes $O(n \log n)$ time. As it was shown in Section 2, the early schedule σ_k for each $k = 1, \ldots, n$ can be constructed in linear time. Thus the total running time of the algorithm is $O(n^2)$.

5   Approximation Ratio

We first present a helpful observation that will be used in evaluating the lengths of the schedules σ_k constructed by algorithm Min_Delay.

Lemma 3. *Let the jobs in J be indexed according to (3). Let σ be a feasible continuous (not necessarily early) schedule with job permutations $\varphi = (1, 2, \ldots, n)$ and $\psi = (1, 2, \ldots, n)$. If some job $j \in J$ in σ is critical, then job n is critical.*

Proof. Let $j \in J$ be a job in σ. Then by the definition of σ

$$C_{\max}(\sigma) \geq \sigma(1, 1) + \sum_{i=1}^{j} a_i + l_j + \sum_{i=j}^{n} a_j = \sigma(1, 1) + a_j + l_j + A.$$

and the inequality holds with equality if and only if j is critical in σ. In view of (3) it follows that

$$C_{\max}(\sigma) \geq \sigma(1, 1) + a_n + l_n + A.$$

Thus if some job j is critical in σ, then $a_j + l_j = a_n + l_n$ and job n is critical in σ as well.

The following lemma establishes a crucial formula expressing the length of the schedule σ_k as a function of the index k and the instance data.

Lemma 4. *For any $k = 1, \ldots, n$,*

$$C_{\max}(\sigma_k) = \max\{X_k, Y_k\}$$

where $X_k = \sum_{j=k+1}^{n} a_j + a_n + l_n$ and $Y_k = A + \sum_{j=k}^{n} a_j + l_k$.

Proof. Consider first the case of $k = n$. Since σ_n is an early schedule, some job $j \in J$ is critical in σ_n. By Lemma 3 it follows that job n is critical in σ_n. Then by the construction of σ_n,

$$C_{\max}(\sigma_n) = \sum_{j=1}^{n} a_j + a_n + l_n = Y_n.$$

Now let $1 \leq k \leq n - 1$. Remind that

$$\sigma_k = [(k+1, \ldots, n, 1, \ldots, k), (1, \ldots, k, k+1, \ldots, n)].$$

By the feasibility of σ_k

$$\sigma_k(1, k) + a_k + l_k \leq \sigma_k(2, k).$$

Since $\sigma_k(1, k) + a_k = \sum_{j=1}^{n} a_j$ and $C_{\max}(\sigma_k) = \sigma_k(2, k) + \sum_{j=k}^{n} a_k$, it follows that

$$C_{\max}(\sigma_k) \geq \sigma_k(1, k) + a_k + l_k + \sum_{j=k}^{n} a_k = A + \sum_{j=k}^{n} a_k + l_k = Y_k.$$

By a similar way it can be shown that $C_{\max}(\sigma_k) \geq X_k$. Thus we have

$$C_{\max}(\sigma_k) \geq \max\{X_k, Y_k\}. \tag{4}$$

Observe now that σ_k consists of two continuous schedules: a schedule τ' of the set of jobs $\{1,\ldots,k\}$ with job permutations $(1,\ldots,k)$ and $(1,\ldots,k)$, and a schedule τ'' of the set of jobs $\{k+1,\ldots,n\}$ with job permutations $(k+1,\ldots,n)$ and $(k+1,\ldots,n)$. Since σ_k is an early schedule, at least one of τ' and τ'' has a critical job. Since both schedules τ' and τ'' satisfy (3), by Lemma 3 either k, or n is a critical job in τ' or τ'', respectively, and therefore at least one of these jobs is critical in σ_k. If job k is critical in σ_k, then

$$C_{\max}(\sigma_k) = \sigma_k(1,k) + a_k + l_k + \sum_{j=k}^{n} a_k = A + \sum_{j=k}^{n} a_k + l_k = Y_k.$$

If job n is critical in σ_k, then a similar computation shows that $C_{\max}(\sigma_k) = X_k$. Thus we arrive at the inequality $C_{\max}(\sigma_k) \leq \max\{X_k, Y_k\}$, which together with (4) proves the lemma.

Theorem 1. *Let I be an instance of the problem and σ be a schedule output by algorithm* Min_Delay*. Then*

$$C_{\max}(\sigma) \leq \frac{3}{2} C^*_{\max}. \tag{5}$$

Proof. For $k = 1,\ldots,n$, set $\theta(k) = X_k - Y_k$. By Lemma 4

$$\theta(k) = a_n + l_n - a_k - l_k - A.$$

Since $a_k + l_k$ is a nondecreasing function of k, it follows that $\theta(k)$ is a nonincreasing function of k. Note that $\theta(n) = -A < 0$. Thus the following two cases are possible: either $Y_k \geq X_k$ for all $k = 1,\ldots,n$, or there exists an index $r \in \{1,\ldots,n-1\}$ such that $Y_r < X_r$ and $Y_k \geq X_k$ for all $k = r+1,\ldots,n$.

CASE 1: $Y_k \geq X_k$ *for all* $k = 1,\ldots,n$. Then by Lemma 4,

$$C_{\max}(\sigma_k) = Y_k$$

for all $k = 1,\ldots,n$. By the construction of σ, it follows that

$$
\begin{aligned}
C_{\max}(\sigma) &\leq \frac{\sum_{k=1}^{n} a_k C_{\max}(\sigma_k)}{A} \\
&= \frac{\sum_{k=1}^{n} a_k Y_k}{A} \\
&= \frac{\sum_{k=1}^{n} a_k (A + \sum_{j=k}^{n} a_j + l_k)}{A} \\
&= A + \frac{\sum_{k=1}^{n} \sum_{j=k}^{n} a_k a_j + \sum_{k=1}^{n} a_k l_k}{A} \\
&= A + \frac{(\sum_{i=1}^{n} a_i)^2 + \sum_{i=1}^{n} a_i^2}{2A} + \frac{\sum_{k=1}^{n} a_k l_k}{A} \\
&= \left(\frac{A}{2} - \frac{\sum_{i=1}^{n} a_i^2}{2A}\right) + \left(A + \frac{\sum_{i=1}^{n} a_i^2}{A} + \frac{\sum_{k=1}^{n} a_k l_k}{A}\right),
\end{aligned}
$$

which by Lemma 2, does not exceed

$$\frac{C_{\max}^*}{2} + C_{\max}^* \le \frac{3}{2}C_{\max}^*.$$

CASE 2: *There exists an index $r \in \{1, \ldots, n-1\}$ such that $Y_r < X_r$ and $Y_k \ge X_k$ for all $k = r+1, \ldots, n$.* By Lemma 4 this implies that $C_{\max}(\sigma_k) = Y_k$ for $k = r+1, \ldots, n$ and $C_{\max}(\sigma_k) = X_k$ for $k = 1, \ldots, r$. Since X_k is a non-increasing function of k and by the construction of σ

$$C_{\max}(\sigma) \le \min\left\{\frac{\sum_{k=r+1}^{n} a_k C_{\max}(\sigma_k)}{\sum_{i=r+1}^{n} a_i}, C_{\max}(\sigma_r)\right\} \le \min\{S_1, S_2\} \tag{6}$$

where

$$S_1 = \frac{\sum_{k=r+1}^{n} a_k Y_k}{\sum_{i=r+1}^{n} a_i} = \frac{\sum_{k=r+1}^{n} a_k(A + \sum_{j=k}^{n} a_j + l_k)}{\sum_{i=r+1}^{n} a_i}, \tag{7}$$

$$S_2 = X_r = \sum_{j=r+1}^{n} a_j + a_n + l_n. \tag{8}$$

Set $A_r = \sum_{i=r+1}^{n} a_i$. By using the identity

$$2 \sum_{k=r+1}^{n} \sum_{j=k}^{n} a_k a_j = A_r^2 + \sum_{i=r+1}^{n} a_i^2$$

rearrange the right-hand side of (7) in the following way:

$$\begin{aligned}
S_1 &= A + \frac{\sum_{k=r+1}^{n} \sum_{i=r+1}^{n} a_k a_j}{A_r} + \frac{\sum_{k=r+1}^{n} a_k l_k}{A_r} \\
&= A + \frac{A_r^2 + \sum_{i=r+1}^{n} a_i^2}{2A_r} + \frac{\sum_{k=r+1}^{n} a_k l_k}{A_r} \\
&= A - \frac{\sum_{i=r+1}^{n} a_i^2}{2A_r} + \frac{A_r}{2} + \frac{\sum_{i=r+1}^{n} a_i^2}{A_r} + \frac{\sum_{k=r+1}^{n} a_k l_k}{A_r} \\
&\le A + \frac{A_r}{2} + \frac{\sum_{i=r+1}^{n} a_i^2}{A_r} + \frac{\sum_{k=r+1}^{n} a_k l_k}{A_r} \\
&= A + \beta\mu + \frac{1}{2}(A_r + \beta) + \beta\left(\frac{1}{2} - \mu\right)
\end{aligned} \tag{9}$$

where

$$\beta = \frac{\sum_{i=r+1}^{n} a_i^2}{A_r} + \frac{\sum_{j=k}^{n} a_j l_k}{A_r},$$

and

$$\mu = \frac{A_r}{A}.$$

By Lemma 2, $A_r + \beta \leq C^*_{\max}$ and

$$A + \beta\mu = A + \frac{\sum_{i=r+1}^{n} a_i^2}{A} + \frac{\sum_{j=k}^{n} a_j l_k}{A}$$

$$\leq A + \frac{\sum_{i=1}^{n} a_i^2}{A} + \frac{\sum_{j=1}^{n} a_j l_k}{A} \leq C^*_{\max}.$$

Thus (9) implies

$$S_1 \leq C_{\max} + \frac{1}{2}C_{\max} + \beta(\frac{1}{2} - \mu). \tag{10}$$

On the other hand, by using the trivial lower bound

$$C^*_{\max} \geq \max\{2a_j + l_j : j \in J\}$$

from (8) we have

$$S_2 = 2a_n + l_n + \sum_{j=r+1}^{n-1} a_j \leq C^*_{\max}(1 + \mu). \tag{11}$$

Now from (10), (11) we see that if $\mu \geq \frac{1}{2}$, then $S_1 \leq \frac{3}{2}C_{\max}$ and if $\mu < \frac{1}{2}$, then $S_2 < \frac{3}{2}C_{\max}$. Taking into account (6) this completes the proof.

References

1. Ageev, A.A., Baburin, A.E.: Approximation algorithms for UET scheduling problems with Exact Delays. Oper. Res. Letters 35, 533–540 (2007)
2. Ageev, A.A., Kononov, A.V.: Approximation algorithms for scheduling problems with exact delays. In: Erlebach, T., Kaklamanis, C. (eds.) WAOA 2006. LNCS, vol. 4368, pp. 1–14. Springer, Heidelberg (2007)
3. Dell'Amico, M.: Shop problems with two machines and time lags. Operations Research 44, 777–787 (1996)
4. Dell'Amico, M., Vaessens, R.J.M.: Flow and open shop scheduling on two machines with transportation times and machine-independent processing times is NP-hard. Materiali di discussione 141, Dipartimento di Economia Politica, Università di Modena (1996)
5. Graham, R.L., Lawler, E.L., Lenstra, J.K., Rinnooy Kan, A.H.G.: Optimization and approximation in deterministic sequencing and scheduling: a survey. Annals of Discrete Mathematics 5, 287–326 (1979)
6. Johnson, S.M.: Optimal two- and three-stage production schedules with setup times included. Naval Research Logistics Quarterly 1, 61–68 (1954)
7. Johnson, S.M.: Discussion: Sequencing n jobs on two machines with arbitrary time lags. Management Science 5, 293–298 (1958)
8. Kern, W., Nawijn, W.M.: Scheduling multi-operation jobs with time lags on a single machine. In: Faigle, U., Hoede, C. (eds.) Proceedings of the 2nd Twente Workshop on Graphs and Combinatorial Optimization, Enschede (1991)
9. Lenstra, J.K.: Private communication (1991)

10. Mitten, L.G.: Sequencing n jobs on two machines with arbitrary time lags. Management Science 5, 293–298 (1958)
11. Strusevich, V.A.: A heuristic for the two-machine open-shop scheduling with transportation times. Discrete Applied Mathematics 93, 287–304 (1999)
12. Yu, W.: The two-machine shop problem with delays and the one-machine total tardiness problem, Ph.D. thesis, Department of Mathematics and Computer Science, Technische Universiteit Eindhoven (1996)
13. Yu, W., Hoogeveen, H., Lenstra, J.K.: Minimizing makespan in a two-machine flow shop with delays and unit-time operations is NP-hard. J. Sched. 7(5), 333–348 (2004)

Online Algorithm for Parallel Job Scheduling and Strip Packing[*]

Johann L. Hurink and Jacob Jan Paulus

University of Twente, P.O. box 217, 7500AE Enschede, The Netherlands,
Fax: +31534894858
j.j.paulus@utwente.nl

Abstract. We consider the online scheduling problem of parallel jobs on parallel machines, $P|\text{online} - \text{list}, m_j|C_{\max}$. For this problem we present a 6.6623-competitive algorithm. This improves the best known 7- competitive algorithm for this problem. The presented algorithm also applies to the special case where machines are ordered on a line and only adjacent machines can be assigned to a job and, therefore, also to online orthogonal strip packing. Since previous results for online orthogonal strip packing assume bounded rectangles, the presented algorithm is the first with a constant competitive ratio.

1 Introduction

Consider the following online machine scheduling problem. Jobs $j = 1, 2, ..., n$ are presented one by one to the decision maker and are characterized by their processing time and the number of machines simultaneously required for processing. Job j has processing time p_j and requires simultaneously m_j out of the available m machines. As soon as a job becomes known, it has to be scheduled irrevocably (i.e. its start time has to be set) without knowledge of successive jobs. Preemption is not allowed and the objective is to minimize the makespan.

Using the three-field notation introduced in [2], this problem is denoted by $P|\text{online} - \text{list}, m_j|C_{\max}$, see also [5,6]. Note that sometimes $size_j$ is used instead of m_j to denote the parallel machine requirement of job j.

The quality of an online algorithm is measured by its competitive ratio. An online algorithm is called ρ-competitive if for any sequence of jobs it produces a schedule with makespan at most ρ times the makespan of the optimal schedule. For background on online scheduling see [6].

The problem $P|\text{online} - \text{list}, m_j|C_{\max}$ gained considerable attention in the last few years. It was pointed out by Johannes [5] that a greedy algorithm which schedules the jobs as early as possible, has a competitive ratio of m. She was also the first to design an online algorithm with a constant competitive ratio, which has a competitive ratio of 12. This result was successively improved by Ye and Zhang, first to an 8 and later to a 7-competitive algorithm [7,8]. For the special case with only 2 machines an greedy algorithm is optimal [4], i.e.

---

[*] Part of this research has been funded by the Dutch BSIK/BRICKS project.

C. Kaklamanis and M. Skutella (Eds.): WAOA 2007, LNCS 4927, pp. 67–74, 2008.
© Springer-Verlag Berlin Heidelberg 2008

no online algorithm for $P2|online - list, m_j|C_{max}$ with competitive ratio strictly less than 2 exists.

Far less is known about lower bounds for the general m machine case. In [4] an ILP formulation is presented to derive lower bounds, and by means of an ILP solver a lower bound of 2.43 is derived. The best analytical lower bound is the bound of 2 from the two machine case.

In the literature also semi-online cases have been studied, e.g. *jobs appear with non-increasing processing times, jobs appear with non-increasing machine requirement* or the *largest processing time is known*. For these semi-online problems the gap between the lower and the upper bound on the competitive ratio is much smaller, see [7,8]. Variations on the scheduling model, where jobs are *malleable* or *preemption* is allowed, or with different online paradigms such as *non-clairvoyance* and *online-time*, are also considered in the literature. For an overview of these various models see [5,6,8].

The problem $P|online - list, m_j|C_{max}$ resembles online orthogonal strip packing. The difference lies in the following. In the scheduling problem any choice of m_j machines for processing job j is allowed, where in strip packing rectangles cannot be split. If the machines were to be ordered on a line and job j requires m_j *adjacent* machines for its processing, the problems become the same. As it turns out, the analysis of the online algorithm presented in this paper also hold in the presence of such a machine ordering and adjacency requirement. Therefore, the presented online algorithm applies to online orthogonal strip packing as well. Till now, the performance ratio of the best online algorithm for online orthogonal strip packing is 6.99, which is due to Baker and Schwarz [1]. It is worthwhile to mention, that the existing bounds for orthogonal strip packing are attained under the assumption that the rectangles have height at most 1. Analogous, the processing time of the jobs is bounded by 1. To the best of our knowledge, the presented algorithm is the first online algorithm with constant competitive ratio for orthogonal strip packing without knowledge of the overall maximum processing time of a job.

The presented approach in this paper leads to a new online algorithm for $P|online - list, m_j|C_{max}$. The algorithm takes two parameters, one parameter defines the borderline between *big* jobs (jobs with large m_j) and *small* jobs, and the second parameter defines *classes of processing times*. Small jobs with processing times of the same class get scheduled in parallel. A proper choice of the two parameters leads to an online algorithm that has a competitive ratio of at most $\frac{7}{2} + \sqrt{10}(\approx 6.6623)$.

In Section 2, we present the online algorithm and prove that it has a competitive ratio of at most 6.6623. In Section 3, we show that the algorithm also applies to the online orthogonal strip packing problem.

2   The Online Algorithm

Before we present the algorithm and the proof of its competitive ratio, we introduce some notation and basic results.

Given a sequence of jobs $\sigma = (1, 2, ..., n)$ we can derive two lower bounds on the makespan of the optimal offline schedule, denoted by $OPT(\sigma)$. On the one hand, the optimal makespan is bounded by the length of the longest job in σ, i.e. $OPT(\sigma) \geq \max_{j=1}^{n}\{p_j\}$. On the other hand, if the work load of a job j is given by $m_j \cdot p_j$, then the total work load divided by m is a lower bound on $OPT(\sigma)$, i.e. $OPT(\sigma) \geq \frac{1}{m}\sum_{j=1}^{n} m_j \cdot p_j$.

Let $\mathcal{S}(\sigma)$ be the schedule created by an online algorithm and denote its makespan by $ON(\sigma)$. For a collection of disjoint intervals X from $[0, ON(\sigma)]$, we denote by $|X|$ the cumulative length of the intervals in X.

The next lemma follows directly from the above presented lower bounds on $OPT(\sigma)$.

Lemma 1. *If $[0, ON(\sigma)]$ can be partitioned in X and Y such that $|X| \leq x \cdot \max_{j=1}^{n}\{p_j\}$ and $|Y| \leq y \cdot \frac{1}{m}\sum_{j=1}^{n} m_j \cdot p_j$, then $ON(\sigma) \leq (x + y) \cdot OPT(\sigma)$.*

In the following, we design an online algorithm for $P|\text{online} - \text{list}, m_j|C_{\max}$ such that the constructed schedules can be partitioned in X and Y as in Lemma 1 such that $x+y$ is small. To do this, we distinguish between two types of jobs; jobs with a large machine requirement and jobs that require only a few machines for processing. A job j is called *big* if it has machine requirement $m_j \geq \lceil \alpha \cdot m \rceil$ with $\alpha \in (0, \frac{1}{2}]$, and called *small* otherwise. This is a generalization of the distinction between big and small jobs found in [5,7,8], where α is a priori fixed to either $\frac{1}{2}$ or $\frac{1}{3}$. Furthermore, the small jobs are classified according to their length. A small job j belongs to job class J_k if $\beta^k \leq p_j < \beta^{k+1}$, where $\beta > 1$ is the second parameter of the algorithm. Note that k may be negative. Similar classifications can be found in *Shelf Algorithms* for Strip Packing [1], which are applied to group rectangles of similar height. The online algorithm to be described in the following, takes α and β as parameters and is denoted by $ON_{\alpha,\beta}$.

In the schedules created by the online algorithm $ON_{\alpha,\beta}$, big jobs are never scheduled in parallel to other jobs, and (where possible) small jobs are put in parallel to other small jobs of the same job class. The intuition behind the online algorithm $ON_{\alpha,\beta}$ is the following. Big jobs have a relative high average load and small jobs are either grouped together to a high average load or there is a small job with a relative long processing time. In the proof of Theorem 1, the intervals with many small jobs, together with the intervals with big jobs will be compared to the *work load bound* for $OPT(\sigma)$ (the Y part for Lemma 1), and the intervals with only a few small jobs are compared to the *longest job bound* for $OPT(\sigma)$ (the X part for Lemma 1).

The following gives a precise description of the algorithm $ON_{\alpha,\beta}$. The algorithm creates schedules with sparse intervals S_k and dense intervals D_k^i for the small jobs of class J_k. With n_k we count the number of dense intervals created for J_k. All small job scheduled in such an interval $[a, b)$ start at a. As a consequence, job j *fits* in interval $[a, b)$ if the machine requirement of the jobs already in $[a, b)$ plus m_j is at most m.

Algorithm $ON_{\alpha,\beta}$:
When scheduling job j and

1. job j is small, i.e. $m_j < \lceil \alpha \cdot m \rceil$, and belongs to job class J_k. Try in the given order:
   - Schedule job j in the first D_k^i where it fits.
   - Schedule job j in S_k.
   - Let $n_k := n_k + 1$ and S_k becomes $D_k^{n_k}$. Create a new interval S_k at the end of the current schedule with length β^{k+1}. Schedule job j in S_k.
2. job j is big, i.e. $m_j \geq \lceil \alpha \cdot m \rceil$. Schedule job j at the end of the current schedule.

The structure of the schedule created by $ON_{\alpha,\beta}$ is illustrated by Fig. 1. Note that at any time for each job class J_k there is at most one sparse interval S_k.

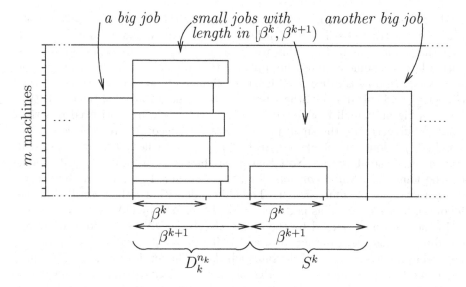

Fig. 1. Part of a schedule created by $ON_{\alpha,\beta}$

In the proof of Theorem 1 we will use the fact that the dense intervals D_k^i contain quite some load, i.e. there is a small job from job class J_k that did not fit in the dense intervals and had to be scheduled in a newly created sparse interval. When considering the length of the dense intervals, we take the load of both the dense and sparse intervals into account. Lemma 2 formalizes this. Slightly abusing notation, we will also refer to S_k and D_k^i as the set of jobs that are scheduled in intervals S_k and D_k^i.

Lemma 2. *For any $\alpha \in (0, \frac{1}{2}]$ and $\beta > 1$, the total work load in the dense and sparse intervals of schedule $\mathbf{S}(\sigma)$ created by $ON_{\alpha,\beta}$, is at least $\frac{2m}{3\beta}$ times the length of all dense intervals.*

Proof. Let σ be an arbitrary list of jobs and let $\mathbf{S}(\sigma)$ be the corresponding schedule constructed by the online algorithm $ON_{\alpha,\beta}$. Consider all dense intervals in $\mathbf{S}(\sigma)$ corresponding to one job class J_k. Since all jobs in D_k^i are scheduled to start at the beginning of interval D_k^i and have a processing time of at least β^k, the interval D_k^i has $\sum_{j \in D_k^i} m_j$ machines in use for at least the first $\frac{1}{\beta}$ fraction of D_k^i. If $\alpha \leq \frac{1}{3}$ this number of machines in use is larger than $\frac{2m}{3}$, and we are done. If $\alpha \in (\frac{1}{3}, \frac{1}{2}]$, we claim that for each job class J_k this number of machine in use is for at most one dense interval less than $\frac{2m}{3}$.

Let $\alpha \in (\frac{1}{3}, \frac{1}{2}]$ and let D_k^l be the first dense interval of job class J_k for which $\sum_{j \in D_k^i} m_j < \frac{2m}{3}$. After the creation of this dense interval, all newly created dense intervals for job class J_k, contain only small jobs with machine requirement $m_j > \frac{m}{3}$ (otherwise these jobs would have been scheduled in D_k^l or in an earlier dense interval). This implies that all successively created dense intervals for job class J_k have at least $\frac{2m}{3}$ machines in use. More precisely, they contain two small jobs with machine requirement $m_j > \frac{m}{3}$. Furthermore, the existence of D_k^l implies that S_k contains at least one job with machine requirement $m_j > \frac{m}{3}$.

So, for each job class J_k there is either one D_k^l with machine usage less than $\frac{2m}{3}$ and S_k contains a job with $m_j > \frac{m}{3}$, or all D_k^i have machine usage of at least $\frac{2m}{3}$. Thus, the total load of the small jobs in job class J_k is at least $\frac{2m}{3\beta}$ times the total length of all dense intervals corresponding to this job class.   □

Next we will prove the upper bound on the performance guarantee of the online algorithm $ON_{\alpha,\beta}$.

Theorem 1. *For any $\alpha \in (0, \frac{1}{2}]$ and $\beta > 1$ the competitive ratio of the online algorithm $ON_{\alpha,\beta}$ for the problem $P|\text{online} - \text{list}, m_j|C_{\max}$ is at most $\max\{\frac{1}{\alpha}, \frac{3\beta}{2}\} + \frac{\beta^2}{\beta-1}$.*

Proof. Let σ be an arbitrary list of jobs and let $\mathbf{S}(\sigma)$ be the corresponding schedule constructed by the online algorithm $ON_{\alpha,\beta}$. We partition $[0, ON_{\alpha,\beta}(\sigma)]$ into three parts: The first part B consists of the intervals in which big jobs are scheduled, the second part D consists of the dense intervals, and finally the third part S contains the sparse intervals.

Since part B contains only jobs with machine requirement $m_j \geq \lceil \alpha \cdot m \rceil$, the total work load in B is at least $\alpha \cdot m \cdot |B|$. According to Lemma 2, the total work load in D and S is at least $\frac{2m}{3\beta} \cdot |D|$. This work load is also in the optimal offline schedule. Therefore, $\min\{\alpha \cdot m, \frac{2m}{3\beta}\} \cdot (|B| + |D|) \leq m \cdot OPT(\sigma)$, or equivalently

$$|B| + |D| \leq \max\{\frac{1}{\alpha}, \frac{3\beta}{2}\} \cdot OPT(\sigma) \ . \tag{1}$$

To simplify the arguments for bounding $|S|$, we normalize the jobs in $\mathbf{S}(\sigma)$ by letting J_0 be the smallest job class, i.e. the smallest processing time of the small jobs is between 1 and β. Then $|S_k| = \beta^{k+1}$. Let $\bar{k}$ be the largest k for which there is a sparse interval in $\mathbf{S}(\sigma)$. Since there is at most one sparse interval for each job class J_k, the length of S is bounded by

$$|S| \leq \sum_{k=0}^{\bar{k}} |S_k| = \sum_{k=0}^{\bar{k}} \beta^{k+1} = \frac{\beta^{\bar{k}+2} - \beta}{\beta - 1} \ .$$

On the other hand, since $S_{\bar{k}}$ is not empty, we know that there is a job in $\mathcal{S}(\sigma)$ with processing time at least $\frac{|S_{\bar{k}}|}{\beta} = \beta^{\bar{k}}$. Thus,

$$|S| \leq \frac{\beta^2}{\beta - 1} \cdot OPT(\sigma) \ . \tag{2}$$

Using Lemma 1, (1) and (2) lead to the following bound on the makespan of the schedule created by online algorithm $ON_{\alpha,\beta}$:

$$ON_{\alpha,\beta}(\sigma) \leq \left(\max\{\frac{1}{\alpha}, \frac{3\beta}{2}\} + \frac{\beta^2}{\beta - 1} \right) \cdot OPT(\sigma) \ .$$

Thus, $ON_{\alpha,\beta}$ has a competitive ratio of at most $\max\{\frac{1}{\alpha}, \frac{3\beta}{2}\} + \frac{\beta^2}{\beta-1}$.   □

To find the best possible performance bound of $ON_{\alpha,\beta}$, we have to find values of α and β which minimize the competitive ratio from Theorem 1.

Corollary 1. *The worst case bound for $ON_{\alpha,\beta}$ is minimal if $\alpha \geq \frac{10}{3(5+\sqrt{10})}$ ($\approx$ 0.4084) and $\beta = 1 + \frac{\sqrt{10}}{5}$ (≈ 1.6325), leading to a competitive ratio of $\frac{7}{2} + \sqrt{10}$ ($\approx$ 6.6623).*

Proof. If $\frac{1}{\alpha} > \frac{3\beta}{2}$ then by increasing the value of α, the value of $\max\{\frac{1}{\alpha}, \frac{3\beta}{2}\}$ can be decreased. Therefore, it is best to choose $\frac{1}{\alpha} \leq \frac{3\beta}{2}$. The competitive ratio then becomes $\frac{3\beta}{2} + \frac{\beta^2}{\beta-1}$. The optimal value for β can be found by differentiating this term.   □

It is interesting to note that there is not just one setting of α and β that gives the best performance guarantee, but for $\beta = 1.6325$ all $\alpha \in [0.4084, 0.5]$ result in 6.6623-competitiveness of $ON_{\alpha,\beta}$.

3   Machines on a Line and Orthogonal Strip Packing

The presented online algorithm also applies to scheduling problems where the machines are ordered on a line and only adjacent machines can be assigned to a specific job. To let the presented algorithm apply to this case, we simply specify that whenever a job j is assigned to some interval, it is scheduled not only at the start of the interval, but also assigned to the *first* m_j machines available (first with respect to the line ordering of the machines). This way we can guarantee that each job j gets assigned to m_j adjacent machines and the algorithm still gives the same schedule as before. To the best of our knowledge the presented online algorithm is the first with constant competitive ratio for this problem. For previous developed online algorithms for $P|$online $-$ list, $m_j|C_{max}$ no such adaptation makes them applicable to this special case.

Since the presented online algorithm also applies to this special case, it applies to the online orthogonal strip packing problem. The online orthogonal strip packing problem is a two-dimensional packing problem. Without rotation rectangles have to be packed on a strip with fixed width and unbounded height. The objective is to minimize the height of the strip used. In the online setting one rectangle is presented after the other and has to be assigned without knowledge of successive rectangles.

To see that these problems are equivalent, let the machines correspond to the width of the strip, and time to the height of the strip. The width of a rectangle j corresponds to the machine requirement of job j and its height to the processing time. Minimizing the height of the strip used is equivalent to minimizing the makespan of the machine scheduling problem.

Although most of the research on online orthogonal strip packing focuses on asymptotic performance ratios, Baker and Schwarz [1] developed a *Shelf Algorithm* that has competitive ratio 6.99 under the assumption that the height of a rectangle is at most 1. So, the presented algorithm not only improves the best known competitive ratio for online orthogonal strip packing, but also does not require the assumption on the bounded height.

4   Conclusions

In this paper we presented a new online algorithm for $P|$online $-$ list, $m_j|C_{max}$ with a competitive ratio of 6.6623. Due to the optimization of the parameters of $ON_{\alpha,\beta}$ a better online algorithm can only be found by employing new ideas, both in the design and analysis. There is room for improvement since the gap with the best lower bound (2.43) is large.

The presented algorithm also applies to the problem where the machines are ordered on a line and to online orthogonal strip packing. It is an interesting open question whether or not the additional requirement of a line ordering will lead to a different competitive ratio of the problem.

Note
In the independent work of Han et al. [3] the same results where obtained in the setting of online orthogonal strip packing.

References

1. Baker, B.S., Schwarz, J.S.: Shelf Algorithms for Two-Dimensional Packing Problems. SIAM Journal on Computing 12(3), 508–525 (1983)
2. Graham, R.L., Lawler, E.L., Lenstra, J.K., Rinnooy Kan, A.H.G.: Optimization and approximation in deterministic sequencing and scheduling: A survey. Annals of Discrete Mathematics 5, 287–326 (1979)

3. Han, X., Ye, D., Zhang, G.: A note on online strip packing. Manuscript (2007)
4. Hurink, J.L., Paulus, J.J.: Online Scheduling of Parallel Jobs on Two Machines is 2-Competitive. Operations Research Letters (to appear) doi:10.1016/j.orl.2007.06.001
5. Johannes, B.: Scheduling parallel jobs to minimize the makespan. Journal of Scheduling 9, 433–452 (2006)
6. Pruhs, K., Sgall, J., Torng, E.: Online Scheduling. In: Leung, J.Y-T. (ed.) Handbook of Scheduling: Algorithms, Models, and Performance Analysis, ch. 15, pp. 15–41. CRC Press, Boca Raton (2004)
7. Ye, D., Zhang, G.: On-line Scheduling of Parallel Jobs. In: Kralovic, R., Sýkora, O. (eds.) SIROCCO 2004. LNCS, vol. 3104, pp. 279–290. Springer, Heidelberg (2004)
8. Ye, D., Zhang, G.: On-line Scheduling of Parallel Jobs in a List. Journal of Scheduling (to appear) doi:10.1007/s10951-007-0032-x

Geometric Spanners with Small Chromatic Number[*]

Prosenjit Bose[1], Paz Carmi[1], Mathieu Couture[1], Anil Maheshwari[1],
Michiel Smid[1], and Norbert Zeh[2]

[1] School of Computer Science, Carleton University
[2] Faculty of Computer Science, Dalhousie University

Abstract. Given an integer $k \geq 2$, we consider the problem of computing the smallest real number $t(k)$ such that for each set P of points in the
plane, there exists a $t(k)$-spanner for P that has chromatic number at
most k. We prove that $t(2) = 3$, $t(3) = 2$, $t(4) = \sqrt{2}$, and give upper and
lower bounds on $t(k)$ for $k > 4$. We also show that for any $\epsilon > 0$, there
exists a $(1 + \epsilon)t(k)$-spanner for P that has $O(|P|)$ edges and chromatic
number at most k. Finally, we consider an on-line variant of the problem
where the points of P are given one after another, and the color of a
point must be assigned at the moment the point is given. In this setting,
we prove that $t(2) = 3$, $t(3) = 1 + \sqrt{3}$, $t(4) = 1 + \sqrt{2}$, and give upper
and lower bounds on $t(k)$ for $k > 4$.

1 Introduction

Let P be a set of n points in the plane. A *geometric graph* with vertex set P is
an undirected graph whose edges are line segments that are weighted by their
Euclidean length. For a real number $t \geq 1$, such a graph G is called a *t-spanner*
if the weight of the shortest path in G between any two vertices p and q does not
exceed $t|pq|$, where $|pq|$ is the Euclidean distance between p and q. The smallest
t having this property is called the *stretch factor* of the graph G. Thus, a graph
with stretch factor t approximates the $\binom{n}{2}$ distances between the points in P
within a factor of t. The problem of constructing t-spanners with $O(n)$ edges for
any given point set has been studied intensively; see the book by Narasimhan
and Smid [6] for an overview.

In this paper, we consider the problem of computing t-spanners whose chromatic number is at most k, for some given value of k. The goal is to minimize
the value of t over all finite sets P of points in the plane. We call a spanner
whose chromatic number is at most k a *k-chromatic spanner*.

Problem 1. Given an integer $k \geq 2$, let $t(k)$ be the infimum of all real numbers t
with the property that for every finite set P of points in the plane, a k-chromatic
t-spanner for P exists. Determine the value of $t(k)$.

[*] Research partially supported by HPCVL, NSERC, MRI, CFI, and MITACS.

C. Kaklamanis and M. Skutella (Eds.): WAOA 2007, LNCS 4927, pp. 75–88, 2008.

Observe that in the definition of $t(k)$, there is no requirement on the number of edges of the chromatic spanner. This is not a restriction, because, as shown by Gudmundsson et al. [5], any t-spanner for P contains a subgraph with $O(n)$ edges which is a $((1 + \epsilon)t)$-spanner for P.

We show how to obtain a 2-chromatic 3-spanner for any point set P, thus showing that $t(2) \leq 3$. We also give an example of a point set P such that any 2-chromatic graph with vertex set P has stretch factor at least three. Thus, we have $t(2) = 3$.

Next, we show how to compute a 3-chromatic 2-spanner of any point set P, thereby proving that $t(3) \leq 2$. We also show, by means of an example, that $t(3) \geq 2$. Thus, we obtain that $t(3) = 2$. For $k = 4$, we show how to compute a 4-chromatic $\sqrt{2}$-spanner of any point set P; thus $t(4) \leq \sqrt{2}$. Again by means of an example, we also show that $t(4) \geq \sqrt{2}$. Therefore, we have $t(4) = \sqrt{2}$.

For $k > 4$, we are not able to obtain the exact value of $t(k)$. Inspired by the *ordered Θ-graph* of Bose et al. [2], we show that $t(k) \leq 1 + 2\sin\frac{\pi}{2(k-1)}$. We also show that the vertex set of the regular $(k + 1)$-gon gives $t(k) \geq 1/\cos\frac{\pi}{k+1}$.

In the second part of the paper, we consider an on-line variant of the problem where the points of P are given one after another, and the color of a point must be assigned at the moment when the point is given; thus, later on, the color of a point cannot be changed. This makes the problem more difficult. Consequently, the bounds are higher, but still tight for $k = 2, 3, 4$. All our bounds are summarized in Table 1.

Problem 2. Given an integer $k \geq 2$, let $t'(k)$ be the infimum of all real numbers t with the property that for every finite set P of points in the plane, which is given on-line, a k-chromatic t-spanner for P exists. Determine the value of $t'(k)$.

A simple variant of the ordered Θ-graph shows that $t'(k) \leq 1 + 2\sin(\pi/k)$. Thus, we have $t'(2) \leq 3$, $t'(3) \leq 1+\sqrt{3}$ and $t'(4) \leq 1+\sqrt{2}$. Since $t'(2) \geq t(2) = 3$, it follows that $t'(2) = 3$. We also give examples showing that $t'(3) \geq 1 + \sqrt{3}$ and $t'(4) \geq 1 + \sqrt{2}$. We finally show that, for $k \geq 5$, $t'(k) \geq 1/\cos\frac{\pi}{k}$.

The rest of this paper is organized as follows: in Section 2, we define the t-ellipse property and show its relationship to our problem. In Section 3, we give upper and lower bounds for the off-line problem (Problem 1). In Section 4, we give give upper and lower bounds for the on-line problem (Problem 2). We conclude in Section 5. In Table 1, we summarize our results. We now motivate our work.

Motivation: In a recent paper, Raman and Chebrolu [7] proposed a new protocol, called 2P, allowing to address rural Internet connectivity in a low-cost manner using off-the-shelf 802.11 hardware. Since their infrastructure uses several directional antennae at one node rather than one single omnidirectional antenna, simultaneous communications are possible at one node. However, due to restrictions inherent in the 802.11 standard, backbone nodes have to communicate with each other using a single channel. While simultaneous transmissions and simultaneous receptions are possible, it is not physically possible for one node to both transmit and receive at the same time. Therefore, backbone nodes

Table 1. Summary of our results

number of colors	$t(k)$ (off-line)		$t'(k)$ (on-line)	
k	lower bound	upper bound	lower bound	upper bound
2	3	3	3	3
3	2	2	$1 + \sqrt{3}$	$1 + \sqrt{3}$
4	$\sqrt{2}$	$\sqrt{2}$	$1 + \sqrt{2}$	$1 + \sqrt{2}$
k	$1/\cos\frac{\pi}{k+1}$	$1 + 2\sin\frac{\pi}{2(k-1)}$	$1/\cos\frac{\pi}{k}$	$1 + 2\sin\frac{\pi}{k}$

have to alternate between the send and receive states. This forces the backbone to be a bipartite graph, i.e., to have chromatic number two.

The backbone creation algorithm of Raman and Chebrolu [7] outputs a tree, which is obviously bipartite. However, the tree structure presents the following disadvantage: it is possible that the path that a message has to follow is much longer than the distance (either Euclidean or in terms of hops) between the originating node and its destination.

Note that the physical constraint preventing nodes to simultaneously receive and transmit can be met even if the graph is not bipartite. In fact, any graph with chromatic number k would meet this requirement: all one has to do is to prevent two nodes that have different colors to transmit simultaneously. A degenerate case is when each node has its own color, in which case at most one node can transmit at any given moment. This case is undesirable, since the amount of time during which a node can transmit decreases as the size of the network increases.

For these reasons, it is desirable to have geometric graphs that have both small chromatic number and small stretch factor.

2   The t-Ellipse Property

In this section, we show that Problem 1, i.e., determining the smallest value of t such that a k-chromatic t-spanner exists for any point set P, is equivalent to minimizing the value of t such that any point set can be colored using k colors in a way that satisfies the so-called t-ellipse property.

Definition 1 (t-ellipse property). *Let $k \geq 2$ be an integer, let P be a finite set of points in the plane and let $c : P \to \{1, \ldots, k\}$ be a k-coloring of P. We say that that the coloring c satisfies the t-ellipse property if, for each pair of distinct points p and q in P with $c(p) = c(q)$, there exists a point $r \in P$ such that $c(r) \neq c(p)$ and $|pr| + |rq| \leq t|pq|$.*

Thus, if p and q have the same color, then the ellipse $\{x \in \mathbb{R}^2 : |px| + |xq| \leq t|pq|\}$ contains a point r of P whose color is different from that of p and q.

Proposition 1. *Let $k \geq 2$, let P be a set of points in the plane, and let G be a k-chromatic t-spanner of P with k-coloring c. Then c satisfies the t-ellipse property.*

Proof. Let $p, q \in P$ be two points with $c(p) = c(q)$. Since G is a t-spanner, there exists a t-spanning path Π in G from p to q. Let r be the point on Π that is adjacent to p. Since the length of Π is at most $t|pq|$, we note that $|pr| + |rq|$ is at most $t|pq|$. Since the edge (p, r) is in G, it follows that $c(p) \neq c(r)$. Therefore, c satisfies the t-ellipse property. $\square$

Proposition 2. *Let $k \geq 2$, let P be a set of points in the plane, and let $c : P \rightarrow \{1, \ldots, k\}$ be a k-coloring of P that satisfies the t-ellipse property. Then, there exists a k-chromatic t-spanner of P.*

Proof. Let $K_c(P)$ be the complete k-partite graph with vertex set P in which there is an edge between two points p and q if and only if $c(p) \neq c(q)$. By definition, $K_c(P)$ is k-colorable. We show that $K_c(P)$ is a t-spanner of P. Let p and q be two distinct points of P such that (p, q) is not an edge in $K_c(P)$. This means that $c(p) = c(q)$. Since c has the t-ellipse property, there exists a point r in P such that $c(r) \neq c(p)$ and $|pr| + |rq| \leq t|pq|$. Since $c(r) \neq c(p)$ (and consequently, $c(r) \neq c(q)$), the edges (p, r) and (r, q) are both in $K_c(P)$. This means that (p, r, q) is a t-spanner path in $K_c(P)$ between p and q. $\square$

From this point on, unless specified otherwise, we define the *stretch factor* of a k-coloring of a point set to be the stretch factor of the complete k-partite graph induced by this coloring. By Propositions 1 and 2, the problem of determining $t(k)$ is equivalent to determining the minimum stretch factor of any k-coloring of any point set.

   We conclude this section by showing why it is sufficient to focus on the coloring problem without worrying about the number of edges in the spanner. The following theorem is due to Gudmundsson *et al.* [5]; its proof is based on the the well-separated pair decomposition of Callahan and Kosaraju [3].

Theorem 1. *[5] Let $\epsilon > 0$ and $t \geq 1$ be constants, let P be a set of n points in the plane, and let G be a t-spanner of P. There exists a subgraph G' of G, such that G' is a $((1 + \epsilon)t)$-spanner of P and G' has $O(n)$ edges.*

Proposition 3. *Let $k \geq 2$, let P be a set of n points in the plane, and let $c : P \rightarrow \{1, \ldots, k\}$ be a k-coloring of P that satisfies the t-ellipse property. Then, for any constant $\epsilon > 0$, there exists a k-chromatic $((1 + \epsilon)t)$-spanner of P that has $O(n)$ edges.*

Proof. By Proposition 2, there exists a k-chromatic t-spanner G of P. By Theorem 1, G contains a subgraph G' with $O(n)$ edges, such that G' is a $((1 + \epsilon)t)$-spanner of P. Since G is k-chromatic, G' is k-chromatic as well. $\square$

3   Upper and Lower Bounds on $t(k)$

The structure of this section is as follows: For $k = 2, 3$, and 4, we give coloring algorithms whose outputs have bounded stretch factor. Then, we show that these stretch factors are tight by providing point sets for which no coloring algorithm can achieve a better stretch factor. Then we present our upper and lower bounds for $t(k)$, when $k > 4$. We now give the coloring algorithm for $k = 2$.

Algorithm 1. Offline 2 Colors

Input: P, a set of points in the plane
Output: c, a 2-coloring of P
 1: Compute a Euclidean minimum spanning tree T of P
 2: $c \leftarrow$ a 2-coloring of T

Lemma 1. *For any point set P, the 2-coloring computed by Algorithm 1 has stretch factor at most 3. Thus, we have $t(2) \leq 3$.*

Proof. It is sufficient to show that the 2-coloring c computed by Algorithm 1 has the 3-ellipse property. Let p and q be two distinct points in P such that $c(p) = c(q)$. Observe that (p, q) is not an edge in the minimum spanning tree T. Let r be the nearest neighbor of p. Since the edge (p, r) is in T, we have $r \neq q$ and $c(r) \neq c(p)$. Since r is closer to p than q, we have

$$|pr| + |rq| \leq |pr| + |rp| + |pq| = 2|pr| + |pq| \leq 2|pq| + |pq| = 3|pq|. \qquad \square$$

Lemma 2. *For every $\epsilon > 0$, there exists a point set P such that every 2-coloring of P has stretch factor at least $3 - \epsilon$. Thus, we have $t(2) \geq 3$.*

Proof. Let n be an odd integer, and let $P = \{p_1, \ldots, p_n\}$ be the set of vertices of a regular n-gon given in counter-clockwise order. Let c be an arbitrary 2-coloring of P. By the pigeonhole principle, there are two points in P which are adjacent on the n-gon and that have the same color. We may assume without loss of generality that these two points are p_1 and p_2. Also, we may assume that $|p_1p_2| = 1$ (see Figure 1, left). Let t be any real number such that c satisfies the t-ellipse property. Then $|p_1p_3| + 1 \leq t$. But $|p_1p_3| = 2\sin((n-2)\pi/2n)$, which tends to 2 as n goes to infinity. $\qquad \square$

We now consider the case when $k = 3$. Our strategy is to construct a graph such that any coloring of that graph has the 2-ellipse property. We then show that this graph is 3-colorable.

Lemma 3. *The graph G computed by Algorithm 2 is triangle-free.*

Proof. Assume that G contains a triangle with vertices p, q, and r. We may assume without loss of generality that (p, r) was the last edge of this triangle that was considered by the algorithm. This means that (p, r) is the longest edge of the triangle. When $e_i = (p, r) = (p_i, q_i)$ in line 4, G already contains the edge (p, q). Since $|p_ip| + |pq_i| = |pp| + |pr| \leq 2|pr|$ and $|p_iq| + |qq_i| = |pq| + |qr| \leq 2|pr|$, the edge (p, r) is not added to G. This is a contradiction and, therefore, G is triangle-free. $\qquad \square$

Lemma 4. *The graph G computed by Algorithm 2 is plane.*

Due to space constrains, we omit the proof of this lemma, which can be found in [1].

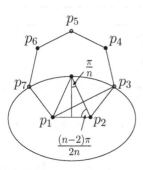
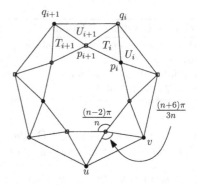

Fig. 1. Left, lower bound of $3 - \epsilon$ for $k = 2$. Right, lower bound of $2 - \epsilon$ for $k = 3$.

Algorithm 2. Offline 3 Colors

Input: P, a set of n points in the plane
Output: c, a 3-coloring of P, and G, a 3-chromatic graph whose vertex set is P
1: Let G be the graph with vertex set P and whose edge set is empty
2: Let $e_1, \ldots, e_{\binom{n}{2}}$ be the pairs of points of P in sorted order of their distances
3: **for** $i = 1$ to $\binom{n}{2}$ **do**
4:    Let $e_i = (p_i, q_i)$
5:    **if** G contains no edge (p, q) where $|p_i p| + |p q_i| \le 2|p_i q_i|$ and $|p_i q| + |q q_i| \le 2|p_i q_i|$
   then
6:       add the edge e_i to G
7:    **end if**
8: **end for**
9: //assertion: G is 3-colorable (see Lemma 5)
10: $c \leftarrow$ a 3-coloring of G

Lemma 5. *The graph G computed by Algorithm 2 is 3-colorable.*

Proof. By Lemmas 3 and Lemma 4, G is plane and triangle-free. It is known that such a graph is 3-colorable; see [4],[8]. □

Lemma 6. *For any point set P, the 3-coloring of P computed by Algorithm 2 has stretch factor at most 2. Thus, we have $t(3) \le 2$.*

Proof. It is sufficient to show that the 3-coloring c produced by Algorithm 2 has the 2-ellipse property. Let p and q be points in P such that $c(p) = c(q)$. Let E be the ellipse whose boundary is the set of points e such that $|pe| + |eq| = 2|pq|$. Since (p, q) is not an edge in G, G must contain an edge (s, t) whose two endpoints are inside E. Since $c(s) \ne c(t)$, at least one of s and t has a different color than p and q. Without loss of generality, s is that point. Since s is inside E, we have that $|ps| + |sq| \le 2|pq|$. □

Lemma 7. *For every $\epsilon > 0$, there exists a point set P such that every 3-coloring of P has stretch factor at least $2 - \epsilon$. Thus, we have $t(3) \ge 2$.*

Proof. Let n be an odd integer, and let $P = \{p_1, \ldots, p_n, q_1, \ldots, q_n\}$ where the p_i's are the vertices of a regular n-gon given in counter-clockwise order, and the q_i's are such that the triangles $T_i = (q_i, p_i, p_{i+1})$ are equilateral with interior lying outside the n-gon (indices are taken modulo n); see Figure 1, right. Now consider the set of triangles $\mathcal{T} = \{T_1, \ldots, T_n, U_1, \ldots, U_n\}$, where $U_i = (q_{i-1}, p_i, q_i)$. A simple parity argument shows that, for any 3-coloring of P, there is at least one triangle of $\mathcal{T}$ that has two vertices u and v that are assigned the same color. If this triangle is a T_i, then the stretch factor between u and v is at least 2. If this triangle is a U_i, then the stretch factor between u and v is at least $1/\sin((n+6)\pi/6n)$, which tends to 2 as n goes to infinity. $\qquad\square$

Next, we consider the case when $k = 4$. For this case, we simply use the Delaunay triangulation to find a 4-coloring. We then show that this coloring satisfies the $\sqrt{2}$-ellipse property.

Algorithm 3. Offline 4 Colors

Input: P, a set of points in the plane
Output: c, a 4-coloring of P
1: Compute the Delaunay triangulation D of P
2: $c \leftarrow$ a 4-coloring of D

Lemma 8. *For any point set P, the 4-coloring of P computed by Algorithm 3 has stretch factor at most $\sqrt{2}$. Thus, we have $t(4) \leq \sqrt{2}$.*

Proof. It is sufficient to show that the coloring c computed by Algorithm 3 has the $\sqrt{2}$-ellipse property. Let p and q be points of P such that $c(p) = c(q)$. Since (p, q) is not an edge in the Delaunay triangulation, the circle C whose diameter is pq contains at least one point of P. For a point r inside C, let $D(r)$ be the circle through p and r whose center is on pq (see Figure 3, left). Let r_0 be the point inside C such that $D(r_0)$ has minimum diameter. Then, $D(r_0)$ is an empty circle with p and r_0 on its boundary, which means that (p, r_0) is a Delaunay edge. Therefore, $c(r_0) \neq c(p)$, and since r_0 is in C, we have $|pr_0| + |r_0q| \leq \sqrt{2}|pq|$. $\quad\square$

Lemma 9. *For every $\epsilon > 0$, there exists a point set P such that every 4-coloring of P has stretch factor at least $\sqrt{2} - \epsilon$. Thus, we have $t(4) \geq \sqrt{2}$.*

Proof. Let n be an odd integer, and let $P = \{p_1, \ldots, p_n, q_1, \ldots, q_n\}$, where the p_i's are the vertices of a regular n-gon, the q_i's are the vertices of a larger regular n-gon with the same center, and $|q_i p_i| = |p_i p_{i+1}|$ for all i; refer to Figure 3, right. Let Q_i be the quadrilateral $(p_i, p_{i+1}, q_{i+1}, q_i)$. A simple parity argument shows that for any 4-coloring of P, there is at least one Q_i that has two vertices u and v that are assigned the same color. The stretch factor between u and v is then at least $1/\sin((n+2)\pi/4n)$, which tends to $\sqrt{2}$ when n goes to infinity. $\quad\square$

Our general algorithm for values $k > 4$ uses ideas from the ordered Θ-graph of Bose *et al.* [2]. We take advantage of the fact that we are in an off-line context,

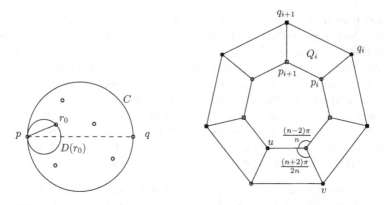

Fig. 2. Left, upper bound of $\sqrt{2}$ for $k = 4$. Right, lower bound of $\sqrt{2} - \epsilon$ for $k = 4$.

so that we can sort the points according to their y-coordinates. We process the points one by one from the lowest to the highest, splitting the half-plane below the current point p being processed into $k-1$ cones of angle $\pi/(k-1)$ and having their apex at p. For each such cone c_j, we take the point r_j in c_j that is closest to p. Then we assign p a color that has not been assigned to any of the r_j's. The fact that this algorithm uses at most k colors is straightforward, since there are at most $k - 1$ such r_j.

Algorithm 4. Offline k Colors

Input: P, a set of points in the plane
Output: c, a k-coloring of P
1: Let $p_1, \ldots, p_n$ be the points of P sorted in non-decreasing order of y-coordinates
2: **for** $i = 1$ to n **do**
3:     partition the half-plane below p_i into $k - 1$ cones of angle $\theta = \pi/(k - 1)$ and apex p_i
4:     for each cone c_j, let r_j be the point in c_j that is closest to p_i
5:     $c(p_i) \leftarrow \min\{l > 0 : \forall r_j, c(r_j) \neq l\}$
6: **end for**

Lemma 10. *For $k > 4$, we have $t(k) \leq 1 + 2\sin(\pi/(2k - 2))$.*

Proof. Let p and q be points of P such that $c(p) = c(q)$. We may assume without loss of generality that $q_y \leq p_y$. Let c be the cone with apex at p that contains q in line 4 of Algorithm 4, let r be the nearest neighbor of p in c, let r' be the intersection between the ray emanating from p through r and the circle centered at p with radius $|pq|$, and let $\alpha = \angle rpq$ (see Figure 3). Then,

$$|pr| + |rq| \leq |pr| + |rr'| + |r'q| = |pq| + |r'q|$$
$$= |pq| + 2\sin\frac{\alpha}{2}|pq| \leq (1 + 2\sin\frac{\pi}{2(k-1)})|pq|.$$

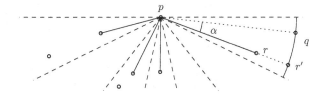

Fig. 3. Upper bound of $1 + 2\sin(\pi/(2k - 2))$ for $k > 4$

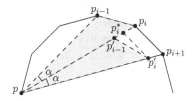

Fig. 4. Illustration of the proof of Lemma 11

It follows that the coloring computed by Algorithm 4 has the $(1 + 2\sin(\pi/(2k - 2)))$-ellipse property. The result follows from the fact that $c(r) \neq c(p)$ and that Algorithm 4 uses at most k colors. $\qquad\square$

Lemma 11. *Let p, q, r be three distinct vertices of a regular $(k + 1)$-gon. Then the ratio $(|pr| + |rq|)/|pq|$ is at least $1/\cos(\frac{\pi}{k+1})$ and this value is achieved when p, r, and q are consecutive vertices.*

Proof. For fixed p and q, the ratio $(|pr| + |rq|)/|pq|$ is minimized when r is adjacent to either p or q. In that case, the angle $\alpha = \angle qpr = \pi/(k+1)$. We show that for a fixed point p and three consecutive vertices p_{i-1}, p_i and p_{i+1} of the regular $(k + 1)$-gon such that $|pp_{i-1}| < |pp_i| < |pp_{i+1}|$ (see Figure 4) the ratio $(|pp_{i-1}| + |p_{i-1}p_i|)/|pp_i|$ is smaller than $(|pp_i| + |p_ip_{i+1}|)/|pp_{i+1}|$ and the result follows.

Without loss of generality, p_{i-1}, p_i and p_{i+1} are in clockwise order. Let p'_{i-1} and p'_i be the rotation of p_{i-1} and p_i around p by a clockwise angle of α. Also, let p_i^* be the intersection of $\overline{pp_i}$ with the parallel line to $\overline{p_ip_{i+1}}$ through p'_i. Triangle $pp_i^*p'_i$ is similar to triangle pp_ip_{i+1}. Therefore,

$$
\begin{aligned}
(|pp_i| + |p_ip_{i+1}|)/|pp_{i+1}| &= (|pp_i^*| + |p_i^*p'_i|)/|pp'_i| \\
&= (|pp'_{i-1}| + |p'_{i-1}p_i^*| + |p_i^*p'_i|)/|pp'_i| \\
&> (|pp'_{i-1}| + |p'_{i-1}p'_i|)/|pp'_i| \\
&= (|pp_{i-1}| + |p_{i-1}p_i|)/|pp_i|.
\end{aligned}
$$

Therefore, the ratio $(|pp_{i-1}| + |p_{i-1}p_i|)/|pp_i|$ is minimized when p_{i-1} is adjacent to p. $\qquad\square$

Lemma 12. *For $k > 4$, we have $t(k) \geq 1/\cos(\frac{\pi}{k+1})$.*

Proof. Let $P = \{p_1, \ldots, p_{k+1}\}$ be the vertex set of a regular $(k + 1)$-gon. By Lemma 11, for any three distinct points p, q, and r in P, the ratio $(|pr| + |rq|)/|pq|$ is at least $1/\cos(\frac{\pi}{k+1})$ and this value is achieved when p, r, and q are consecutive vertices.

By the pigeonhole principle, any k-coloring of P has to assign the same color to at least two points, say p and q. By the argument above, the stretch factor between p and q is at least $1/\cos(\frac{\pi}{k+1})$. □

The constructions we have shown in this section use a quadratic number of edges since we consider the complete k-partite graph induced by the coloring of the points. To reduce this to a linear number of edges we apply Proposition 3, which slightly increases the stretch factor, giving us the following:

Theorem 2. *The following are true:*

1. *For any point set P in the plane, the complete k-partite graph induced by the k-coloring of P computed by the above algorithms has a stretch factor at most 3, 2, $\sqrt{2}$, and $1 + 2\sin\frac{\pi}{2(k+1)}$ for $k = 2$, $k = 3$, $k = 4$, $k > 4$, respectively.*
2. *For any $\epsilon > 0$, there exist point sets such that no coloring algorithm can compute a k-coloring that has the t-ellipse property for t smaller than $3 - \epsilon$, $2 - \epsilon$, $\sqrt{2} - \epsilon$, and $1/\cos\frac{\pi}{k+1}$ for $k = 2$, $k = 3$, $k = 4$, $k > 4$, respectively.*
3. *Thus, we have $t(2) = 3$, $t(3) = 2$, $t(4) = \sqrt{2}$, and $1 + 2\sin\frac{\pi}{2(k+1)} \geq t(k) \geq 1/\cos\frac{\pi}{k+1}$ for $k > 4$.*
4. *It is possible to obtain a $((1 + \epsilon)t(k))$-spanner that has $O(|P|)$ edges, from the coloring computed by the above algorithms.*

4   Upper and Lower Bounds on $t'(k)$

Recall that in the on-line setting, the algorithm receives the points of P one at a time and assigns a color to a point as soon as it receives it. It cannot change the color of a point after this assignment. Naturally, this setting is more difficult which is reflected by higher bounds for $t'(3)$ and $t'(4)$. However, we are still able to give the exact value of $t'(k)$ for $k = 2, 3, 4$ and provide upper and lower bounds when $k > 4$. In the online setting, we actually provide a general algorithm that is the same for all values of $k \geq 2$. Although it is similar to Algorithm 4, there are at least two important differences. First, since we are in an on-line setting, we cannot process the points in the order of their y-coordinates. Therefore, we have to use cones with an angle greater than $\pi/(k - 1)$. If we choose the cones a priori as we do in Algorithm 4, we obtain cones whose angle is $2\pi/(k - 1)$. However, by aligning the cone's bisectors on the points that are chosen to be neighbors, we can get a slightly better stretch factor, since in this case, the angle is reduced to $2\pi/k$.

Lemma 13. *For $k \geq 2$, Algorithm 5 computes a k-coloring that has the t-ellipse property for $t = 1 + 2\sin(\pi/k)$. Thus, we have $t'(k) \leq 1 + 2\sin(\pi/k)$.*

Algorithm 5. Online k Colors

Input: P, an arbitrarily ordered list of points in the plane
Output: c, a k-coloring of P
1: Let $p_1, \ldots, p_n$ be the points of P in the given order
2: **for** $i = 1$ to n **do**
3:     $P_i \leftarrow \{p_1, \ldots, p_{i-1}\}$
4:     $j \leftarrow 0$
5:     **while** $P_i \neq \emptyset$ **do**
6:         $j \leftarrow j + 1$
7:         $r_j \leftarrow$ a nearest neighbor of p_i in P_i
8:         $P_i \leftarrow P_i \setminus \{r_j\}$
9:         **for** each $q \in P_i$ **do**
10:             **if** $\angle qp_ir_j \leq 2\pi/k$ **then**
11:                 $P_i \leftarrow P_i \setminus \{q\}$
12:             **end if**
13:         **end for**
14:     **end while**
15:     $c(p_i) \leftarrow \min\{l > 0 \mid \forall r_j, c(r_j) \neq l\}$
16: **end for**

Proof. Algorithm 5 produces a k-coloring, because each p_i selects at most $k - 1$ points r_j. If there were more than $k - 1$ such points, then two of them would form an angle of $2\pi/k$ or less around p_i. However, this situation cannot occur because of lines 10 and 11. The proof on the stretch factor is identical to the one given in Lemma 10.   □

Corollary 1. *We have* $t'(2) \leq 3$, $t'(3) \leq 1 + \sqrt{3}$ *and* $t'(4) \leq 1 + \sqrt{2}$.

Since an off-line lower bound also provides an on-line lower bound, we have $t'(2) \geq t(2) = 3$. It follows that $t'(2) = 3$. We now prove that Algorithm 5 is also optimal for $k = 3$ and 4.

Lemma 14. *Let $\mathcal{A}$ be an arbitrary on-line coloring algorithm that guarantees a 3-coloring that has the t-ellipse property. Then its stretch factor, t, is at least $1 + \sqrt{3}$.*

Proof. The proof is by an adversarial argument, where the adversary forces a stretch factor of at least $1 + \sqrt{3}$. The main objective of the adversary is to force $\mathcal{A}$ to assign different colors to the vertices of an equilateral triangle. Then, the next point is placed in the center of this triangle (see Figure 5(a)). This results in a stretch factor of $1 + \sqrt{3}$.

Consider Figure 5(b), where the points are numbered by the order of insertion. Up to symmetry, there is only one way to assign colors to points p_1 to p_6 such that $t < 1 + \sqrt{3}$ (e.g., $c(p_1) = red, c(p_2) = blue, c(p_3) = red, c(p_4) = green, c(p_5) = green, c(p_6) = green$). The key property is that the points p_3, p_4 and p_5 must be assigned the same color that is different from the colors assigned to the first three points. If any of these conditions is violated, then the spanning ratio is at least $1 + \sqrt{3}$.

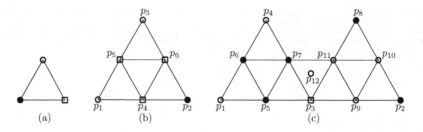

Fig. 5. Online lower bound of $1 + \sqrt{3}$ for $k = 3$

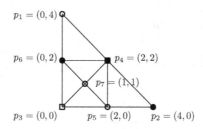

Fig. 6. Online lower bound of $1 + \sqrt{2}$ for $k = 4$

Next, consider Figure 5(c), where the point set of Figure 5(b) is reproduced twice. Consider triangles $\triangle(p_3, p_5, p_7)$, $\triangle(p_3, p9, p_{11})$ and $\triangle(p_3, p_7, p_{11})$ after the insertion of p_{11}. At least one of these triangles has to be assigned three different colors, otherwise, the stretch factor would already be $1 + \sqrt{3}$. Assume w.l.o.g. that triangle $\triangle(p_3, p_7, p_{11})$ is assigned three different colors then by the insertion of point p_{12} in the center of the triangle, we force a spanning ratio of $1 + \sqrt{3}$, as required.   □

Lemma 15. *Let $\mathcal{A}$ be an arbitrary on-line coloring algorithm that guarantees a 4-coloring that has the t-ellipse property. Then the stretch factor, t is at least $1 + \sqrt{2}$.*

Proof. Consider the point set depicted in Figure 6. $\mathcal{A}$ must assign different colors to p_3, p_4, p_5 and p_6, otherwise the stretch factor will already be greater than $1 + \sqrt{2}$. Upon introduction of p_7, $\mathcal{A}$ must assign it the same color as one of p_3, p_4, p_5 or p_6. The stretch factor between p_7 and that point is $1 + \sqrt{2}$.   □

Lemma 16. *Let $\mathcal{A}$ be an arbitrary on-line coloring algorithm that guarantees a k-coloring that has the t-ellipse property. Then the stretch factor, t, is at least $1/\cos(\frac{\pi}{k})$.*

Proof. Let $P = \{p_1, \ldots, p_k, q\}$, where the p_i' are the vertices of a regular k-gon K and q is the center of the circumcircle of K. If, after processing p_1 to p_k, $\mathcal{A}$ assigned the same color to two points, then as in Lemma 12, the stretch factor is $1/\cos(\frac{\pi}{k})$. Otherwise, all p_i are assigned different colors. When q is introduced, the color $\mathcal{A}$ assigns to it has already been assigned to some other point p. In that case, the stretch factor for the edge (q, p) is $1 + 4\sin(\pi/2k) > 1/\cos(\frac{\pi}{k})$. $\qquad\square$

The constructions we have shown in this section use a quadratic number of edges since we consider the complete k-partite graph induced by the coloring of the points. To reduce this to a linear number of edges we apply Proposition 3, which slightly increases the stretch factor, giving us the following:

Theorem 3. *The following are true:*

1. *For any sequence P of points in the plane, the complete k-partite graph induced by the on-line k-coloring of P computed by the above algorithms has a stretch factor at most 3, $1 + \sqrt{3}$, $1 + \sqrt{2}$, and $1 + 2\sin\frac{\pi}{k}$ for $k = 2$, $k = 3$, $k = 4$, $k > 4$, respectively.*
2. *For any $\epsilon > 0$, there exist point sets such that no on-line coloring algorithm can compute an on-line k-coloring that has the t-ellipse property for t smaller than $3 - \epsilon$, $1 + \sqrt{3} - \epsilon$, $1 + \sqrt{2} - \epsilon$, and $1/\cos\frac{\pi}{k}$ for $k = 2$, $k = 3$, $k = 4$, $k > 4$, respectively.*
3. *Thus, we have $t'(2) = 3$, $t'(3) = 1 + \sqrt{3}$, $t'(4) = 1 + \sqrt{2}$, and $1 + 2\sin\frac{\pi}{k} \geq t'(k) \geq 1/\cos\frac{\pi}{k}$ for $k > 4$.*
4. *It is possible to obtain a $((1 + \epsilon)t'(k))$-spanner that has $O(|P|)$ edges, from the coloring computed by the above algorithms.*

5   Conclusion

In this paper, we investigated the problem of computing a spanner of a point set that has chromatic number k. For small values of k ($k \leq 4$), we provided tight upper and lower bounds on the smallest possible stretch factor of such spanners. For larger values of k, we provided general upper and lower bounds which, unfortunately, are not tight. Our construction algorithms show how to color a point set with k colors such that the complete k-partite graph induced by this coloring has the stated stretch factor. The number of edges in these graphs can be reduced from quadratic to linear with a slight increase in the spanning ratio by applying the general technique of Gudmundsson et al. [5]. An interesting open problem in this setting of the problem is to find tight upper and lower bounds when $k > 4$.

We also considered an on-line variant of this problem where the points are presented sequentially and our algorithm assigns a color to each point upon reception such that the complete k-partite graph induced by the coloring is a constant spanner. Again, for small values of k ($k \leq 4$), we provided tight upper and lower bounds on the smallest possible stretch factor of such spanners and for $k > 4$, we provided general upper and lower bounds that are not tight. A linear-sized spanner can be constructed after all the points have been colored by applying the technique of Gudmundsson et al. [5]. However, in this case, our algorithm for computing the linear-sized constant spanner is *not* on-line. Therefore, there are two open problems in the on-line setting. First, to close the gap between the upper and lower bound for $k > 4$. Second, provide an on-line algorithm that computes the linear-sized constant spanner.

References

1. Bose, P., Carmi, P., Couture, M., Maheshwari, A., Smid, M., Zeh, N.: Geometric spanners with small chromatic number. Technical Report 0711.0114v1 (2007), http://arxiv.org/abs/0711.0114v1
2. Bose, P., Gudmundsson, J., Morin, P.: Ordered theta graphs. Comput. Geom. Theory Appl. 28(1), 11–18 (2004)
3. Callahan, P.B., Kosaraju, S.R.: A decomposition of multidimensional point sets with applications to k-nearest-neighbors and n-body potential fields. Journal of the ACM 42, 67–90 (1995)
4. Grötzsch, H.: Ein Dreifarbensatz für dreikreisfreie Netze auf der Kugel. Wiss. Z. Martin-Luther-Univ. Halle-Wittenberg Math.-Natur. Reihe 8, 109–120 (1959)
5. Gudmundsson, J., Levcopoulos, C., Narasimhan, G., Smid, M.: Approximate distance oracles for geometric graphs. In: Proceedings of the 13th ACM-SIAM Symposium on Discrete Algorithms, pp. 828–837 (2002)
6. Narasimhan, G., Smid, M.: Geometric Spanner Networks. Cambridge University Press, New York (2007)
7. Raman, B., Chebrolu, K.: Design and evaluation of a new mac protocol for long-distance 802.11 mesh networks. In: MobiCom 2005, pp. 156–169. ACM Press, New York (2005)
8. Thomassen, C.: A short list color proof of Grotzsch's theorem. Journal of Combinatorial Theory B 88, 189–192 (2003)

Approximating Largest Convex Hulls for Imprecise Points*

Maarten Löffler and Marc van Kreveld

Department of Information and Computing Sciences
Utrecht University, the Netherlands
{loffler,marc}@cs.uu.nl

Abstract. Assume that a set of imprecise points is given, where each point is specified by a region in which the point will lie. Such a region can be modelled as a circle, square, line segment, etc. We study the problem of maximising the area of the convex hull of such a set. We prove NP-hardness when the imprecise points are modelled as line segments, and give linear time approximation schemes for a variety of models, based on the *core-set* paradigm.

1 Introduction

In computational geometry, many fundamental problems take a point set as input on which some computation is done, for example to determine the convex hull, the Voronoi diagram, or a travelling sales route. These problems have been studied for decades. The vast majority of research assumes the locations of the input points to be known exactly. In practice, however, this is often not the case. Coordinates of the points may have been obtained from the real world, using equipment that has some error interval, or they may have been stored as floating points with a limited number of decimals. In real applications, it is important to be able to deal with such imprecise points.

When considering imprecise points, various interesting questions arise. Sometimes it is sufficient to know just some possible solution, which can be achieved by applying existing algorithms to some point set that is possibly the true point set. More information about the outcome can be obtained by computing a probability distribution over all possibilities, for example using Monte Carlo methods. In many applications it is also important to know concrete lower and upper bounds on some measure on the outcome, given concrete bounds on the input.

There are a number of basic geometric measures on point sets, such as the diameter, the size of the smallest enclosing circle, the area of the convex hull, etc. For most of these measures the lower and upper bounds can be computed exactly in an efficient way [15], as is summarised in Table 1. However, there are a few problems for which no efficient exact algorithm is known.

* This research was partially supported by the Netherlands Organisation for Scientific Research (NWO) under BRICKS/FOCUS grant number 642.065.503 and through the project GOGO.

C. Kaklamanis and M. Skutella (Eds.): WAOA 2007, LNCS 4927, pp. 89–102, 2008.

Table 1. Exact algorithms for basic geometric measures on imprecise point sets, when the imprecision of a point is modelled as a square region. The $O(n^7)$ result for the largest area convex hull only applies for disjoint squares, see Table 2.

structure	smallest		largest	
diameter	$O(n \log n)$	[15]	$O(n \log n)$	[15]
closest pair	$O(n \log n)$	[15]	NP-hard	[6]
smallest enclosing circle	$O(n)$	[11]	$O(n)$	[15]
convex hull (area)	$O(n^2)$	[14]	$O(n^7)$	[14]
minimum spanning tree	NP-hard	[14]	open	

Table 2. New and previous results for maximising convex hull area of a set of imprecise points

model	restrictions	exact algorithm		approximation scheme
line segments	parallel	$O(n^3)$	[14]	$O(n + \eta^3)$
line segments	-	$O(2^n n \log n)$ (NP-hard)		$O(n) + 2^{O(\eta^2)}$
squares	disjoint	$O(n^7)$	[14]	$O(n + \eta^{14})$
squares	equal size	$O(n^5)$	[14]	$O(n + \eta^{12})$
squares	disjoint and equal size	$O(n^3)$	[14]	$O(n + \eta^{12})$
squares	-	$O(4^n n \log n)$		$O(n) + 2^{O(\eta^2)}$
k-gons	disjoint or equal size	$n^{O(k)}$		$O(n) + 2^{O(k \log \eta)}$
k-gons	-	$O(k^n n \log n)$		$O(n) + 2^{O(\eta^2 \log k)}$
circles	disjoint or equal size			$O(n) + 2^{O(\sqrt{\eta} \log \eta)}$
circles	-			$O(n) + 2^{O(\eta^2 \log \eta)}$

In [13,14], we studied a wide range of problems concerning the convex hull of imprecise points. We varied the imprecision model (line segment, square, circle), the objective function (area, perimeter), the goal (maximisation, minimisation), and the restrictions on the input (equal size, disjoint, no restrictions). It appeared that the maximisation of area variant (see Figure 1(a)) was one of the hardest, where we found many polynomial time algorithms of rather high degree, and were unable to find any polynomial time algorithm for several variants.

Here we present linear time approximation schemes for all variants of the largest area convex hull problem. The algorithms are all of the form $O(n + f(\eta))$, where n is the input size and $\frac{1}{\eta} = \varepsilon$ is the required precision of the answer. The dependence on n is linear, provided that the ceiling operation can be performed in constant time. The dependence on η is polynomial if we have a polynomial time exact algorithm to solve the problem, and superpolynomial otherwise. We also prove NP-hardness of one of the problems for which no polynomial time exact algorithm is known. Our previous exact results and new approximate results are summarised in Table 2.

A lot of research about imprecision in computational geometry is directed at computational imprecision [16]. Recently, however, interest in exact approaches to deal with data imprecision is growing [1,4,9,12]. A more extensive overview of related work in this area is given in [14].

The *core-set* framework, introduced by Agarwal and Har-Peled [2], is a powerful way to obtain approximation algorithms, and still an active research topic [3,7,8]. In this framework, a point set P is given, and the problem is to maximise some measure $\mu(P)$. To do this, one constructs a *core-set* $P' \subset P$, such that $\mu(P') > (1 - \varepsilon)\mu(P)$. The size of the core-set must only depend on ε, and not on n (or sublinearly on n, depending on the application). Now the total running time of the algorithm is the time it takes to construct P', and the time it takes to compute $\mu(P')$, where the second step does not depend on n. If the first part can be done in linear time, one obtains a *linear time approximation scheme* (LTAS) [5]. A good survey on core-sets is provided by Agarwal *et al.* [3].

We generalise the concept of core-sets to sets of imprecise points. We are not given a set of points, but a set of regions. A core-set of such a set is still a subset of bounded size that guarantees a good solution. However, the criteria to include a region in the core-set become more elaborate; in some cases they depend on the size and shape of a region as well as its location, rather than only on its location, as is usually the case. For the classical (precise) convex hull problem it is well known that a small core-set always exists [3]. This immediately implies that a core-set for the imprecise convex hull problem exists as well: take a core-set for the optimal solution. In the remainder of this paper we show how to compute such core-sets efficiently.

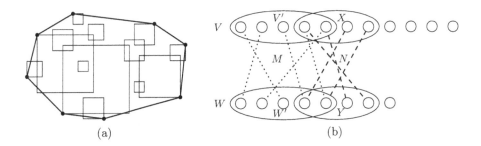

(a)   (b)

Fig. 1. (a) The maximum area convex hull for a set of imprecise points modelled as squares. (b) There are vertices in $X - V'$, and from these vertices there is an augmenting path that ends in either $V' - X$ or $Y - W'$.

2   Preliminaries

Before treating the main results, we first give some small results that are independent of the rest of the text, but are needed in some of the proofs.

Let G be a bipartite graph with two sets of vertices V and W, and a set of edges $E \subset V \times W$. Let $M \subset E$ be a maximum matching of G, and let $V' \subset V$ and

$W' \subset W$ be the two vertex sets that are used by M. A matching between two subsets $A \subset V$ and $B \subset W$ is called *perfect* if it consists of exactly $\min(|A|, |B|)$ edges.

Lemma 1. *For every subset $Y \subset W$, if there is a perfect matching between V and Y then there is also a perfect matching between V' and Y.*

Proof. Suppose the theorem is false. Let $Y \subset W$ be a subset of W such that there exists a perfect matching between V and Y, but no perfect matching between V' and Y. Let $N \subset E$ be the matching among all perfect matchings between V and Y that uses the largest number of vertices of V'. Let $X \subset V$ be the set of vertices used by N, apart from Y. Then $X \not\subset V'$, so there is a vertex $x \in X$ with $x \notin V'$, see Figure 1(b).

Now start an augmenting path from x that uses only edges of $M \cup N$. This path takes alternating edges from N and from M, since no two from the same set can use the same vertex. Therefore, this path ends either in a vertex $v \in V' - X$ or in a vertex $w \in Y - W'$. In the first case, we have a perfect matching between $X - \{x\} \cup \{v\}$ and Y, which is in contradiction with the choice of N. In the second case, we have a perfect matching between $V' \cup \{x\}$ and $W' \cup \{w\}$, which contradicts the maximality of M.

Lemma 2. *Let P be a set of n points. The diameter and width of P can be approximated within a factor $\sqrt{2}$ in $O(n)$ time.*

Proof. The proof is elementary, and can be found in the full paper.

3   NP-Hardness

Problem 1. *Given a set of line segments, choose a point on each segment such that the area of the convex hull of the resulting point set is as large as possible.*

This problem is NP-hard, and we prove this by reduction from SAT. The full proof can be found in [13]; here we merely sketch the basic idea. We start with a large circle, and divide it into enough arcs, see Figure 2(a). We separate these

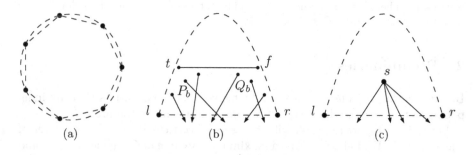

(a)   (b)   (c)

Fig. 2. The division of the circle into independent arcs. (b) A variable. (c) A clause.

arcs by precise points. The solution will contain at least the convex hull of these precise points. We will make sure never to place any line segments outside this circle, so maximising the area of the convex hull is now equal to maximising the sum of the areas within the arcs.

For each Boolean variable b in the SAT instance, we add the configuration of Figure 2(b) inside an empty arc. We can only use one of the points t and f, so for a maximal area we need either all points P_b or all points Q_b. For each clause we add a single point s inside an empty arc, see Figure 2(c). We include an edge between s and every variable in this clause. Now an assignment to the variables to satisfy the SAT instance can be made if and only if a solution to the convex hull maximisation problem of maximal area exists. All points can be chosen with rational coordinates of polynomial complexity.

Theorem 1. *Given a set of n arbitrarily oriented, possibly intersecting line segments, the problem of choosing a point on each segment such that the area of the convex hull of the resulting point set is as large as possible is NP-hard. The decision version of the problem is NP-complete.*

4   Approximation

We study the problem of finding the largest convex hull for a set of imprecise points. We are given a set $\mathcal{L} \subset \mathcal{P}(\mathbb{R}^2)$ of imprecise points; that is, $\mathcal{L}$ is a set of subsets of $\mathbb{R}^2$. We want to find a core-set $\mathcal{L}' \subset \mathcal{L}$ with respect to the measure μ, where μ measures the area of the largest possible convex hull. We model the imprecise points as line segments, squares and circles. We are always looking for a $(1 - \varepsilon)$-approximation, and we also denote $\eta = \varepsilon^{-1}$.

4.1   Parallel Line Segments

Problem 2. *Given a set of parallel line segments, choose a point on each segment such that the area of the convex hull of the resulting point set is as large as possible.*

We can solve this problem exactly in $O(n^3)$ time, using a dynamic programming solution [14].

Core-Set Construction. Assume that there are a point p_l on the leftmost segment and a point p_r on the rightmost segment that have the same y-coordinate. We can do this without loss of generality, because we can freely skew the problem without changing any areas.

Let $\mathcal{L}$ be the set of input segments. We will select a core-set $\mathcal{L}' \subset \mathcal{L}$ of at most 16η segments. Let w be the difference in x-coordinates between p_l and p_r, and let h be the maximal difference in y-coordinate between any two vertices of the optimal solution. Let $\delta = \frac{1}{4}\varepsilon \cdot w$. We will now divide the plane into 4η vertical strips of width δ, see Figure 3(a). In each strip, we take the two topmost and the two bottommost endpoints and add the segments they belong to to $\mathcal{L}'$.

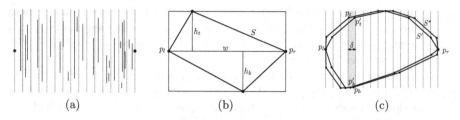

Fig. 3. (a) A set of parallel line segments divided in vertical strips. (b) There is a solution S with area $\frac{1}{2}wh$, so S^* is at least as large. (c) In one strip, the horizontal difference between the points in S^* and S' is at most δ.

Proof of $\mathcal{L}'$ being a Core-Set. Let S^* be the optimal solution for $\mathcal{L}$, the original input, and let S'^* be the optimal solution for $\mathcal{L}'$. The area of S^* is at least $\frac{1}{2}wh$, see Figure 3(b). We prove that the area difference between S^* and S'^* is a fraction of this area, dependent on ε.

Lemma 3. *There exists a solution S' for $\mathcal{L}'$ with a difference to S^* of at most $2\delta h$.*

Proof. For each vertical strip, let p_t be the topmost vertex of S^* within that strip and p_b the bottommost vertex. Since p_t and p_b cannot be endpoints of the same segment, there are points p'_t and p'_b in $\mathcal{L}'$ such that p'_t is equal to or above p_t, and p'_b is equal to or below p_b, and they are not endpoints of the same segment, see Figure 3(c). Use these points in S'. If there are no vertices of S^* in the strip, we just skip it. We know that S' is a valid solution for $\mathcal{L}$, so $S' \leq S^*$. On the other hand, because of the above, S^* can never be larger than S' with a strip of horizontal width δ around it: the Minkowski sum of S' and the horizontal line segment from $(-\delta, 0)$ to $(\delta, 0)$. So $S^* \leq S' + 2\delta h$.

Putting it all together, we have $S'^* \geq S' \geq S^* - 2\delta h = S^* - \frac{1}{2}\varepsilon wh \geq S^* - \varepsilon S^* = (1 - \varepsilon)S^*$.

As mentioned earlier, we assume that the ceiling operation can be performed in constant time. This is necessary to put the segments into the correct strips in linear time. Without this assumption, the algorithm can be made to run in $O(n \log \eta)$ time.

Theorem 2. *We can compute a core-set of size $O(\eta)$ for Problem 2 in $O(n)$ time.*

This problem can be solved exactly in $O(n^3)$, and therefore approximated in $O(n + \eta^3)$ time.

4.2 Arbitrary Line Segments

Problem 3. *Given a set of line segments, choose a point on each segment such that the area of the convex hull of the resulting point set is as large as possible.*

As we proved in Section 3, this problem is NP-hard. However, for the core-set approach, we do need an exact algorithm. There exists an optimal solution that has every point on an endpoint of its segment. Therefore we can solve the problem in $O(2^n n \log n)$ time by computing the convex hull of every possible set of endpoints. Of course this can be improved slightly.

Core-Set Construction. For technical reasons, we need to scale the input to ensure that the width and diameter of the set of endpoints are not too different (the input is not too narrow). Note that we can freely do this as it does not influence the relative area of any solution. We can ensure that the ratio between the width and diameter is at most $4\sqrt{2}$.

Let l_{max} be the longest segment in $\mathcal{L}$. We put l_{max} in $\mathcal{L}'$. Let p and q be two points of the vertex set of $\mathcal{L} - \{l_{max}\}$ that approximate its diameter within a factor 2. Call the direction from p to q $\boldsymbol{e}_1$, and the direction perpendicular to this $\boldsymbol{e}_2$. Determine the axis-parallel bounding box B of $\mathcal{L} - \{l_{max}\}$ in the $(\boldsymbol{e}_1, \boldsymbol{e}_2)$ axis system. Let w be the width (the maximum extent in the $\boldsymbol{e}_1$ direction) of B, and h the height (the maximum extent in the $\boldsymbol{e}_2$ direction) of B. Assume that $w \geq h$, and exchange axes otherwise.

Divide B into 1024η by 1024η grid cells, see Figure 4(a). The cells are $\delta_1 = \frac{\varepsilon}{1024} w$ long in the $\boldsymbol{e}_1$ direction, and $\delta_2 = \frac{\varepsilon}{1024} h$ long in the $\boldsymbol{e}_2$ direction. Consider the bipartite graph where one set of nodes corresponds to the set of line segments $\mathcal{L} - \{l_{max}\}$, and the other set corresponds to the cells of the grid. There is an edge between segment l and cell c if one of the endpoints of l is in c. Let M be a maximum matching of this graph, and add all segments that occur in M to $\mathcal{L}'$.

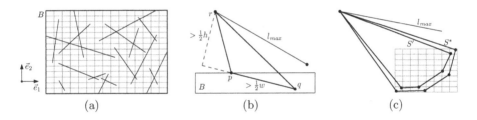

(a)   (b)   (c)

Fig. 4. (a) A set of line segments divided according to their cells. (b) In the narrow case, there is a solution with area at least $\frac{1}{8}wh$. (c) In one cell, the horizontal difference between the points in S^* and S' is at most δ_1, and the vertical difference is at most δ_2.

Proof of $\mathcal{L}'$ being a Core-Set. Let S^* be the optimal solution for $\mathcal{L}$, the original input, and let S'^* be the optimal solution for $\mathcal{L}'$, the core-set. Let A denote an amount of area: $A = area(B) + g \cdot w$, where g is the distance from the centre of B to the furthest endpoint of l_{max}. We will first show that the area of S^* is at least a constant fraction of A. Then we will show that the difference in area between S^* and S'^* is only an ε-dependent fraction of A.

Lemma 4. *The area of S^* is at least $\frac{1}{128}$ times the area of B.*

Proof. We distinguish two cases. If B is narrow, that is, $h < \frac{1}{16}w$, see Figure 4(b), then there exist two points p and q among the endpoints of $\mathcal{L} - \{l_{max}\}$ such that the distance between them is at least $\frac{1}{2}w$, and they are not endpoints of the same segment. There also exists a point r that is an endpoint of l_{max} that is at least $(2\sqrt{2}-1)h$ away from B, since we know that the width of the original input was at least $\frac{1}{8}\sqrt{2}$ times the diameter. Now r has to be at least $(2\sqrt{2}-2)h > \frac{1}{2}h$ away from the line extending $\overline{pq}$, so the area of $\triangle pqr$ is at least $\frac{1}{8}wh$.

If B is not narrow, that is, $h \geq \frac{1}{16}w$, there exist three points p, q and r among the endpoints of $\mathcal{L} - \{l_{max}\}$ such that the distance between any pair is at least h, otherwise there would be a smaller bounding box. If they are all endpoint of different segments, $\triangle pqr$ is a valid solution of area at least $\frac{1}{4}h^2 > \frac{1}{64}wh$. Otherwise, one of the segments has length at least h, and therefore also l_{max} has length at least h. Suppose p and q are endpoints of the same segment. Now there has to be an endpoint s of l_{max} such that s is at least $\frac{1}{2}h$ away from either the line extending $\overline{pr}$ or the line extending $\overline{qr}$. This means that either $\triangle prs$ or $\triangle qrs$ (or both) has area at least $\frac{1}{8}h^2 > \frac{1}{128}wh$.

In both cases we have a valid solution with an area of at least $\frac{1}{128}wh$, so the area of the optimal solution will also be at least $\frac{1}{128}wh$.

Lemma 5. *The area of S^* is at least $\frac{1}{128}gw$.*

Proof. There must be at least two points belonging to different line segments in B that are half the diameter apart. Because the original input was scaled to be not narrow, there must be one endpoint of l_{max} far enough away in the direction perpendicular to the diameter of B. These three points form a triangle with the required area. The complete proof is in the full paper.

As a consequence of these lemmata, we now know that the area of S^* is at least a constant fraction of A: $area(S^*) \geq \frac{1}{256}A$.

Lemma 6. *There exists a solution S' for $\mathcal{L}'$ such that the difference between the areas of S' and S^* is at most $\frac{\varepsilon}{256}$ times A.*

Proof. Let Y be the set of grid cells used by the optimal solution S^*. There exists a perfect matching between $\mathcal{L}$ and Y, otherwise S^* would not be possible. By Lemma 1, we know that there is also a perfect matching between $\mathcal{L}'$ and Y. Let S' be the convex hull of the point set that realises this matching, and uses the same endpoint of l_{max} as S^*. Then for each vertex of S^* there is a point of S' in the same grid cell. Going from S^* to S', all vertices can move a distance of δ_1 in the e_1 direction, and δ_2 in the e_2 direction, see Figure 4(c). In the worst case, the transformed solution has a complete 'band' around it. The area of such a band is composed of two triangles incident to l_{max}, which together are smaller than $(\delta_1 + \delta_2)d$, and a part that lies completely within B, which is smaller than $2\delta_1 h + 2\delta_2 w$. In total this is smaller than $\frac{\varepsilon}{256}A$.

We need that the optimal solution of $\mathcal{L}'$ is at least $(1 - \varepsilon)$ times as large as the optimal solution of $\mathcal{L}$. By Lemma 6, there is a solution S' of $\mathcal{L}'$ with an area of at most $\frac{\varepsilon}{256}A$ away from the area of S^*. Furthermore, by Lemmata 4 and 5 we know that the area of S^* is at least $\frac{1}{256}A$. Therefore we have $S'^* \geq S' \geq S^* - \frac{\varepsilon}{256}A \geq S^* - \frac{\varepsilon}{256}256S^* = (1 - \varepsilon)S^*$.

Running Time Analysis. To ensure that the input $\mathcal{L}$ is not too skinny, we need to compute an approximate bounding box of the endpoints, which can be done in linear time, according to Lemma 2. Again, we need to perform the ceiling operation to allocate the endpoints of the segments to the right cells of the grid.

To compute a maximum matching, we can use the algorithm by Hopcroft and Karp [10], which runs in $O(\sqrt{|V|}|E|)$ time. In our case, we have $n-1$ nodes on the left side and $2^{20}\eta^2$ nodes on the right side, and every left node has degree 2. When there are more than two left nodes that are connected to the same two right nodes, we will never use more than two of them, so we can reduce the number of left nodes to at most $2 \cdot 2^{40}\eta^4$ by using radix sort. The number of edges is twice the number of left nodes. Now we can compute a maximum matching in $O(\eta^6)$ time. In total this takes $O(n+\eta^6)$ time.

Theorem 3. *We can compute a core-set of size $O(\eta^2)$ for Problem 3 in $O(n+\eta^6)$ time.*

The problem is NP-hard, so unless P=NP there is no polynomial time exact algorithm. A trivial approach takes $O(2^n n \log n)$ time. Using this, we achieve an approximation running time of $O(n) + 2^{O(\eta^2)}$.

4.3 Squares

Problem 4. *Given a set of axis-parallel squares, choose a point in each square such that the area of the convex hull of the resulting point set is as large as possible.*

The status of the general version of this problem is open. In the optimal solution, every point has to be chosen on a corner of its square. Therefore we can solve the problem in $O(4^n n \log n)$ time by computing the convex hull of every possible set of corners.

Under certain conditions, the problem can be solved more efficiently. If the squares are disjoint, we can solve it exactly in $O(n^7)$ time. If the squares all have the same size, we can solve it in $O(n^5)$ time. If the squares are both disjoint and of the same size, we only need $O(n^3)$ time. All of these results can be found in [14].

Core-Set Construction. Let s_{max} be the largest square in $\mathcal{L}$, and s_{max2} the second largest square. Put s_{max} and s_{max2} in $\mathcal{L}'$. Let p and q be two points that approximate the diameter d of the vertices of $\mathcal{L} - \{s_{max}, s_{max2}\}$ by a factor 2. Let e_1 be the direction from p to q, and e_2 the direction perpendicular to this. Let B be the smallest bounding box of $\mathcal{L} - \{s_{max}, s_{max2}\}$ in the (e_1, e_2) axis system, and let w be its width and h its height.

Divide B into $2^{14}\eta$ by $2^{14}\eta$ grid cells, see Figure 5(a). The cells will be $\delta_1 = 2^{-14}\varepsilon w$ long in the e_1 direction, and $\delta_2 = 2^{-14}\varepsilon h$ long in the e_2 direction. Consider the bipartite graph where one set of nodes corresponds to the set of squares $\mathcal{L} - \{s_{max}, s_{max2}\}$, and the other set of nodes corresponds to the cells of the grid. There is an edge between square s and cell c if one of the corners of s is in c. Let M be a maximum matching of this graph, and add all squares that occur in M to $\mathcal{L}'$.

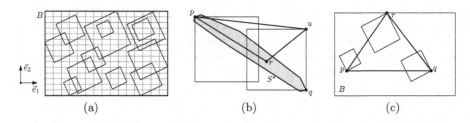

Fig. 5. (a) A set of squares divided according to their cells. (b) Triangle $\triangle pru$ has a larger area than S^*. (c) If all squares are small, the triangle $\triangle pqr$ has a large area.

Proof of $\mathcal{L}'$ being a Core-Set. Let S^* be the optimal solution for $\mathcal{L}$, the original input, and let S'^* be the optimal solution for $\mathcal{L}'$, the core-set. First we show that the area of S^* is bounded from below by a constant factor of the area of B. Then we prove that the difference in area between S^* and S'^* is only a fraction of the area of B, dependent on ε.

Lemma 7. *If $n \geq 3$, then the width of S^* is at least $\frac{1}{8}$ times the side length of s_{max2}.*

Proof. Let b be the side length of s_{max2}, and let w^* be the width of S^*. Suppose the lemma is not true, so $w^* < \frac{1}{8}b$. Let p and q be the vertices of S^* that define the diameter d of S^*. Then we know that the area of S^* is $a^* \leq dw^* < \frac{1}{8}db$. Suppose either p or q is not a corner of one of the two largest squares. Then one of the two largest squares has a corner u that is at least $\frac{1}{2}b$ away from the line extending $\overline{pq}$, and there exists a solution of area $\frac{1}{4}db > \frac{1}{8}db$, so in this case S^* would not be optimal, a contradiction. Now suppose that both p and q are corners of the largest two squares, see Figure 5(b). Let $r \neq p, q$ be an arbitrary vertex of S^*. Now r is at least $\frac{1}{2}d$ away from either p or q, say p without loss of generality. Now the square that has q as a corner has another corner u that is at least $\frac{1}{2}b$ away from the line extending $\overline{pr}$, and there exists a solution of area $\frac{1}{8}db$, so in this case S^* would not be optimal either. Therefore the assumption is false, and the lemma is true.

This lemma implies that the area of S^* is at least 2^{-8} times the area of s_{max2}.

Lemma 8. *The area of S^* is at least 2^{-12} times the area of B.*

Proof. Let b be the side length of s_{max2}. If $b \geq \frac{1}{4}w$, then this square has area at least $2^{-4}wh$. The optimal solution has area at least 2^{-8} times the second largest square, so at least $2^{-12}wh$. Next, assume that $b < \frac{1}{4}w$. If p and q, approximating the diameter of the vertices of $\mathcal{L} - \{s_{max}, s_{max2}\}$, would be corners of the same square, then the width of this square would be at least $\frac{1}{2}d \geq \frac{1}{2}w$, which is larger than the diameter of s_{max2}. So p and q are corners of different squares. Let r be the point in P furthest from the line extending $\overline{pq}$, see Figure 5(c). If r is a corner of yet another square, then the solution pqr has an area of at least $\frac{1}{8}wh$, and so the optimal solution also has at least that area. If r is a corner of the same square as either p or q, say p, then this means that the width of this square is larger than $\frac{1}{4}\sqrt{2}h$, so $b > \frac{1}{4}h$. The optimal solution S^* uses some corner p'_l of the same square as p_l, the leftmost point in the e_1 direction, and some corner p'_r of the same square as p_r, the rightmost point in the e_1 direction and we know that the distance between p'_l and p'_r is at least $w - 3b > \frac{1}{4}w$. We also know that S^* has a width of at least $\frac{1}{8}b > \frac{1}{32}h$, so the area of S^* is at least $2^{-6}wh$.

Lemma 9. *There exists a solution S' for $\mathcal{L}'$ such that the difference between the areas of S' and S^* is at most $2^{-12}\varepsilon$ times the area of B.*

Proof. The proof of this lemma is omitted, and very similar to the proof of Lemma 6.

This time we have $S'^* \geq S' \geq S^* - 2^{-12}\varepsilon wh \geq S^* - 2^{-12}\varepsilon 2^{12}S^* = (1 - \varepsilon)S^*$.

Running Time Analysis. The computation of B takes linear time, by Lemma 2. Again, we need to perform the ceiling operation to allocate the corners of the squares to the right cells of the grid. We can compute a maximum matching in $O(n + \eta^{12})$ time, since we now have four edges per left node.

Theorem 4. *We can compute a core-set of size $O(\eta^2)$ for Problem 4 in $O(n+\eta^{12})$ time.*

For arbitrary squares, we can solve the problem exactly in $O(4^n n \log n)$ time; therefore we can approximate it in $O(n) + 4^{O(\eta^2)}$ time. For unit size squares, we have an $O(n^5)$ exact algorithm so we get a strong linear time approximation scheme that runs in $O(n + \eta^{12} + \eta^{10}) = O(n + \eta^{12})$. We can solve disjoint squares exactly in $O(n^7)$ so we get an $O(n + \eta^{14})$ LTAS. For squares that are both unit size and disjoint, we have an $O(n^3)$ exact algorithm, but this gives no better result than the general unit size case since the running time is dominated by the term η^{12}.

4.4   Circles

Our exact solution to the convex hull problem for square regions makes use of the four extreme points in the cardinal directions, which makes it impossible to extend to circular regions. A second difficulty is of an algebraic kind. Even if we know which circles have points that contribute to the largest area convex hull, it is not easy to determine where on the circles the points should be. These difficulties remain even for disjoint unit size regions [13].

When we model the points as circular regions (discs), we only need to consider the boundaries, since no vertex of an optimal solution need be chosen in the interior of a region.

Problem 5. *Given a set of circles, choose a point in each circle such that the area of the convex hull of the resulting point set is as large as possible.*

Approximate Circles by k-gons. Let ε be given. Let $\mathcal{C}$ be the set of circles, and $\mathcal{C}'$ the set of circles with the same centres but radii a factor $(1 - \delta)$ smaller, where $\delta = \frac{1}{8}\varepsilon$.

Lemma 10. *The area of the optimal solution for C' is at least $(1-\varepsilon)$ times the area of the optimal solution of C.*

Proof. The proof is omitted due to space constraints and can be found in the full paper.

We will now approximate the circular imprecise points by k-gons that lie completely within the band between the original circle and the circle with a factor $(1-\delta)$ smaller radius. A k-gon fits inside this band when $2k\arccos(1-\delta) \geq 2\pi$, and this can be estimated by $k \geq 2\pi\sqrt{\eta}$. Let $k = \lceil 2\pi\sqrt{\eta}\rceil$, and $\mathcal{G}$ the set of k-gons (with the same orientation) that have their corners on the circles of C.

Theorem 5. *The area of the optimal solution for $\mathcal{G}$ is a $(1-\varepsilon)$ approximation for the optimal solution for C.*

Exact Algorithms. Again, we have a trivial exponential algorithm that runs in $O(k^n n \log n)$ time. As in the case of squares, we can achieve a better algorithm under certain constraints. If the k-gons are either disjoint or unit size, we can solve the problem in $n^{O(k)}$ instead of $k^{O(n)}$ time. We can adapt the algorithm described in [13] in a mostly straightforward manner to the k-gon case. We will briefly discuss the main differences and new ideas that are needed to make these algorithms work.

For both algorithms, we need to know the k extreme points of the solution. These are the vertices of the solution that lie furthest in one of the k directions that are perpendicular to the edges of a k-gon. Trying all possibilities gives a factor $O(n^k)$.

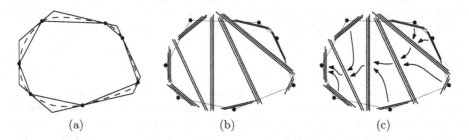

(a)   (b)   (c)

Fig. 6. (a) The division of the plane for $k = 7$. (b) There are 11 groups of parallel line segments. (c) The order in which the groups can be combined.

Suppose the k-gons are disjoint. The k extreme points divide the plane into k triangular regions, see Figure 6(a). For each k-gon, we only need to consider the endpoints that are within their respective triangle. Since the k-gons are disjoint, there can be at most $k-2$ k-gons that intersect more than two of these triangles. For these k-gons, we try every possible combination of their candidate endpoints. This gives a factor $O(k^k)$.

The remaining k-gons can now be represented as line segments. There are at most $2k-3$ groups of line segments, see Figure 6(b). We can solve the problem in this situation in $O(kn^3)$ time, using a dynamic programming approach as described in [13]. We start with two consecutive groups that pass over only one extreme point, for which there is no group between them. For these two groups, we compute the optimal solution for every pair of points. Then we combine them with the group that passes over both extreme points. This process is repeated until we have found the optimal solution, see Figure 6(c).

Now suppose the k-gons have equal sizes (but are not necessarily disjoint). The algorithm described in [13] still works exactly as described there, only with k chains instead of four. This gives a running time of $O(n^{k+1})$ instead of $O(n^5)$.

Core-Set Construction. A core-set of a set of regular k-gons can be computed in exactly the same way as with squares, as long as $k \geq 4$. The same proof also applies.

Running Time Analysis. Constructing a core-set of size $O(\eta^2)$ takes $O(\sqrt{|V|}|E|)$ time. In our case, we have $O(\eta^{2k})$ nodes at the left side after removing doubles, and $O(\eta^2)$ nodes at the right side, and each left node has exactly k edges, so $|V| = O(\eta^{2k} + \eta^2)$ and $|E| = O(\eta^{2k+\frac{1}{2}})$. This means that the core-set selection algorithm runs in $O(n + \eta^{3k+\frac{1}{2}}) = O(n) + 2^{O(\sqrt{\eta}\log\eta)}$ time. Again, provided that the ceiling operation takes constant time.

Theorem 6. *We can compute a core-set of size $O(\eta^2)$ for Problem 5 in $O(n) + 2^{O(\sqrt{\eta}\log\eta)}$ time.*

The general problem can be solved exactly in $O(k^n n \log n)$ time. The approximation algorithm then takes $O(n) + 2^{O(\sqrt{\eta}\log\eta)} + O(k^{\eta^2}\eta^2 \log\eta) = O(n) + 2^{O(\eta^2\log\eta)}$ time in total.

Under the assumption that the circles are either disjoint or unit size, we have a better exact algorithm, which runs in $n^{O(k)}$ time. The approximation algorithm then takes $O(n) + 2^{O(\sqrt{\eta}\log\eta)} + \eta^{O(\sqrt{\eta})} = O(n) + 2^{O(\sqrt{\eta}\log\eta)}$ time.

5 Conclusions

The *core-set* paradigm has been successfully applied to sets of imprecise points, to obtain approximation algorithms for computationally hard problems. The dependence of the running time on the input size is linear and does not multiply with the dependence on ε, which makes the algorithms suitable for very large sets of imprecise points. On the other hand, the dependence on ε is often highly polynomial or exponential, which limits the achievable precision.

Acknowledgements. The authors would like to thank anonymous referees for their detailed and helpful comments.

References

1. Abellanas, M., Hurtado, F., Ramos, P.A.: Structural tolerance and Delaunay triangulation. Inf. Proc. Lett. 71, 221–227 (1999)
2. Agarwal, P.K., Har-Peled, S.: Approximating extent measures of points. In: Proc. 12th Annual ACM-SIAM Symposium on Discrete Algorithms, pp. 148–157 (2001)

3. Agarwal, P.K., Har-Peled, S., Varadarajan, K.R.: Geometric approximation via coresets. In: Goodman, J.E., Pach, J., Welzl, E. (eds.) Combinatorial and Computational Geometry, MSRI Publications, vol. 52, Cambridge University Press, Cambridge (2005)
4. Bandyopadhyay, D., Snoeyink, J.: Almost-Delaunay simplices: Nearest neighbour relations for imprecise points. In: Proc. 15th ACM-SIAM Symposium on Discrete Algorithms, pp. 410–419 (2004)
5. Chan, T.M.: Approximating the diameter, width, smallest enclosing cylinder, and minimum-width annulus. Int. J. Comput. Geometry Appl. 12(1-2), 67–85 (2002)
6. Fiala, J., Kratochvil, J., Proskurowski, A.: Systems of distant representatives. Discrete Applied Mathematics 145, 306–316 (2005)
7. Frahling, G., Sohler, C.: A fast k-means implementation using coresets. In: Proc. 22nd Annual ACM Symposium on Computational Geometry, pp. 135–143 (2006)
8. Gao, J., Langberg, M., Schulman, L.J.: Analysis of incomplete data and an intrinsic dimension Helly theorem. In: Proc. 17th Symposium on Discrete Algorithms, pp. 464–473 (2006)
9. Guibas, L.J., Salesin, D., Stolfi, J.: Constructing strongly convex approximate hulls with inaccurate primitives. Algorithmica 9, 534–560 (1993)
10. Hopcroft, J.E., Karp, R.M.: An $n^{\frac{5}{2}}$ algorithm for maximum matching in bipartite graphs. SIAM Journal on Computing 4, 225–231 (1973)
11. Jadhav, S., Mukhopadhyay, A., Bhattacharya, B.K.: An optimal algorithm for the intersection radius of a set of convex polygons. J. Algorithms 20, 244–267 (1996)
12. Khanban, A.A., Edalat, A.: Computing Delaunay triangulation with imprecise input data. In: Proc. 15th Canad. Conf. on Comput. Geom., pp. 94–97 (2003)
13. Löffler, M., van Kreveld, M.: Largest and smallest convex hulls for imprecise points. Technical Report UU-CS-2006-019, Utrecht University, Institute of Information and Computing Sciences (May 2006)
14. Löffler, M., van Kreveld, M.: Largest and smallest tours and convex hulls for imprecise points. In: Arge, L., Freivalds, R. (eds.) SWAT 2006. LNCS, vol. 4059, pp. 375–387. Springer, Heidelberg (2006)
15. van Kreveld, M., Löffler, M.: Largest bounding box, smallest diameter, and related problems on imprecise points. In: Proc. 10th Workshop on Algorithms and Data Structures, LNCS, vol. 4619, pp. 447–458 (2007)
16. Yap, C.-K.: Robust geometric computation. In: Goodman, J.E., O'Rourke, J. (eds.) Handbook of Discrete and Computational Geometry, ch. 41, pp. 927–952. Chapman & Hall/CRC (2004)

A 2-Approximation Algorithm for the Metric 2-Peripatetic Salesman Problem

Alexander A. Ageev[*] and Artem V. Pyatkin[**]

Sobolev Institute of Mathematics, pr. Koptyuga 4, Novosibirsk, Russia
{ageev,artem}@math.nsc.ru

Abstract. In the m-peripatetic traveling salesman problem (m-PSP), given an n-vertex complete undirected edge-weighted graph, it is required to find m edge disjoint Hamiltonian cycles of minimum total weight. The problem was introduced by Krarup (1974) and has network design and scheduling applications. It is known that 2-PSP is NP-hard even in the metric case and does not admit any constant-factor approximation in the general case. Baburin, Gimadi, and Korkishko (2004) designed a $(9/4 + \varepsilon)$-approximation algorithm for the metric case of 2-PSP, based on solving the traveling salesman problem. In this paper we present an improved 2-approximation algorithm with running time $O(n^2 \log n)$ for the metric 2-PSP. Our algorithm exploits the fact that the problem of finding two edge disjoint spanning trees of minimum total weight is polynomially solvable.

1 Introduction

In the m-peripatetic traveling salesman problem (m-PSP), we are given a complete undirected graph $G = (V, E)$ on n vertices with nonnegative edge weight function $w : E \to \mathbb{R}_+$. It is required to find m edge disjoint Hamiltonian cycles $C_1, \ldots, C_m \subset E$ minimizing $\sum_{k=1}^{m} \sum_{e \in C_k} w(e)$. The 1-PSP coincides with the Traveling Salesman Problem (TSP). Applications of m-PSP include the design of watchman tours [12] where it is often important to assign a set of edge disjoint rounds to the watchman in order to avoid repeating the same tour and thus enhance security. De Kort [8] cites a network design application where several edges-disjoint cycles must be determined in order to protect the network from link failure. De Kort also mentions a scheduling application of the 2-PSP where each job must be processed twice by the same machine but technological constraints prevent the repetition of identical job sequences.

Related results. De Kort [8] proved that the 2-PSP is NP-hard by constructing a polynomial-time reduction from the Hamiltonian Path Problem. By similar

---

[*] Supported by the Russian Foundation for Basic Research (project codes 06-01-00255, 05-01-00960).

[**] Supported by the Russian Foundation for Basic Research (project code 05-01-00395) and INTAS (project code 04–77–7173).

arguments one can show that m-PSP is NP-hard for each $m > 2$. De Brey and Volgenant [4] identified several polynomially solvable cases of 2-PSP. De Kort [6,7,8] designed and analyzed lower and upper bounds for 2-PSP as possible ingredients of branch-and-bound algorithms. Duchenne, Laporte, and Semet [5] discussed a polyhedral approach for solving m-PSP.

It is easy to show (by using the reduction in [8] and essentially the same argument that is used for TSP) that 2-PSP admits no constant-factor approximation algorithm in the case of general edge weights. For the case when the weights satisfy the triangle inequality (the metric 2-PSP) Baburin, Gimadi, and Korkishko [2] designed a $(\frac{9}{4} + \varepsilon)$-approximation algorithm.

Like that of TSP, the maximization version of 2-PSP admits constant factor approximations even in the general case. The currently best result is due Ageev, Baburin, and Gimadi [1] who presented a 3/4-approximation algorithm for the problem.

Our result. In this paper we present a 2-approximation algorithm for solving the metric case of 2-PSP. The algorithm runs in time $O(n^2 \log n)$.

2   Algorithm: A General Scheme

Since 2-PSP has no feasible solution for $n < 5$, we further assume that $n \geq 5$.

Below we present a general scheme of our algorithm. Recall that a graph G is *outerplanar* if it can be drawn at the plane in such a way that no two edges meet in a point other than a common vertex and all vertices of G lie in the outer face.

Algorithm. Disj_Ham_Cycles

Phase 0. By using the algorithm of Roskind and Tarjan [10] find two disjoint spanning trees T_1^* and T_2^* of total minimum weight.

Phase 1. Find a Hamiltonian cycle C_1 and two disjoint spanning trees T_1 and T_2 such that
1. $T_1 \cup T_2 = T_1^* \cup T_2^*$;
2. $T_2 \cap C_1 = \emptyset$;
3. The graph $C_1 \cup T_1$ is outerplanar with the outer face C_1.

Phase 2. Find a Hamiltonian cycle C_2 such that
1. $C_1 \cap C_2 = \emptyset$;
2. The graph $C_2 \cup T_2$ is outerplanar with the outer face C_2.

Output C_1 and C_2.

At Phase 0 we use the algorithm of Roskind and Tarjan [10] that finds k edge disjoint spanning trees of minimum total weight in time $O(n^2 \log n + k^2 n^2)$. Thus Phase 0 can be implemented in time $O(n^2 \log n)$.

The detailed descriptions of Phases 1 and 2 with the related theoretical background and running time bounds are given in the next two sections.

The general description of Algorithm `Disj_Ham_Cycles` is sufficient to establish the bound on its approximation ratio:

Lemma 1. *The algorithm `Disj_Ham_Cycles` outputs a feasible solution of 2-PSP whose weight is at most twice the weight of the optimal solution.*

Proof. Clearly, C_1 and C_2 are edge disjoint. We show that $w(C_1) + w(C_2) \leq 2(w(C_1^*) + w(C_2^*))$ where C_1^*, C_2^* is an optimal solution of 2-PSP. Clearly, $w(C_1^*) + w(C_2^*) \geq w(T_1^*) + w(T_2^*)$. Now let us evaluate $w(C_i)$. Let $i \in \{1, 2\}$ and let e be an edge of the cycle C_i that does not belong to T_i. Denote by G_i the outerplanar graph $C_i \cup T_i$ and by $F(e)$, the edges of the inner face containing e. Due to the triangle inequality, the weight of the edge e can be bounded by the sum of the weights of the edges from the unique path in T_i connecting the endpoints of e. Since G_i is an outerplanar graph and T_i is a spanning tree, each face of G_i contains exactly one edge not from T_i. Hence, the path in T_i connecting the endpoints of e consists of all edges of $F(e) \setminus \{e\}$. Each chord of G_i belongs to two inner faces, so its weight is counted twice. Each edge in $C_i \cap T_i$ lies in one inner face; so, its weight is counted once for the path, and once for itself (as an edge of C_i). Therefore $w(C_1) + w(C_2) \leq 2(w(T_1) + w(T_2)) = 2(w(T_1^*) + w(T_2^*)) \leq 2(w(C_1^*) + w(C_2^*))$. □

3   Implementation of Phase 1

The general idea is the following: we first split T_1^* into a few relatively small subtrees, then for each of them find an appropriate cycle (exchanging edges between T_1^* and T_2^* if necessary), and finally construct a desired cycle C_1 from these cycles.

We first show by a graph theoretical argument that a cycle C_1 with the desired properties does exist and then present a linear-time algorithm that follows from the proof.

Let H be an undirected graph. By $V(H)$ we denote the vertex set of H. For a subset $V' \subseteq V(H)$, denote by $H[V']$ a subgraph of H induced by V'.

Let T_1 and T_2 be edge disjoint spanning trees of the input graph $G = (V, E)$. Suppose that $V' \subseteq V$ is such that $T = T_1[V']$ is a spanning tree of $G[V']$, $F = T_2[V']$ is a forest in $G[V']$, and T and F are edge disjoint. We say that T is *F-walkable* if there exist a spanning tree T', a forest F' and a Hamiltonian cycle C such that:

1) $T' \cup F' = T \cup F$;
2) $(T_1 \setminus T) \cup T'$ and $(T_2 \setminus F) \cup F'$ are spanning trees of G;
3) $F' \cap C = \emptyset$
4) $T' \cup C$ is an outerplanar graph whose outer face is bounded by C.

The tree T is called *walkable* if it is F-walkable for every forest F such that F and T can be extended to edge disjoint spanning trees T_1 and T_2.

The following theorem is the main result of this section.

Theorem 1. *Every tree on at least 5 vertices is walkable.*

The algorithmic proof of this theorem is based on the following lemmas.

Lemma 2. *Let D_1 and D_2 be two walkable trees with disjoint vertex sets. Let $v_1 \in D_1$ and $v_2 \in D_2$ be arbitrary vertices of these trees. Let D be a tree obtained from D_1 and D_2 by identifying v_1 and v_2 into the vertex v (see Fig. 1a). Then D is walkable.*

Proof. Let $V = V[D]$, $V_i = V[D_i]$, $i = 1, 2$ and let F be an arbitrary forest on V such that D and F can be extended to edge disjoint spanning trees T_1 and T_2 of G. We show that D is F-walkable. Denote by F_1 the restriction of F to V_1. By the assumption, there exist a tree D_1', a forest F_1', and a cycle C_1 satisfying 1)–4) in the definition of walkable trees. By 2), $(F \setminus F_1) \cup F_1'$ is a forest and $(D \setminus D_1) \cup D_1'$ is a spanning tree in $V[D]$ and they can be extended to edge disjoint spanning trees $T_1' = (T_1 \setminus D_1) \cup D_1'$ and $T_2' = (T_2 \setminus F_1) \cup F_1'$. Denote by F_2 the restriction of F to V_1. Since D_2 is walkable, there exist a tree D_2', a forest F_2', and a cycle C_2 satisfying 1)–4) in the definition of walkable trees with T_1' and T_2' standing for T_1 and T_2 respectively. Let $F' = (F \setminus (F_1 \cup F_2)) \cup F_1' \cup F_2'$ and $D' = (D \setminus (D_1 \cup D_2)) \cup D_1' \cup D_2'$. Note that D' and F' satisfy 1) and 2). Now construct the desired cycle C. Let x_i and y_i be the neighbors of v in C_i, $i = 1, 2$. Then at least one of the edges $x_1 x_2, x_1 y_2, y_1 y_2$, and $y_1 x_2$ must be not in F'. By symmetry, we may assume that $x_1 x_2 \notin F$. Let $C = (C_1 \cup C_2 \cup x_1 x_2) \setminus \{x_1 v, x_2 v\}$. Then $C \cup F' = \emptyset$ and $C \cup D'$ is outerplanar and C is the boundary of the outer face of $C \cup D'$, i. e. D is F-walkable. Since F is an arbitrary forest, D is walkable. $\square$

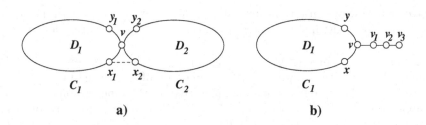

Fig. 1. Induction steps in Lemmas 2 and 3

Lemma 3. *Let D_1 be a walkable tree, $v \in D_1$ and D be a tree obtained from D_1 by adding the vertices v_1, v_2, v_3 and edges $vv_1, v_1 v_2, v_2 v_3$ (see Fig. 1b). Then D is walkable.*

Proof. Let $V_1 = V(D_1)$ and $V = V(D)$. Denote the path $vv_1 v_2 v_3$ by P. Consider an arbitrary forest F on V (such that D and F can be extended to edge disjoint spanning trees T_1 and T_2 of G) and denote by F_1 its restriction to V_1 and by F_2 the restriction to $V(P)$. If $v_3 v \notin F$ then P is clearly F_2-walkable (just take $C = P \cup v_3v$) and by the same arguments as were used in Lemma 2, D is F-walkable. Assume that $v_3v \in F$. Since D_1 is walkable, there exist a tree D_1', a forest F_1' and a cycle C_1 satisfying 1)–4). Let $F' = (F \setminus F_1) \cup F_1'$ and $D' = (D \setminus D_1) \cup D_1'$. Denote by x and y the neighbors of v in C_1. If $v_3x \notin F$ then let $C = (C_1 \cup P \cup v_3x) \setminus vx$. Clearly, F', D' and C satisfy 1)–4). If $v_3y \notin F$ we proceed in a similar way. So $v_3x, v_3y \in F$. Then at least one of the edges v_2x, v_2y cannot be in F (say, $v_2x \notin F$). If $v_3v_1 \notin F$ then we construct C from C_1 by removing the edge vx and adding the path $vv_1v_3v_2x$. Again, F', D' and C satisfy 1)–4). Finally, assume that v_3 is adjacent to x, y, v, and v_1 in F. Then $v_1x \notin F$. Let $D'' = (D' \cup v_3v) \setminus v_1v$, $F'' = (F' \cup v_1v) \setminus v_3v$, and $C = (C_1 \cup vv_3v_2v_1x) \setminus vx$. It is clear that D'', F'', and C satisfy 1),3), and 4). In order to see that they satisfy 2), note that the vertices v_3 and v lie in different components of $T_1 \setminus vv_1$; so the tree $T_1 \cup v_3v \setminus v_1v$ is spanning. Note also that the vertices v, v_1, and v_3 form the unique cycle in $T_2 \cup v_1v$; thus $T_2 \cup v_1v \setminus v_3v$ is a spanning tree as well. So, D is F-walkable. Since F is an arbitrary forest, D is walkable.   □

For a tree T, a forest F and edges $e \in T$ and $f \in F$, by the *exchange* we mean the operation of removing this edges from T and F and adding them to F and T, respectively (i. e. $T' = (T \cup f) \setminus e$ and $F' = (F \cup e) \setminus f$). The exchange is *correct* if T' and F' satisfy 2). The following observation helps to check that the exchanges used below are correct.

Note 1. An exchange is correct if f lies on a cycle in $F \cup e$ and the endpoints of f are in different components of $T \setminus e$.

Proposition 1. *Every tree on 5 vertices is walkable.*

Proof. There are three nonisomorphic trees on 5 vertices. So, three cases arise. Let F be an arbitrary forest on 5 vertices.

Case 1. T is a path of length 4 (see Fig. 2a). If $v_1v_5 \notin F$ then let $C = T \cup v_1v_5$, and T is walkable (for $T' = T, F' = F$). If F contains neither v_1v_3 nor v_3v_5 then we can apply Lemma 2 for trees D_1, D_2 induced by the sets $\{v_1, v_2, v_3\}$ and $\{v_3, v_4, v_5\}$ respectively. So, we may assume that $v_1v_3, v_1v_5 \in F$ but $v_3v_5 \notin F$. If $v_1v_4 \notin F$ then we may take the cycle $v_1v_2v_3v_5v_4$ as C, and T is walkable. Finally, if $v_1v_4 \in F$ then let $T' = (T \cup v_1v_4) \setminus v_3v_4$, $F' = (F \cup v_3v_4) \setminus v_1v_4$ and $C = v_1v_2v_3v_5v_4$. Clearly, T', F', and C satisfy 1)–4).

Case 2. T is as in Fig. 2b. If $v_4v_5 \in F$ then either $(T_2^* \cup v_3v_4) \setminus v_4v_5$ or $(T_2^* \cup v_3v_5) \setminus v_4v_5$ is a spanning tree. So, we can exchange the edge v_4v_5 with one of the edges v_3v_4 or v_4v_5 and reduce the problem to the previous case. So, $v_4v_5 \notin F$. If $v_1v_4 \notin F$ then we may take $C = v_1v_2v_3v_5v_4$; if $v_1v_3 \notin F$ then the subtrees $T \setminus \{v_1, v_2\}$ and $T \setminus \{v_4, v_5\}$ are both F-walkable, and by Lemma 2, T is walkable. So, we may assume that $v_1v_3, v_1v_4 \in F$. Then exchange v_1v_4 with v_3v_4 and reduce the problem to Case 1.

Case 3. T is the star $K_{1,4}$. If none of the edges connecting the leaves of the star belong to F then T is clearly walkable (an arbitrary cycle can be taken as C). If some edge e connecting two leaves lies in F then we can exchange this

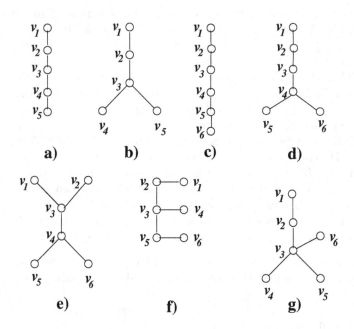

Fig. 2. Trees on 5 and 6 vertices

edge with one of the edges connecting an endpoint of e with the center of the star, thus reducing the problem to the Case 2.   □

Proposition 2. *Every tree on 6 vertices is walkable.*

Proof. There are six nonisomorphic trees on 6 vertices. Thus six different cases arise. Let F be an arbitrary forest on 6 vertices.

Case 1. T is a path of length 5 (see Fig. 2c). If $v_1v_6 \notin F$ then T, F, and $C = T \cup v_1v_6$ satisfy 1)–4). If F does not contain v_1v_3 then the subtree $T \setminus \{v_4, v_5, v_6\}$ is clearly F-walkable and by Lemma 3, T is also walkable. So $v_1v_3 \in F$, and analogously, $v_4v_6 \in F$. But then $v_1v_4, v_3v_6 \notin F$ and we can take $C = v_1v_2v_3v_6v_5v_4$.

Case 2. T is as in Fig. 2d. If $v_5v_6 \in F$ then, like in Proposition 1, we can exchange the edge v_5v_6 with one of the edges v_4v_5 or v_4v_6 and reduce the problem to the previous case. If $v_5v_6 \notin F$, then the subtree $T \setminus \{v_1, v_2, v_3\}$ is F-walkable and by Lemma 3, T is also walkable.

Case 3. T is as in Fig. 2e. At least on of the edges connecting $\{v_1, v_2\}$ with $\{v_5, v_6\}$ must be missing in F. We may assume that $v_1v_5 \notin F$. If either $v_1v_2 \in F$ or $v_5v_6 \in F$ then we reduce the problem to Case 2 in the same way as Case 2 was reduced to Case 1. Otherwise, take $C = v_1v_2v_3v_4v_6v_5$.

Case 4. T is as in Fig. 2f. Consider two subcases.

a) Either $v_1v_4 \notin F$ or $v_4v_6 \notin F$ (assume that the second alternative holds). If $v_1v_3 \notin F$ then both trees $T \setminus \{v_1, v_2\}$ and $T \setminus \{v_4, v_5, v_6\}$ are F-walkable, and by Lemma 2, T is also walkable. If $v_1v_5 \notin F$ then T, F, and $C = v_1v_2v_3v_4v_6v_5$ satisfy 1)–4). Finally, if $v_1v_3, v_1v_5 \in F$ we can exchange v_3v_5 with v_1v_5, arriving at Case 1.

b) Both v_1v_4 and v_4v_6 are in F. Clearly, $v_1v_6 \notin F$. If $v_2v_4 \notin F (v_5v_4 \notin F)$ then T, F, and $C = v_1v_2v_4v_3v_5v_6 (C = v_1v_2v_3v_4v_5v_6)$ satisfy 1)–4). If $v_2v_4, v_5v_4 \in F$ then $v_1v_5, v_2v_5 \notin F$. Exchange v_5v_6 with v_4v_6 and let $C = v_1v_6v_4v_3v_2v_5$.

Case 5. T is as in Fig. 2g. If the set $\{v_4, v_5, v_6\}$ contains at least one edge from F, then this tree can be reduced to the previous case by the same way as Case 2 was reduced to Case 1. Otherwise, trees $T \setminus \{v_1, v_2\}$ and $T \setminus \{v_1, v_2, v_4\}$ are F-walkable. Then by Lemma 2, $v_1v_3 \in F$ and $v_1v_4 \in F$. Now we can exchange v_3v_4 with v_1v_4 arriving at Case 2.

Case 6. If T is the star $K_{1,5}$, then it can be reduced to Case 5 by exactly the same arguments as were used in Case 3 of Proposition 1.   □

Let v be a vertex of a tree T. Denote by $A_1, A_2, \ldots, A_k$ the components of $T \setminus v$ and let $a_i = |A_i|, i = 1, 2, \ldots, k$. We may assume that $a_1 \geq a_2 \geq \ldots \geq a_k$. The vertex v is called a *center* of T if a_1 achieves minimum for all vertices of T. It is easy to see that if v is a center, then the inequality

$$a_1 \leq a_2 + a_3 + \ldots + a_k + 1 \tag{1}$$

holds. In particular, $a_1 \leq \lfloor n/2 \rfloor$.

We say that a vertex $v \in V(T)$ is a *bud* if it is adjacent to at least two leaves.

Proof of Theorem 1. Suppose that the theorem is false. Then consider counterexamples to it with the minimum number of vertices n. Among them choose one with the minimum number of leaves adjacent to buds. Denote it by T. By Propositions 1 and 2, $n \geq 7$. Let F be an arbitrary forest nonintersecting with T.

Claim 1. T has no buds.

Indeed, if u is a bud adjacent to leaves v and w then the tree $T \setminus \{v, w\}$ has at least 5 vertices. By the minimality of T, it is walkable. If $vw \notin F$ then the tree induced by u, v, and w is F-walkable, and by Lemma 2, T is also walkable, a contradiction. Otherwise, we can exchange the edge vw with either uv or uw, obtaining a tree T' with a smaller number of leaves adjacent to buds. By the choice of T, we have that T' is walkable, and hence T is walkable. Claim 1 is proved.

Let v be a center of T. By Claim 1, $a_{k-1} \geq 2$.

Claim 2. There is no $I \subset \{1, 2, \ldots, k\}$ such that $\sum_{i \in I} a_i \geq 4$ and $\sum_{i \notin I} a_i \geq 4$.

Indeed, otherwise both trees induced by $\cup_{i \in I} A_i \cup \{v\}$ and $\cup_{i \notin I} A_i \cup \{v\}$ have at least 5 vertices. By minimality of T both of them are walkable. But then by Lemma 2, T is also walkable. Claim 2 is proved.

In particular, if $a_1 = 4$ then $a_2 + a_3 + \ldots + a_k \leq 3$. Note that by (1) and Claim 2, $a_1 \leq 4$.

Claim 3. If $a_i = 3$ for some i then $n = 7$.

Indeed, by Claim 1, the tree induced by $A_i \cup \{v\}$ must be a path of length 3. If $n > 7$ then $|T \setminus A_i| \geq 5$ and by the minimality of T, the tree $T \setminus A_i$ is walkable. But then by Lemma 3, T is also walkable, a contradiction. Claim 3 is proved.

Only the following special trees satisfy the properties stated in Claims 1–3.

1. Two trees with parameters $a_1 = 4, a_2 = 2, a_3 = 1$ are depicted in Fig. 3d and Fig. 3e

2. The tree with parameters $a_1 = 3, a_2 = 3$ is depicted in Fig. 3a.

3. The tree with parameters $a_1 = 3, a_2 = 2, a_3 = 1$ is depicted in Fig. 3b.

4. The tree with parameters $a_1 = 2, a_2 = 2, a_3 = 2$ is depicted in Fig. 3c.

5. The tree with parameters $a_1 = 2, a_2 = 2, a_3 = 2, a_4 = 1$ is depicted in Fig. 3f.

Consider each of the special trees separately (the name of the case corresponds to the name of the graph in Fig. 3).

a) If $T = P_6$ then by Lemma 3, $v_1 v_4, v_4 v_7 \in F$. However, then $v_1 v_7 \notin F$ and we can take $C = v_1 v_2 v_3 v_4 v_5 v_6 v_7$.

b) We have $v_6 v_7 \in F$ by Lemma 3 and $v_4 v_6 \in F$ by Lemma 2 and choice of T. Then we exchange $v_4 v_7$ with $v_6 v_7$ reducing the problem to the previous case.

c) By Lemma 2 and choice of T, $v_1 v_3, v_3 v_5, v_3 v_7 \in F$. Then $v_1 v_7, v_5 v_7 \notin F$ and at least one of the edges $v_1 v_6, v_5 v_6$ is not in F (say, $v_5 v_6 \notin F$). Then T, F, and $C = v_1 v_2 v_3 v_4 v_5 v_6 v_7$ satisfy 1)–4).

d) By the minimality of T, subtree $T \setminus \{v_5, v_6, v_7\}$ is walkable. Then by Lemma 3, T is also walkable.

e) By Lemma 2 and choice of T, the trees $T \setminus \{v_5, v_6\}$ and $T \setminus \{v_5, v_6, v_7\}$ are walkable; so, $v_4 v_6, v_6 v_7 \in F$. Then exchange $v_4 v_7$ with $v_6 v_7$ and reduce the problem to the previous case.

f) Like in case c), $v_1 v_3, v_3 v_5 \in F$. But then $v_1 v_8 \notin F$ or $v_5 v_8 \notin F$. In any case, T is walkable by Lemma 2. $\qquad\square$

It is easy to see that the proof of Theorem 2 in fact contains an algorithm for finding the cycle C_1. The algorithm **Phase_1** is recursive and can be described as follows.

Algorithm Phase_1.

Set $T_1 := T_1^*$ and $T_2 := T_2^*$.

Step 1. If T_1 has 5 or 6 vertices or T_1 is a graph shown in Fig. 3, then the cycle C_1 can be found straightforwardly as described in the proofs of Propositions of 4.4, 4.5 and in the final part of the proof of Theorem 4.1.

Otherwise, T_1 does not satisfy a conclusion of one of the Claims 1–3 in the proof of Theorem 4.1.

Step 2. Suppose that T_1 has a bud x with leafs u, v (i. e., T_1 does not satisfy the conclusion of Claim 1).

If $u, v \in T_2$ then transform T_1 and T_2 by the exchange described in the proof of Claim 1 into trees T_1' and T_2', where T_1' has less number of leaves adjacent to buds than T_1. Set $T_1 := T_1', T_2 := T_2'$ and go to Step 1.

If $u, v \notin T_2$ then applying Phase_1 to the subtree $T_1 \setminus \{u, v\}$ find the cycle C', and construct a cycle C_1 for T_1 from the cycles C' and $C'' = xuv$ using the procedure described in the proof of Lemma 4.2.

Step 3. Find a center v of T_1 (indeed, it is enough to find any vertex satisfying (1)).

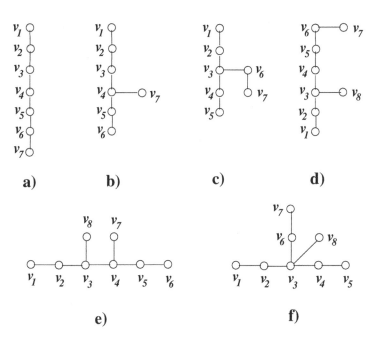

Fig. 3. Special trees

Step 4. If $n > 7$ and $a_i = 3$ for some i (i. e., T_1 does not satisfy the conclusion of Claim 3), then applying Phase_1 to the tree $T_1 \setminus A_i$ find a cycle C' for these tree. Using the procedure described in Lemma 4.3, construct the desired cycle C_1 for T_1

Step 5. Find a set $S \subset \{1, 2, \ldots, k\}$ such that $\sum_{i \in S} a_i \geq 4$ and $\sum_{i \notin S} a_i \geq 4$. (Such a set must exist since T_1 does not satisfy the conclusion of Claim 2.) If $n = 7$, then S is found by the complete enumeration. Otherwise, either $a_1 \geq 4$ and then set $S = \{1\}$ or $a_1 = 2$ and then set $S = \{1, 2\}$.

By applying Phase_1 to the subtrees induced by $\cup_{i \in I} A_i \cup \{v\}$ and $\cup_{i \notin I} A_i \cup \{v\}$, find the cycles C' and C'' for these trees and using the procedure in the proof of Lemma 4.2, construct a cycle C_1 for T_1 from these cycles.

It is clear that the running of Phase 1 is dominated by the running time of Phase 0 (in fact it can be easily verified that Phase 1 can be implemented in linear time).

4   Implementation of Phase 2

After the first step, we have a Hamiltonian cycle C_1 and spanning tree T_2 non-intersecting with C_1. We need to find a cycle C_2 non-intersecting with C_1 such that $C_2 \cup T_2$ would be outerplanar with the outer face C_2. We use the similar idea as in previous section (splitting T_2 into subtrees), but here we, obviously, cannot exchange edges between T_2 and C_1. On the other hand, we can use the fact that the forbidden graph C_1 has maximum degree 2.

By *partial tour* in this section we mean either a Hamiltonian cycle or a graph whose connected components are disjoint paths. Let T be a spanning tree. We say that T is *passable* if for every partial tour F there exists a cycle C such that

1) $F \cap C = \emptyset$;
2) $T \cup C$ is an outerplanar graph with the outer face C.

We call a vertex of the tree T a *branch vertex* if all except one of its neighbors are leaves. The subtree induced by a branch vertex and all leaves adjacent to it is called the *branch*.

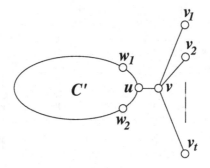

Fig. 4. Induction step in Lemma 4

Lemma 4.   *Let T be a spanning tree and B be a branch. If $T' = T \setminus B$ is passable, then T is passable.*

Proof. . Let v be a branch vertex of B and $v_1, v_2, \ldots, v_t$ be the leaves adjacent to it. Consider an arbitrary partial tour F. Let C' be a Hamiltonian cycle for T' and F' where F' is a restriction of F to the vertex set of T'. Let u be a non-leaf neighbor of v in T and denote by w_1, w_2 the neighbors of u in C' (see Fig. 4). By the definition of partial tour, if $t \geq 2$ then at least one of the edges $v_1 w_1, v_1 w_2, v_t w_1, v_t w_2$ must be not in F. We may assume that $v_t w_1 \notin F$. Consider four cases.

1. If $t \geq 4$ then we can reorder the vertices of $B \setminus v$ in such a way that $v_i v_{i+1} \notin F$ for all $i = 1, 2, \ldots, t-1$. Then remove the edge $u w_1$ from C' and add a path $u v v_1 v_2 \ldots v_t w_1$ instead. The obtained cycle C satisfies 1)–2).

2. If $t = 3$ then we may assume that $v_1 v_2, v_2 v_3 \in F$ (otherwise we do the same as in the previous case). Then $v_2 u \notin F$ and $v_1 v_3 \notin F$ since F is a partial

tour. We remove the edge uw_1 from C' and add a path $uv_2vv_1v_3w_1$ instead. The obtained cycle C satisfies 1)–2).

3. If $t = 2$ then we may assume that $v_1v_2 \in F$ (otherwise, do as in the first case). If $uv_1 \notin F$ then we remove the edge uw_1 from C' and add a path $uv_1vv_2w_1$ instead. Otherwise, $uv_2, v_1w_1 \notin F$ since F is a partial tour. Then in C', we replace the edge uw_1 by the path $uv_2vv_1w_1$, and again obtain a cycle C satisfying 1)–2).

4. If $t = 1$ and $v_1w_i \notin F$ for some $i = 1, 2$ the we substitute the path uvv_1w_i for the edge uw_i in C'. Otherwise, $uv_1 \notin F$ and $vw_i \notin F$ for some $i = 1, 2$ and in C', we replace the edge uw_i by the path uv_1vw_i.   □

Now we can prove the main theorem of this section.

Theorem 2. *Every tree on at least 5 vertices is passable.*

Proof. Suppose that the theorem is false and choose a counterexample T with the minimum number of vertices $n \geq 5$. Let F be an arbitrary partial tour. If T is the star $K_{1,n-1}$ then since $n - 1 \geq 4$, its leaves can be reordered in such a way that $v_iv_{i+1} \notin F$ for all $i = 1, 2, \ldots, n - 2$. Then we can take $C = vv_1v_2 \ldots v_{n-1}$ as a desired cycle (v is a center of the star here). If T is not a star then it has branch vertices. If there is a branch B in T such that $|T \setminus B| \geq 5$ then by the minimality of T and Lemma 4, T is passable. So, for every branch B, the tree $T \setminus B$ has at most 4 vertices. It is straightforward to verify that there are exactly 10 such trees (see Fig. 5). We will consider each of these trees separately (the name of the case corresponds to the name of the graph in Fig. 5).

a) We may assume that $v_4v_5 \in F$ (otherwise we can add the branch $B = T(\{v_1, v_2\})$ to the cycle $C' = v_3v_4v_5$ in the same way as in Lemma 4). If $v_1v_5 \in F$ then $v_1v_4, v_2v_5 \notin F$ by the definition of partial tour and we can take $C = v_1v_4v_3v_5v_2$. If $v_2v_5 \in F$ then, analogously, $C = v_2v_4v_3v_5v_1$. Assume that $v_1v_5, v_2v_5 \notin F$. At least one of the edges v_2v_4, v_2v_5 (say, v_2v_4) is also not in F. Then we can take $C = v_1v_2v_4v_3v_5$.

b) As in the previous case, we may assume that $v_4v_5, v_5v_6 \in F$. Then $v_2v_5, v_4v_6 \notin F$ and v_1 is not adjacent to either v_4 or v_6 in F. Then we can take $C = v_1v_2v_5v_3v_6v_4$ or $C = v_1v_2v_5v_3v_4v_6$ respectively.

c) We may assume that $v_1v_2, v_5v_6 \in F$ (otherwise we can apply the technique of Lemma 4). Then there is at most one edge connecting the sets $\{v_1, v_2\}$ and $\{v_5, v_6\}$ in F. So, we may assume that $v_1v_5, v_2v_6 \notin F$ and let $C = v_1v_3v_2v_6v_4v_5$.

d) We can assume that $v_1v_2, v_5v_6, v_6v_7 \in F$ and $v_1v_5, v_2v_6, v_5v_7 \notin F$. Then $C = v_1v_3v_2v_6v_4v_7v_5$ satisfies 1)–2).

e) Since $v_1v_2, v_2v_3, v_6v_7, v_7v_8 \in F$, we can take $C = v_1v_3v_4v_2v_8v_6v_5v_7$.

f) If $v_3v_5 \notin F$ then we add the branch $B = T \setminus \{v_3, v_4, v_5\}$ to the cycle $C' = v_3v_4v_5$ in the same way as in Lemma 4. So, $v_3v_5 \in F$. Analogously, $v_1v_3 \in F$. Then $v_1v_5 \notin F$, and we can take $C = v_1v_2v_3v_4v_5$.

g) As in the previous case, $v_1v_3 \in F$ and F must contain two edges from the set $\{v_3v_5, v_3v_6, v_5v_6\}$. Since the degree of v_3 is at most 2 in F, we may assume that $v_3v_5, v_5v_6 \in F$. Then $v_1v_5, v_3v_6 \notin F$ and we take $C = v_1v_2v_3v_6v_4v_5$.

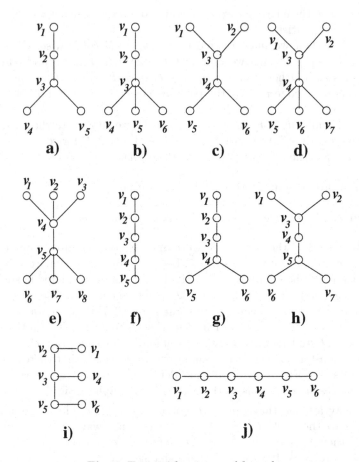

Fig. 5. Trees without a good branch

h) As in the previous case, we may assume that $v_1v_2, v_2v_4, v_4v_6, v_6v_7 \in F$. Then $v_1v_4, v_4v_7, v_2v_6 \notin F$. So, the cycle $C = v_1v_3v_2v_6v_5v_7v_4$ satisfies 1)–2).

i) If $v_1v_4 \notin F$ then we add the branch $B = T(\{v_5, v_6\})$ to the cycle $C' = v_1v_2v_3v_4$ in the same way as in Lemma 4. So, $v_1v_4 \in F$. Analogously, $v_4v_6 \in F$. Then $v_1v_6, v_4v_5 \notin F$ and we can take $C = v_1v_2v_3v_4v_5v_6$.

j) As in case f), $v_1v_4, v_3v_6 \in F$. If $v_1v_6 \notin F$ then $C = v_1v_2v_3v_4v_5v_6$. Otherwise, $v_1v_5, v_4v_6 \notin F$, and so the cycle $C = v_1v_2v_3v_4v_6v_5$ satisfies 1)–2).   □

The proof of Theorem 2 can be easily converted into an algorithm for finding the cycle C_2.

Algorithm **Phase_2**

Set $T_2 := T_2^*$.

Step 1. If T_2 is isomorphic to the star $K_{1,n-1}$ or to one of the trees shown in Fig. 3, then construct the desired cycle C_2 straightforwardly by using the procedures described in the proof of Theorem 2.

Step 2. If T_2 has a branch B in T such that $|T \setminus B| \geq 5$ then by applying Phase_2 to the tree $T_2' = T_2 \setminus B$ find a cycle C' for these tree. By using the procedure described in the proof of Lemma 4 construct the desired cycle C_2 from C'.

Again, it is clear that the running time of Phase_2 is dominated by the running time of Phase 0.

So the description of Disj_Ham_Cycles completed and we have the following

Theorem 3. *Algorithm Disj_Ham_Cycles finds a feasible solution of 2-PSP whose weight is at most twice the weight of the optimum in time $O(n^2 \log n)$.* □

References

1. Ageev, A.A., Baburin, A.E., Gimadi, E.K.: A polynomial algorithm with an accuracy estimate of 3/4 for finding two nonintersecting Hamiltonian cycles of maximum weight (in Russian). Diskretn. Anal. Issled. Oper. Ser. 13(2), 11–20 (2006)
2. Baburin, A.E., Gimadi, E.K., Korkishko, N.M.: Approximate algorithms for finding two edge-disjoint Hamiltonian cycles of minimal weight (in Russian), Diskretn. Anal. Issled. Oper. Ser. 2. 11(1), 11–25 (2004)
3. Christofides, N.: Worst-case analysis of a new heuristic for the traveling salesman problem, Technical Report CS-93-13, Carnegie Mellon University (1976)
4. De Brey, M.J.D., Volgenant, A.: Well-solved cases of the 2-peripatetic salesman problem. Optimization 39(3), 275–293 (1997)
5. Duchenne, E., Laporte, G., Semet, F.: Branch-and-cut algorithms for the undirected m-peripatetic salesman problem. European J. Oper. Res. 162(3), 700–712 (2005)
6. De Kort, J.B.J.M.: Lower bounds for symmetric K-peripatetic salesman problems. Optimization 22(1), 113–122 (1991)
7. De Kort, J.B.J.M.: Upper bounds for the symmetric 2-peripatetic salesman problem. Optimization 23(4), 357–367 (1992)
8. De Kort, J.B.J.M.: A branch and bound algorithm for symmetric 2-peripatetic salesman problems. European J. of Oper. Res. 70, 229–243 (1993)
9. Krarup, J.: The peripatetic salesman and some related unsolved problems, Combinatorial programming: methods and applications. In: Proc. NATO Advanced Study Inst., Versailles, 1974, pp. 173–178 (1975)
10. Roskind, J., Tarjan, R.E.: A note on finding minimum-cost edge-disjoint spanning trees. Math. Oper. Res. 10(4), 701–708 (1985)
11. Serdjukov, A.I.: Some extremal bypasses in graphs (in Russian). Upravlyaemye Sistemy 17(89), 76–79 (1978)
12. Wolfter, C.R., Cordone, R.: A Heuristic Approach to the Overnight Security Service Problem. Computers & Operations Research 30, 1269–1287 (2003)

Covering the Edges of Bipartite Graphs Using $K_{2,2}$ Graphs

Dorit S. Hochbaum[1,*] and Asaf Levin[2]

[1] Department of Industrial Engineering and Operations Research and Walter A. Haas
School of Business, University of California, Berkeley
hochbaum@ieor.berkeley.edu
[2] Department of Statistics, The Hebrew University, Jerusalem, Israel
levinas@mscc.huji.ac.il

Abstract. We consider an optimization problem arising in the design of optical networks. We are given a bipartite graph $G = (L, R, E)$ over the node set $L \cup R$ where the edge set is $E \subseteq \{[u, v] : u \in L, v \in R\}$, and implicitly a collection of all four-nodes cycles in the complete graph over V. The goal is to find a minimum size sub-collection of graphs $G_1, G_2, \ldots, G_p$ where for each i G_i is isomorphic to a cycle over four nodes, and such that the edge set E is contained in the union (over all i) of the edge sets of G_i. Noting that every four edge cycle can be a part of the solution, this covering problem is a special case of the unweighted 4-set cover problem. This specialization allows us to obtain an improved approximation guarantee. Whereas the currently best known approximation algorithm for the general unweighted 4-set cover problem has an approximation ratio of $H_4 - \frac{196}{390} \approx 1.58077$ (where H_p denotes the p-th harmonic number), we show that for every $\epsilon > 0$ there is a polynomial time $(\frac{13}{10} + \epsilon)$-approximation algorithm for our problem. Our analysis of the greedy algorithm shows that when applied to covering a bipartite graph using copies of $K_{q,q}$ bicliques, it returns a feasible solution whose cost is at most $(H_{q^2} - H_q + 1)OPT + 1$ where OPT denotes the optimal cost, thus improving the approximation bound by a factor of almost 2.

Keywords: Approximation algorithms, network design, set cover.

1 Introduction

In the area of designing optical networks, one of the issues is how to pack the demands on each link into optical channels. At each node there are digital routers limited in their capabilities and they can only pack demands on a link together if they arrive from up to q different directions, or go to up to q different directions. For edge $e = [u, v]$ the demands going through this edge are described in terms of the paths they follow through the edge, such as $\{a, e, c\}$ or $\{b, e, d\}$, as in Figure 1. Due to technical limitations of the optical routers this value of q is typically 2 or 3. The problem is to route all demands through an edge with minimum number of optical routers.

* Research supported in part by NSF award No. DMI-0620677 and CBET-0736232.

C. Kaklamanis and M. Skutella (Eds.): WAOA 2007, LNCS 4927, pp. 116–127, 2008.
© Springer-Verlag Berlin Heidelberg 2008

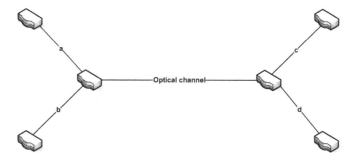

Fig. 1. A demonstration of a valid packing of demands on one edge in the optical network design problem with $q = 2$

Consider an abstraction of this problem for an edge $e = [u, v]$ with a bipartite graph, $B = (V_1 \cup V_2, E)$ that has the set of nodes V_1 each representing all the edges incoming (adjacent) to u except for edge e, and the set of nodes V_2 representing all the edges adjacent to v except for edge e. Each demand going through edge e is of the form $\{v_1, e, v_2\}$ with $v_1 \in V_1$ and $v_2 \in V_2$. In this bipartite graph, a valid channel packing of demands corresponds to a $K_{q,q}$ biclique (where $K_{q,q}$ denotes the complete bipartite graph with q nodes in each side of the bipartition). The problem of packing all demands using the minimum number of optical channels is then the problem of covering all the edges of B with a minimum number of $K_{q,q}$ bicliques. We call this problem the BIPARTITE $K_{q,q}$-COVERING PROBLEM. We note that the $K_{q,q}$ bicliques need not be subgraph of B.

The problem of covering a bipartite graph with bicliques is also prominent within the subject of *biclustering* and *gene expression*, [2,16,17]. Biclustering was defined by Mirkin [16] as the simultaneous clustering of both row and column sets in a "data matrix". The term *balanced biclustering* refers to finding a (large) biclique that corresponds to a square submatrix, or a $K_{q,q}$ biclique. An application of balanced biclique covering has been used in [18] to identify leukocyte-serum immunological reaction matrices.

As shown here, the BIPARTITE $K_{q,q}$-COVERING PROBLEM is NP-hard, even for $q = 2$. For $q = 1$ the problem is trivial – it is to cover the edges of the bipartite graph with singleton edges. The fact that the BIPARTITE $K_{2,2}$-COVERING PROBLEM (BK$_{2,2}$C) is NP-hard motivates our search for approximation algorithms. An α-approximation algorithm for a minimization problem is a polynomial time algorithm that always returns a feasible solution whose cost is at most α times the cost of an optimal solution, and α is called the approximation ratio, or approximation bound, or the performance guarantee of the algorithm. Our focus here is on BK$_{2,2}$C, showing it is hard, and devising a $(1.3 + \epsilon)$-approximation algorithm for the problem.

The general problem of bipartite $K_{q,q}$-covering is formulated here as a set cover problem. For the set cover problem there is an approximation algorithm with an approximation bound of $H_d = \sum_{i=1}^{d} \frac{1}{d} \approx \log d$, [3], where d is the largest number of elements covered by a set. Since the formulation of the bipartite $K_{q,q}$-covering as an instance of the unweighted set cover has each set with up to q^2 elements (the number of edges in $K_{q,q}$ biclique) this approximation ratio for the greedy algorithm for this problem is H_{q^2}. Using the special structure of the problem we show that the greedy algorithm for this set cover problem returns a feasible solution whose cost is at most $(H_{q^2} - H_q + 1)OPT + 1$ where OPT denotes the optimal cost. This is an improvement of a factor of (almost) 2 in the approximation bound.

The problem $BK_{2,2}C$ is defined as follows. The input to the problem is a bipartite graph $G = (L \cup R, E)$ with the bipartition of the nodes to L and R (i.e., each edge in E connects a node from L and a node from R). The problem is to cover the edges in E using $K_{2,2}$ bicliques. In other words, $BK_{2,2}C$ is to find a collection $\{G_1, G_2 \ldots, G_p\}$ of subgraphs of G, each a biclique $K_{2,2}$ intersection with the edges of G, where the union of the edge sets of all these subgraphs is E. The goal is to find a minimum size collection of such subgraphs that covers E, i.e., to minimize p.

A H-*decomposition* of a graph $G = (V, E)$ is a partition of E into subgraphs isomorphic to H. For a fixed graph H the H-DECOMPOSITION PROBLEM is to determine whether an input graph G admits a H-decomposition. Holyer [9] proved that H-decomposition problem is NP-complete for (H) a complete graph on at least three nodes, and also for (H) a cycle on at least four nodes. Since then a stronger result was proved by Dor and Tarsi [5] showing that if H is connected with at least three edges, then the H-decomposition problem is NP-complete.

The reduction of Holyer for H-decomposition where H is a four nodes cycle creates a bipartite graph. Therefore, H-decomposition where H is a four nodes cycle is NP-complete even when restricted to bipartite graphs. Also, the H-decomposition problem defined on a bipartite graph, where H is the cycle over four nodes, is reducible to $BK_{2,2}C$ by checking whether the optimal cost for $BK_{2,2}C$ equals $\frac{|E|}{4}$. We conclude that $BK_{2,2}C$ is also NP-hard. We do not know whether $BK_{2,2}C$ is APX-hard.

In the WEIGHTED SET-COVER PROBLEM we are given a set of elements $E = \{e_1, e_2, \ldots, e_n\}$ and a collection $\mathcal{F}$ of subsets of E, where $\cup_{S \in \mathcal{F}} S = E$ and each $S \in \mathcal{F}$ has a positive cost c_S. The goal is to compute a sub-collection $SOL \subseteq \mathcal{F}$ such that $\cup_{S \in SOL} S = E$ and its cost $\sum_{S \in SOL} c_S$ is minimum. Such a sub-collection of subsets is called a *cover*. When we consider instances of the WEIGHTED SET-COVER with each S_j having at most k elements ($|S| \leq k$ for all $S \in \mathcal{F}$), we obtain the WEIGHTED k-SET COVER PROBLEM. The UNWEIGHTED SET COVER PROBLEM and the UNWEIGHTED k-SET COVER PROBLEM are special cases of the WEIGHTED SET COVER and of WEIGHTED k-SET COVER, respectively, where $c_S = 1 \ \forall S \in \mathcal{F}$. Problem $BK_{2,2}C$ can thus be viewed as an instance of the unweighted 4-set cover problem, by considering the element set to be the edge set E of the input graph, and the collection $\mathcal{F}$ to be the set of all four-edge cycles over nodes of G. Thus, $BK_{2,2}C$ is precisely the resulting instance of the unweighted 4-set cover problem.

Chvátal, in [3], established that a greedy algorithm is a H_k-approximation algorithm for the weighted k-set cover. This greedy algorithm works by choosing iteratively a in the cover that maximizes the ratio of the number of remaining elements it covers to its cost. The k-th harmonic number bound is tight for the greedy algorithm even for the unweighted k-set cover problem (see, [11,15]). On the other hand, the unweighted k-set cover problem is known to be NP-complete [12] and MAX SNP-hard for all $k \geq 3$ [4,13,19].

Goldschmidt, Hochbaum and Yu [7] modified the greedy algorithm for the unweighted k-set cover and showed that the resulting algorithm has a performance guarantee of $H_k - \frac{1}{6}$. Halldórsson [8] presented an algorithm based on local search that has an approximation ratio of $H_k - \frac{1}{3}$ for the unweighted k-set cover, and a $(1.4+\epsilon)$-approximation algorithm for the unweighted 3-set cover. Duh and Fürer [6] further improved this result and presented a $(H_k - \frac{1}{2})$-approximation algorithm for the unweighted k-set cover. The current best approximation guarantee for the unweighted k-set cover problem is $H_k - \frac{196}{390}$ (for all $k \geq 4$) [14] (see [1] for some improvement of this for values of $k \geq 6$). Therefore, prior to this study the best known approximation ratio for problem $BK_{2,2}C$ is $H_4 - \frac{196}{390} \approx 1.58077$. This best known previous result is significantly improved here for problem $BK_{2,2}C$. The algorithm of [7] as well as all the other known improvements of the greedy approximation algorithm [8,6,14] are not greedy algorithms, and require much higher running times, though still polynomial.

To motivate our improvement we show in Section 2 that the greedy algorithm for the set cover problem has a better (asymptotic) performance guarantee when it is applied to problem $BK_{2,2}C$ ($H_4 - \frac{1}{2}$ instead of H_4). For the general bipartite $K_{q,q}$-covering problem we show that the greedy algorithm has an asymptotic performance guarantee of $H_{q^2} - H_q + 1$ instead of H_{q^2}. Then, in Section 3, we show our improved $(\frac{13}{10} + \epsilon)$-approximation algorithm for $BK_{2,2}C$.

Our results. We show that the greedy algorithm when applied to covering a bipartite graph using copies of $K_{q,q}$ bicliques, returns a feasible solution whose cost is at most $(H_{q^2} - H_q + 1)OPT + 1$ where OPT denotes the optimal cost. We also present an improved $(\frac{13}{10} + \epsilon)$-approximation algorithm for $BK_{2,2}C$.

2   The Approximation Ratio of the Greedy Algorithm

We show here that the greedy algorithm is a $(H_4 - \frac{1}{2})$-approximation algorithm for $BK_{2,2}C$. In fact the main result shown in this section is more general – it is a $(H_4 - \frac{1}{2})$-approximation algorithm for the problem of covering the edges of *any* graph by a 4-cycle, C_4. Since the running time of the greedy algorithm is much faster than the algorithms of [6,14] as well as the algorithm of the next section, the result of this section presents an improvement over the other results in either its approximation ratio or its time complexity. The key idea in the improved approximation ratio of the greedy algorithm for this problem is that greedy uses singletons (sets that cover exactly one new previously uncovered element) at most once, as shown next.

We consider an unweighted set cover instance $E = \{e_1, e_2, \ldots, e_n\}$ and $\mathcal{F} \subseteq 2^E$ so that each set $S \in \mathcal{F}$ has at most p elements, and $\mathcal{F}$ contains all q-subsets of E (for some integers $q \leq p$). We call such an instance for the unweighted set cover, a (q,p)-*uniform unweighted set cover problem*.

The greedy algorithm starts with an empty collection of subsets in the solution and no element being covered. Then it repeats the following procedure until all elements of E are covered:

Let n_S be the number of elements that are still uncovered in a set $S \in \mathcal{F}$, and the current *ratio of* S is $r_S = \frac{1}{n_S}$. Let S^* be a set such that r_{S^*} is minimized. The algorithm adds S^* to the collection of subsets of the solution, updates the status of the elements of S^* as covered, and assigns a *price* of r_{S^*} to all the elements newly covered in this iteration (i.e., the elements of E that were first covered by S^*).

Theorem 1. *The greedy algorithm for (q,p)-uniform unweighted set cover problem returns a feasible solution whose cost is at most $(H_p - H_q + 1)OPT + 1 - \frac{1}{q}$, where OPT is the cost of the optimal solution.*

Proof. We modify Chvátal's proof of the Harmonic bound, [3], as follows. First, note that the cost of the greedy solution equals the sum of prices assigned to the elements of E. Let OPT be an optimal solution of value OPT, and consider a subset S that belongs to OPT (S has at most p elements). Then, OPT pays 1 for S. When the i-th element of S is covered by the greedy algorithm, the algorithm could select S as a feasible set with a current ratio of $\frac{1}{|S|-i+1}$. Therefore, the price assigned to this item is at most $\frac{1}{|S|-i+1}$. Thus we have established that the total price assigned to the elements of S is at most H_p where the q last elements of S to be covered by the algorithm have at most H_q units of price. We next argue that this bound can be improved for all S in OPT except perhaps for one such set. This is so because each of the last q elements of S is covered by a set with at least q (previously uncovered) elements, and hence its price is at most $\frac{1}{q}$. This argument can be applied to every subset in OPT except perhaps one subset where the last q uncovered elements may have larger price, and may be assigned a total price of at most one. It follows that the total price assigned to the elements of S (for all S in OPT except possibly one set) is at most $\sum_{i=1}^{|S|-q} \frac{1}{|S|-i+1} + q \cdot \frac{1}{q} = \sum_{i'=1}^{|S|} \frac{1}{i'} - \sum_{i=1}^{q} \frac{1}{i} + 1 \leq H_p - H_q + 1$. We note that the last set S pays an additional price of at most $1 - \frac{1}{q}$ price units (there are at most $q - 1$ elements that do not fit into the q-subsets selected by the algorithm), and hence the claim follows. □

We note that the unweighted set cover instances resulting from bipartite $K_{q,q}$-covering problem are (q, q^2)-uniform, and hence we establish the following proposition.

Proposition 1. *1. When the greedy algorithm is applied to the problem of minimum cover for the edges of a graph with 4-cycles, it returns a feasible solution whose cost is at most $(H_4 - \frac{1}{2}) \cdot OPT + \frac{1}{2}$ where OPT is the cost of an optimal solution.*

2. *When the greedy algorithm is applied for the bipartite $K_{q,q}$-covering problem, it returns a feasible solution whose cost is at most $(H_{q^2} - H_q + 1) \cdot OPT + \frac{q-1}{q}$ where OPT is the cost of an optimal solution.*

3   Improved Approximation Algorithm

In this section we present our $(\frac{13}{10} + \epsilon)$-approximation algorithm for problem $BK_{2,2}C$. The analysis of this improved algorithm makes use of the fact that the input graph is bipartite.

Let $\{G_1, G_2 \ldots, G_t\}$ be a collection of subgraphs, each a biclique $K_{2,2}$ intersection with the edges of G, forming a feasible solution to $BK_{2,2}C$. We associate each edge $e \in E$ of G with the first subgraph G_i on this ordered list that contains it. We call the subgraphs G_i that have four associated edges *cycles of G*. Other subgraphs that have three associated edges we call *3-paths of G*. The remaining subgraphs G_i have only one or two associated edges, such subgraphs are not referred to as 3-paths. The *end-nodes* of a 3-path G_i of G are the two nodes adjacent to the edge of the biclique of G_i that is not associated with G_i (this edge is either not in G or it is associated with another subgraph G_j). So the end-nodes of a 3-path G_i are the end-nodes of the subgraph resulting from G_i by removing the edge that is not associated with G_i (see Figure 2 for an illustration).

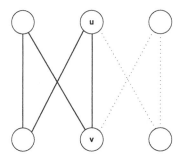

Fig. 2. The solid edges are associated with G_1 whereas the dotted edges are associated with G_2. In this figure G_1 is a cycle and G_2 is a 3-path. The end-nodes of G_2 are u and v.

A 3-path of G is called a *good 3-path* if both its end-nodes have odd degrees in G. A set S of subgraphs $G_1, G_2 \ldots, G_i$ is a *good disjoint collection of subgraphs* if the following three conditions hold: 1. All of these subgraphs are edge-disjoint, 2. Each of them is either a cycle of G or a good 3-path of G, and 3. All the end-nodes of the good 3-paths of G in S are disjoint.

The first step of our algorithm is a pre-processing step that removes an approximate maximum size good disjoint collection of subgraphs[1]. This is done by applying the local-search based algorithm for packing problems of Hurkens and Schrijver [10]. This pre-processing step is referred in the sequel as the *local-search phase*. The algorithm of [10] has an integer parameter t, and when it is applied to approximate the maximum size good disjoint collection of subgraphs, it maintains a current collection that is a good disjoint collection of subgraphs. It starts with an empty set of subgraphs as the current collection (since an empty collection is clearly a good disjoint collection of subgraphs we can start with this initial collection). At each step the algorithm tries to delete t subgraphs from the current collection and to add $t + 1$ subgraphs to the collection, while enforcing the property that the resulting set of subgraphs is a good disjoint collection of subgraphs. If the process cannot increment the current collection (i.e., it is a local-maximum size good disjoint collection of subgraphs), then the algorithm returns the current collection.

The approximation ratio and the time complexity of the local-search algorithm both depend on t. When $t = 2r$ (for even values of t) the approximation ratio is $\frac{(2(k-1)^r-2)}{(k(k-1)^r-2)}$ and for $t = 2r - 1$ the approximation ratio is $\frac{(2(k-1)^r-k)}{(k(k-1)^r-k)}$, where k is the maximum number of items in an input set. In our case an item can be either an edge or an end-node. Therefore, each selected graph can have at most five items (either four associated edges from G, or three associated edges and two end-nodes). Hence, the approximation ratio of the Hurkens and Schrijver's algorithm is $\frac{2}{5} - \epsilon$ where ϵ is $O\left(\frac{1}{4^{t/2}}\right)$.

The good collection of subgraphs that we found in the local-search phase is part of our cover. Additional bicliques are added next to attain a feasible solution. Denoting by $\tilde{E}$ the set of edges of G that are not covered by the selected good disjoint collection of subgraphs, we partition the edges of the graph $\tilde{G} = (V, \tilde{E})$ into two parts as follows. The first part is a subgraph of $\tilde{G}$ such that each of its connected component is Eulerian (i.e., a subgraph of G where the degree of each node is an even number) denoted as $G_e = (V, E_e)$, and the other parts are paths $P_1, P_2, \ldots, P_k$ each of them connects two odd-degree nodes (where P_i has an arbitrary number of edges). Each of these parts is an edge induced subgraph of G. The partition is chosen so that it has the additional property that the node set induced by the paths $P_1, P_2, \ldots, P_k$ is disjoint to the node set induced by E_e. To find such a partition we apply the following procedure. We add to $\tilde{G}$ a set of *fake edges* that is a matching over the odd-degree nodes of $\tilde{G}$. Then, in the resulting graph the degree of each node is even, and for each connected component of the resulting graph we find an Eulerian tour traversing all its edges (fake edges or regular edges that belong to $\tilde{G}$). We next remove all the fake edges, and by doing so some of the Eulerian tours are partitioned into a set of paths that we select to the partition. The other connected components are Eulerian in

---

[1] In our algorithm two pairs of odd-degree nodes may result in a subgraph G_i with only two associated edges, and hence we would like to decrease the number of odd-degree nodes in the subgraph after the pre-processing.

$\tilde{G}$, and these are node disjoint to the selected paths. We denote these Eulerian connected components by $C_1, C_2, \ldots, C_l$, and we let G_e to be their union.

Note that each C_i is an Eulerian tour in a bipartite graph, and hence has an even number of edges. Our algorithm traverses each of $C_1, C_2, \ldots, C_l$, as well as each of $P_1, P_2, \ldots, P_k$, and partitions them into a set of 3-paths of G each of which has three consecutive edges along the Eulerian tour or along P_i, and a remainder of at most two edges called *remaining edges* from each of $C_1, \ldots, C_l, P_1, \ldots, P_k$. Each 3-path of G that we find is a part of the output. The remaining edges are paired up arbitrarily, and each such pair of edges belongs to a common biclique in the solution returned by the algorithm. If there is an unpaired edge then we add one biclique to the output that covers this edge.

To see that our algorithm returns a feasible solution we note that any pair of edges of G can be covered using one copy of $K_{2,2}$, and any three edges of G that form a 3-path of G can be covered using one copy of $K_{2,2}$ (together with the edge between the two end-nodes of the 3-path of G). Therefore, our algorithm returns a feasible solution. It runs in polynomial time because given the graph G, we can find G_e in polynomial time as described above. Therefore, we establish the following lemma.

Lemma 1. *The approximation algorithm runs in polynomial time and returns a feasible solution.*

The time complexity of the algorithm consists of the preprocessing step and a linear time for the rest of the algorithm. Recall that the time complexity of the preprocessing step is exponential in $\frac{1}{\epsilon}$.

In the rest of this section we analyze the approximation ratio of the algorithm.

Consider a fixed optimal solution denoted by OPT. Denote by CY the number of cycles of G in OPT, and denote by CH_3 the number of 3-paths of G in OPT where CH stands for chains and 3 stands for the number of edges associated with such a 3-path. We find a collection of subgraphs in OPT that is maximal with the property that each node in G has an even degree in the selected collection (0 is obviously even). This collection forms an Eulerian subgraph. The other subgraphs in OPT that are 3-paths of G are partitioned as follows: as long as the following succeeds we identify a set of subgraphs of G such that the set of associated edges from these subgraphs forms a path between two odd-degree (in G) nodes. Such a path made of a collection of 3-paths that is not a single good 3-path is called *a superchain of OPT*. We denote by CH_G the number of good 3-paths in OPT, by CH_S the number of 3-paths of G that belong to superchains of OPT, and by CH_E the number of 3-paths of G that belong to the Eulerian subgraph: note that since CH_G, CH_S and CH_E count disjoint sets of 3-paths of G, we conclude that $CH_3 \geq CH_G + CH_S + CH_E$ but this inequality may be strict inequality). Denote by CH_2 the number of subgraphs of OPT with two associated edges, and by CH_1 the number of subgraphs of OPT with exactly one associated edges. We denote by n_o the number of odd-degree nodes in G.

Let the solution APP that the algorithm returns be of cost APP. Denote by A the number of cycles of G found by the local-search phase and by B the number of good 3-paths of G found by the local search phase. Then, by the performance guarantee of the algorithm of [10] we conclude that $A+B \geq (\frac{2}{5}-\epsilon) \cdot (CY+CH_G)$.

We allocate next *APP-prices* to the elements of G. Each edge is assigned an APP-price of $\frac{3}{8}$ and each odd-degree node is assigned an APP-price of $\frac{3}{16}$. We associate the odd-degree nodes with the different subgraphs of APP in the following way. For an odd-degree node v such that there is a good 3-path (in APP) with end-node v, we associate v with this good 3-path. In APP, after the removal of the good 3-paths of G during the local-search phase there are $2k$ odd-degree nodes, and APP has at most $2k$ remaining edges that do not belong to the Eulerian subgraph. We associate one odd-degree node with each such remaining edge. Other odd-degree nodes are ignored when we lower bound the total *APP*-price of all items (i.e., we modify the *APP*-price of such a node to zero). We next upper bound the cost of APP using the total APP-price.

Lemma 2. *The total APP-prices is at least* $APP + \frac{1}{2} \cdot (A+B) + \frac{k}{8} - \frac{1}{2}$. *I.e., APP is at most the total APP-prices minus* $\frac{1}{2} \cdot (A+B) + \frac{k}{8} - \frac{1}{2}$.

Proof. Consider a cycle that was removed during the local-search phase. It has four edges, each of them has an APP-price of $\frac{3}{8}$, and therefore the total APP-price is $\frac{12}{8} = \frac{3}{2}$. Similarly, a good 3-path that was removed during the local-search phase has a total APP-price of $\frac{9}{8} + \frac{6}{16} = \frac{3}{2}$. Note that APP pays one unit for each of these subgraphs (cycle or good 3-path), and therefore the sum of the total cost of APP for the removed cycles and good paths plus $\frac{1}{2} \cdot (A+B)$ is the total APP-price paid for the elements of the removed cycles and good paths of APP.

Next, consider a connected component C_i of the Eulerian subgraph. Since C_i is simple (without parallel edges) and bipartite, it has at least six edges (if C_i has only four edges, then this contradicts the local optimality of the good disjoint collection of subgraphs that we find in the local-search phase). Denoting the number of edges of C_i by c_i, if c_i is even such that $c_i \geq 6$ and $c_i \equiv 0 \pmod 3$ or $c_i \equiv 2 \pmod 3$, then APP pays $\lceil \frac{c_i}{3} \rceil$ and $\frac{3}{8} \cdot c_i \geq \lceil \frac{c_i}{3} \rceil$. For every even number c_i such that $c_i \equiv 1 \pmod 3$ then APP pays $\lfloor \frac{c_i}{3} \rfloor + \frac{1}{2}$ and since $c_i \geq 6$ then $\frac{3}{8} \cdot c_i \geq \lfloor \frac{c_i}{3} \rfloor + \frac{1}{2}$.

Next, consider a 3-path of APP that is not part of the Eulerian graph, then its total APP-price is at least $\frac{9}{8}$ that is greater than 1. I.e., for such a 3-path APP pays less than the total APP-price of the elements of the 3-path.

It remains to consider the remaining edges that do not belong to the Eulerian subgraph. There are at most $2k$ such edges, and for each of these edges there is at least one associated odd-degree node, so that each of these remaining edges have (together with the odd node) a total APP-price of at least $\frac{3}{8} + \frac{3}{16} = \frac{1}{2} + \frac{1}{16}$. Since APP pays for such a remaining edge $\frac{1}{2}$ (except perhaps the last edge that is charged one unit), we have an extra of at least $\frac{1}{16}$ units of APP-price with respect to the cost APP for each odd-degree node that is left after we removed the good 3-paths of G during the local-search phase. Since $\frac{2k}{16} = \frac{k}{8}$ and the last remaining edge might be charged one unit instead of $\frac{1}{2}$, the claim follows. $\square$

We allocate next *OPT-prices* to the elements of G. Each edge that belongs to either a cycle of OPT or a good path of OPT is assigned an OPT-price of $\frac{3}{8} - \frac{1-\epsilon}{20}$. Other edges are assigned an OPT-price of $\frac{3}{8}$. We associate the odd-degree nodes with the different subgraphs of OPT in the following way. For an odd-degree node v which is an end-node of a good 3-path (in OPT) we associate v with this good 3-path and let its OPT-price be $\frac{3}{16} - \frac{1-\epsilon}{40}$. If a node v is not an end-node of a good 3-path, then its OPT-price is $\frac{1}{8}$. In that case v is either an end-node of a 3-path of G (where this 3-path is not part of the Eulerian subgraph) or of a singleton remaining edge e. We then associate v with either the 3-path subgraph for which it is an end-node or with the subgraph G_i of OPT that covers e.

We next show that the total *OPT*-price of the elements of each subgraph of OPT is at most $\frac{13}{10} + O(\epsilon)$. Summing over all subgraphs of OPT we will conclude that the total $O\hat{P}T$-price of all subgraphs of OPT is at most $(\frac{13}{10} + \frac{\epsilon}{5})OPT + \frac{1}{2}$.

Lemma 3. *Consider a subgraph G_i of* OPT, *then the total OPT-price that is assigned to G_i is at most $\frac{13}{10} + \frac{\epsilon}{5}$.*

Proof. The proof is via case analysis of the different types of subgraphs in OPT.

- Assume that G_i is a cycle. Then it has four associated edges and an OPT-price of $4 \cdot \left(\frac{3}{8} - \frac{1-\epsilon}{20}\right) = \frac{13}{10} + \frac{\epsilon}{5}$.
- Assume that G_i is a 3-path in the Eulerian subgraph of OPT. Then, G_i has three associated edges and does not have an associated odd-degree nodes. Therefore, its OPT-price is $3 \cdot \frac{3}{8} = \frac{9}{8} < \frac{13}{10}$.
- Assume that G_i is a good 3-path of OPT. Then, the OPT-price of G_i is $3 \cdot \left(\frac{3}{8} - \frac{1-\epsilon}{20}\right) + 2 \cdot \left(\frac{3}{16} - \frac{1-\epsilon}{40}\right) = \frac{13}{10} + \frac{\epsilon}{5}$.
- Assume that G_i is a 3-path in G that is not a good path and also it is not a part of the Eulerian subgraph. Such G_i has at most one associated odd-degree node. Note that such odd-degree node that is assigned to G_i is not an end-node of a good 3-path of OPT, and therefore it has an OPT-price of $\frac{1}{8}$. Therefore, the OPT-price of G_i is at most $3 \cdot \frac{3}{8} + \frac{1}{8} = \frac{5}{4} < \frac{13}{10}$.
- Otherwise, G_i has at most two associated edges and two associated odd-degree nodes. Again, the associated odd-degree nodes are not end-nodes of good 3-paths, and therefore the OPT-price of each such odd-degree node is $\frac{1}{8}$. Therefore, G_i has an OPT-price of at most $2 \cdot \frac{3}{8} + 2 \cdot \frac{1}{8} = 1$. □

Corollary 1. $APP \le (\frac{13}{10} + \frac{\epsilon}{5}) \cdot OPT + \frac{1}{2}$.

Proof. By Lemma 2, APP is at most the total APP-prices minus $\frac{1}{2} \cdot (A + B) + \frac{k}{8} - \frac{1}{2}$. By Lemma 3, the total OPT-prices is at most $\left(\frac{13}{10} + \frac{\epsilon}{5}\right) \cdot OPT$. Next, we argue that the total APP-price is larger than the total OPT-price assigned to the elements of G by at most $\frac{A+B}{2} + \frac{k}{8}$. To see this note firstly that the total OPT-price of the edges and end-nodes of a cycle or a good path of OPT is $\frac{1-\epsilon}{5}$ lower than the total APP-prices of these elements. Secondly, from the performance guarantee of the local-search phase $A + B \ge \frac{2}{5}(CY + CH_G)$. Thirdly, there are exactly $2k$ odd-degree nodes that are assigned APP-price of $\frac{3}{16}$ (and the other odd-degree nodes have zero APP-price), and each of these nodes have an OPT-price of at least $\frac{1}{8}$. Therefore, the difference between the contribution of odd-degree nodes to the total APP-price and the total OPT-price is at most $\frac{k}{8}$. □

We note that if $OPT \leq \frac{1}{\epsilon}$, then we can enumerate all partitions into at most $\frac{1}{\epsilon}$ subgraphs, and for each of them we test the feasibility of the partition (as a solution to $BK_{2,2}C$) and we pick the cheapest feasible solution that we find. Therefore, we can assume without loss of generality that $OPT > \frac{1}{\epsilon}$, and therefore $\frac{1}{2} < \frac{\epsilon}{2} \cdot OPT$. Hence, we establish the following theorem.

Theorem 2. *For every $\epsilon > 0$, there is an approximation algorithm for $BK_{2,2}C$ that returns a feasible solution whose cost is at most $\left(\frac{13}{10} + \epsilon\right) \cdot OPT$.*

References

1. Athanassopoulos, S., Caragiannis, I., Kaklamanis, C.: Analysis of approximation algorithms for k-set cover using factor-revealing linear programs. In: FCT 2007. Proceedings of the 16th International Symposium on Fundamentals of Computation Theory (to appear, 2007)
2. Cheng, Y., Church, G.M.: Biclustering of expression data. In: Proceedings of the Eighth International Conference on Intelligent Systems for Molecular Biology (ISMB), pp. 93–103 (2000)
3. Chvátal, V.: A greedy heuristic for the set-covering problem. Mathematics of Operations Research 4, 233–235 (1979)
4. Crescenzi, P., Kann, V.: A compendium of NP optimization problems (1995), http://www.nada.kth.se/theory/problemlist.html
5. Dor, D., Tarsi, M.: Graph decomposition is NP-complete: a complete proof of Holyer's conjecture. SIAM J. Comput. 26, 1166–1187 (1997)
6. Duh, R., Fürer, M.: Approximation of k-set cover by semi local optimization. In: Proc. STOC 1997, pp. 256–264 (1997)
7. Goldschmidt, O., Hochbaum, D.S., Yu, G.: A Modified Greedy Heuristic for the Set Covering Problem with Improved Worst Case Bound. Information Processing Letters 48, 305–310 (1993)
8. Halldórsson, M.M.: Approximating k set cover and complementary graph coloring. In: Cunningham, W.H., Queyranne, M., McCormick, S.T. (eds.) IPCO 1996. LNCS, vol. 1084, pp. 118–131. Springer, Heidelberg (1996)
9. Holyer, I.: The NP-completeness of some edge-partition problems. SIAM J. Comput. 10, 713–717 (1981)
10. Hurkens, C.A.J., Schrijver, A.: On the size of systems of sets every t of which have an SDR, with an application to the worst-case ratio of heuristics for packing problems. SIAM Journal on Discrete Mathematics 2, 68–72 (1989)
11. Johnson, D.S.: Approximation algorithms for combinatorial problems. Journal of Computer and System Sciences 9, 256–278 (1974)
12. Karp, R.M.: Reducibility among combinatorial problems. In: Miller, R.E., Thatcher, J.W. (eds.) Complexity of computer computations, pp. 85–103. Plenum Press, New-York (1972)
13. Khanna, S., Motwani, R., Sudan, M., Vazirani, U.V.: On syntactic versus computational views of approximability. SIAM Journal on Computing 28, 164–191 (1998)

14. Levin, A.: Approximating the unweighted k-set cover problem: greedy meets local search. In: Erlebach, T., Kaklamanis, C. (eds.) WAOA 2006. LNCS, vol. 4368, pp. 290–301. Springer, Heidelberg (2007)
15. Lovász, L.: On the ratio of optimal integral and fractional covers. Discrete Mathematics 13, 383–390 (1975)
16. Mirkin, B.: Mathematical Classification and Clustering. Kluwer, Dordrecht (1996)
17. Morgan, J.N., Sonquist, J.A.: Problems in the analysis of survey data, and a proposal. Journal of The American Statistical Association 58, 415–434 (1963)
18. Nau, D.S., Markowsky, G., Woodbury, M.A., Amos, D.B.: A mathematical analysis of human leukocyte antigen serology. Math. Biosci. 40, 243–270 (1978)
19. Papadimitriou, C.H., Yannakakis, M.: Optimization, approximation and complexity classes. Journal of Computer System Sciences 43, 425–440 (1991)

On Min-Max r-Gatherings

Amitai Armon

School of Computer Science,
Tel-Aviv University, Tel-Aviv 69978, Israel
armon@tau.ac.il

Abstract. We consider a *min-max* version of the previously studied *r-gathering* problem with unit-demands. The problem we consider is a metric facility-location problem, in which each open facility must serve at least r customers, and the *maximum* of all the facility and connection costs should be minimized (rather than their *sum*). This problem is motivated by scenarios in which r customers are required for a facility to be worth opening, and the costs represent the time until the facility/connection will be available (*i.e.*, we want to have the complete solution ready as soon as possible).

We present a 3-approximation algorithm for this problem, and prove that it cannot be approximated better (assuming $P \neq NP$). Next we consider this problem with the additional natural requirement that each customer will be assigned to a nearest open facility, and present a 9-approximation algorithm. We further consider previously introduced special cases and variants, and obtain improved algorithmic and hardness results.

1 Introduction

Facility-location has been studied in many forms over the past decades (see, *e.g.*, [3,4,6,9,11,12,13,14,16,18,19]). In the classic *metric facility-location problem*, we are given a set of customer locations S and a set of potential locations of facilities F (which may intersect S). Each location $f_i \in F$ is associated with a cost $p(f_i)$ for opening a facility there. For every $s_i \in S$ and $f_j \in F$, there is a cost $d(s_i, f_j)$ for connecting a customer in s_i to a facility in f_j. These costs are equivalent to the distances, and satisfy the symmetry and triangle-inequality requirements. The goal is to open facilities and assign each customer to a facility, such that the total cost is minimized (*i.e.*, the sum of the facility opening-costs and the connection-costs should be minimal).

The metric facility-location problem models many realistic scenarios, in which service-posts of a certain type should be opened to serve a set of customers. Applications range from classic power-plants or warehouses location problems to locating servers in computer-networks (see, *e.g.*, [6,18] for surveys). The current best approximation algorithm for metric facility-location achieves an approximation-ratio of 1.5 [4]. On the other hand, this problem cannot be approximated within less than a factor of 1.463, assuming $P \neq NP$ [11].

One of the interesting recent variants of metric facility-location is the *r-gathering* problem, introduced in parallel by Karger and Minkoff [13] and by

Guha et al. [12] (who called it *load-balanced facility-location*). The basic additional requirement in the *r-gathering* problem is that each facility will be assigned at least r customers (customers are not necessarily assigned to the nearest open facility in this problem). This variant captures the idea that opening a facility is economically justified when it serves at least a certain amount of demand (and this constraint may even be more natural than facility costs in some settings). Furthermore, in various settings there is an inherent lower bound on the number of customers in each facility. For example, in *secret-sharing* schemes (see [15]), at least r shares are needed to uncover a secret. We may need to locate servers in the network, to which clients will connect in order to uncover the secret, and we may want this process to be as fast or as cheap as possible.

Both papers [12,13] considered the generalization of *r-gathering* in which customers have different demands, the connection-costs are the product of the demand and distance, and each facility must serve customers having a total of at least r demand [12,13]. They both presented a $(\frac{1+\alpha}{1-\alpha}\beta, \alpha)$ bicriteria approximation, for any $\alpha < 1$, where β is the approximation-ratio of the metric facility-location problem (currently 1.5 for the classic problem [4] and 1.582 for the generalization in which customers may have different demands [16]). Namely, their algorithm guarantees that each open facility in the solution will serve at least αr demand, and the cost will be at most $\frac{1+\alpha}{1-\alpha}\beta$ times the optimal cost of the *r-gathering* problem. Choosing $\alpha = \frac{r-1}{r} + \epsilon$ for the case of unit-demands provides a $1.5(2r - 1 + \epsilon)$-approximate feasible solution. Note that we cannot hope for a significant improvement in the approximation-ratio due to improvement of β, since β is lower-bounded by 1.463 [11].

Although the first papers considered minimizing the *sum* of costs [12,13], a natural variant is to minimize the *maximal* cost (in the spirit of the *k-center* problem [9]). This may model, for example, the time until all the facilities and connections will be available (if each cost represents the time until the corresponding facility/connection will be ready). A special case of the *min-max* version of this problem with unit-demands, called "*r-gather clustering*", has been recently considered by Aggrawal et al. [1]. In their special case, motivated by a clustering application, all the facility costs are zero and all the locations of customers are included in the set of optional facility locations $(S \subseteq F)$[1]. Their paper presented a 2-approximation algorithm for this case, and proved that it cannot be approximated better, for any $r \geq 7$ (assuming $P \neq NP$). They also considered a generalization called (r, ϵ)-*gather clustering*, in which the solution can ignore ϵn of the customers ("outlier points"), and stated that this problem can be approximated within a factor of 3 if facilities (cluster-centers) can only be located at customer (input points) locations [1]. We note that unlike the algorithm of [12,13], the algorithm of [1] does not guarantee that each customer will be assigned to a nearest open facility.

For the basic special case of $r = 2$, a recent paper of Anshelevich and Karagiozova [2] proves that both *min-sum 2-gathering* without facility-costs and *min-max 2-gathering* can be solved in polynomial time.

Demaine et al. [5] have recently introduced another problem related to *min-max 2-gathering*, which they called *min-max minimum-movement facility location*. In our terminology, there are two types of customers in that problem: Customers from type A ("clients") must be assigned to a facility having at least one customer from type B ("server") assigned to it, while customers from type B do not have to be assigned. Also, $S \subseteq F$ and there are no facility costs. Demaine et al. [5] asked whether this problem can be approximated within less than a factor of 2. We prove that the answer is negative, assuming $P \neq NP$.

In this paper we focus on *min-max r-gathering* in the basic case of unit-demands - our results refer to this problem unless stated otherwise. In addition to the basic *r-gathering problem*, we consider the version in which there is an additional *proximity requirement*: Each customer in the solution must be assigned to the nearest open facility. This is clearly a plausible quality of a solution in many facility-location settings, and also in clustering scenarios (*e.g.*, in geographic data-mining, see [10]). We manage to obtain a constant-factor approximation for this problem as well.

1.1   Our Results

We start by presenting a simple 3-approximation algorithm for *min-max r-gathering*. On the other hand, we prove that this problem cannot be approximated within less than a factor of 3 (assuming $P \neq NP$), for any $r \geq 3$. By using a similar reduction, we also show that *r-gather clustering* cannot be approximated within less than a factor of 2 for any $r \geq 3$, thus improving the hardness result of [1].

The same approximation algorithm extends to provide a 3-approximate solution for a generalization considered by [12,13], in which each $f_i \in F$ has a different lower-bound r_i on the number of customers required. Furthermore, it extends to provide the same approximation-ratio for the generalization in which there are several types of customers, and each open facility f_i must have at least r_{i_j} customers of type j (this may be useful for example for achieving "*p-Sensitive k-Anonymity*"[17] in publishing information from databases, similarly to the use of *r-gather clustering* for achieving "*k-Anonymity*"[1]).

By using another extension of this algorithm, we provide a 3-approximation for the generalization of *min-max r-gathering* in which an ϵ-fraction of the customers can be ignored. We thus match the approximation-ratio stated in [1] for the special case of (r, ϵ)-*gather clustering*.

Interestingly, practically the same algorithm also provides a $2r$ approximation for the *min-sum* version of the problem, if there are no facility costs. For this case, this improves upon the $1.5(2r - 1) + \epsilon$ approximation implied by the bicriteria algorithm of [12,13].

Next we consider the *proximity requirement*, and present a 9-approximation algorithm for *min-max r-gathering* which satisfies it (*i.e.*, each customer is assigned to a nearest open facility). For the special case of *r-gather clustering*, our technique provides a 6-approximation algorithm. In addition, we provide a 2-approximation algorithm for *2-gather clustering* which satisfies the proximity requirement. We show that this approximation factor cannot be improved: An algorithm for r-gather clustering which guarantees the proximity requirement cannot guarantee an approximation-ratio smaller than 2.

Finally, we show that although min-max 2-gathering is polynomial [2], the related min-max minimum-movement facility-location [5] is NP-hard and cannot be approximated within less than a factor of 2 (assuming $P \neq NP$). This resolves the open-question recently posed by Demaine et al. [5].

All our algorithms are based on discrete combinatorial techniques. Our hardness results use reductions from Exact-k-cover and SAT.

The rest of this paper is organized as follows. In Section 2 we present formal problem definitions and notations. Section 3 presents our simple approximation algorithm for min-max r-gathering, and analyzes its use for other versions. Section 4 considers the requirement of assigning each customer to a nearest open facility. The hardness results are provided in Section 5. We conclude with some concluding remarks and open problems.

2   Problem Definitions and Notations

We now formally state the basic problems we consider and introduce some of the notations we use. (we use slightly different notations from those of [12,13]).

The input for an r-gathering problem consists of a set of customer-locations $S = \{s_1, ..., s_n\}$, a set of potential facility-locations $F = \{f_1, ..., f_m\}$ with opening costs $p : F \rightarrow R^+ \bigcup \{0\}$, and distances (connection-costs) $d : (S \bigcup F)$ x $(S \bigcup F) \rightarrow R^+ \bigcup \{0\}$. The input also consists of a positive integer $r > 1$.

A solution is an assignment of the n customers to (not necessarily distinct) facilities, $t_1, ..., t_n$, which are considered open, such that customer i is assigned to facility $t_i \in F$, and the number of customers assigned to each open facility is at least r. In the min-max version of the problem, the goal is to minimize $\max_{1 \leq i \leq n}(\max(d(s_i, t_i), p(t_i)))$ (we refer to this as the cost of the solution). In the min-sum version, the goal is to minimize $\sum_{i=1}^{n} d(s_i, t_i) + \sum_{f_i \in \{t_1, ..., t_n\}} p(f_i)$ (each cost of an open facility is considered once in this sum).

A special case of min-max r-gathering is r-gather clustering [1], where $S \subseteq F$, and there are no facility costs ($p(f_i) = 0$, for $1 \leq i \leq m$).

3   Approximating Min-Max r-Gathering

Definition 1. *The "**min-cost**" of customer i, denoted $c(i)$, is the minimum cost of assigning r customers, including customer i, to a single facility (considering both the facility-cost and the customers' connection-costs). The location of this min-cost assignment, $g_i \in F$, is called "**the best facility**" of customer i. The "**partners**" of customer i are the r-1 customers, other than customer i, who participate in this min-cost assignment. (If there are several options we arbitrarily prefer locations and customers with smaller indices).*

We now provide a simple approximation algorithm for the problem, **Best-or-Rest** (see Figure 1).

Algorithm Best-or-Rest

1. For each customer, find his *min-cost, best facility* and *partners*.
2. Sort the customers in non-decreasing *min-cost* order.
3. For each customer i in this sorted order:
   If customer i and all his *partners* have not been assigned yet - assign them to the *best facility* of customer i (open this facility if it is not open yet). Otherwise, do nothing and continue to the next customer.
4. Assign any unassigned customer to the nearest open facility. (In case of a tie, arbitrarily choose the location with smallest index).

Fig. 1. A 3-approximation algorithm for *min-max r-gathering*

Lemma 1. *The cost of the solution found by algorithm* **Best-or-Rest** *for* min-max r-gathering *is at most thrice the maximal* min-cost.

Proof. First, observe that the cost of a customers' assignment at stage (3) is the min-cost of one of the customers assigned at this stage (customer i), which is at most the maximal min-cost of any of the n customers.

Now consider a customer i assigned at stage (4). This customer was not assigned at stage (3), which means that when customer i was considered at stage (3), at least one of his partners, say customer j, had already been assigned to another facility, $t_j = g_k$ (the best facility of some customer $k \neq i$). Customer i can also be assigned to t_j, with a cost of $d(s_i, t_j)$. Clearly, $d(s_i, t_j) \leq d(s_i, s_j) + d(s_j, t_j)$. Observe that $d(s_i, s_j) \leq 2c(i)$, since $d(s_i, g_i) \leq c(i)$ and $d(g_i, s_j) \leq c(i)$ (as j is one of the *partners* of customer i and g_i is the best-facility of customer i). Also, $d(s_j, t_j) = d(s_j, g_k) \leq c(k)$, since customer j is one of the partners of customer k. Since we performed stage (3) in a non-decreasing order of min-cost, $c(k) \leq c(i)$. So taken together, for each customer assigned at stage (4), $d(s_i, t_i) \leq 3c(i)$ (the customer is assigned to a nearest open facility, and we saw that there exists an open facility which satisfies this). This yields the required result.

Theorem 1. *Algorithm* **Best-or-Rest** *finds a 3-approximate solution for* min-max r-gathering, *and can be implemented to run in* $O(n(m+r) + (n+mr)\log n)$ *time.*

Proof. The cost of an optimal solution for the problem is clearly at least the maximal *min-cost* (since there is a customer whose assignment requires at least that cost in any solution). Therefore, the previous lemma proves that the algorithm finds a 3-approximate solution.

For implementing the first stage efficiently, we can first find for each $t \in F$ the sorted list of r customers closest to t. This can be done in $O(n + r\log n)$ time for each facility (using a binary heap). Let D_t be the distance from t of the r-th customer in that (non-descending) list. Thus, for each customer i, the minimal cost of assigning him along with r-1 other customers to location t is $max(D_t, d(s_i, t), p(t))$. So computing these costs for each customer and for each $t \in F$ takes an overall time of $O(m(n + r \log n))$. We now find the *best facility* of each customer according to these costs (in an overall time of $O(mn)$). The *partners* of customer i are clearly the first other r-1 customers in the list of his best facility, and noting them for each customer requires a total of $O(nr)$ time (nr may be higher than $m(n + r \log n)$). Thus, stage (1) can be implemented to run in $O(nr + m(n + r \log n))$ time. Stage (2) clearly requires $O(n \log n)$ time (which may be higher than $O(mr \log n)$). The next stages are less time-consuming than the first one, and thus the total running time is as stated.

Algorithm **Best-or-Rest** can also be used for the generalization in which an ϵ-fraction of the customers may be ignored (ϵ is specified in the input). We can simply ignore the ϵn customers with highest min-costs (in case of ties we ignore only those whose min-cost is strictly higher than the min-cost of $(1 - \epsilon)n$ other customers), and then run this algorithm. This guarantees an approximation-ratio of 3 for this generalization of the problem, since the optimal cost must be at least the highest min-cost of the customers we considered (note that the customers we ignored are not partners of customers we haven't ignored, since their min-cost is higher). As mentioned in the Introduction, this matches the approximation-ratio stated in [1] for a special case of this generalization.

It is also easy to see that algorithm **Best-or-Rest** can be used to achieve the same approximation-ratio even if there is a different lower-bound r_i on the number of customers for each facility $f_i \in F$, a generalization considered by [12,13]. This should simply be taken into account in the definitions of *min-cost*, *best-facility* and *partners*, and the first stage of the algorithm will change accordingly (and will be similarly implemented). Furthermore, it can be used to achieve the same approximation-ratio for the generalization in which there are several types of customers, and each open facility f_i must have at least r_{i_j} customers of type j (again, this should simply be taken into account in Definition 1, changing the first stage of the algorithm accordingly).

We next prove that algorithm **Best-or-Rest** can be used to provide a $2r$ approximation for *min-sum r-gathering* (with unit demands), in the basic case introduced by [13] where there are no facility costs. We call this case *unweighted min-sum r-gathering*. This improves upon the ratio of $1.5(2r - 1) + \epsilon$ implied by the algorithm of [12,13] for this case of the problem.

We define the *min-cost*, *best-facility* and *partners* in the corresponding way for the *min-sum* problem (the cost of an assignment to a facility is the *sum* of the connection-costs of the customers rather than their *maximum*).

Lemma 2. *The cost of the solution found by algorithm **Best-or-Rest** for un-weighted min-sum r-gathering is at most twice the sum of the min-costs of all the customers.*

Proof. The proof is similar to Lemma 1, and is omitted from this version.

Lemma 3. *The cost of an optimal solution for unweighted min-sum r-gathering is at least a $(1/r)$-fraction of the sum of the min-costs of all the customers.*

Proof. Consider a facility $t \in F$ opened by an optimal solution OPT. Let $x = yr + z$ be the number of customers assigned to t in this solution (where y, z are integers such that $y \geq 1$, $r > z \geq 0$). Now divide these customers into $(y + 1)$ sets in the following way. For each customer a assigned to t, calculate $d(s_a, t)/c(a)$, *i.e.*, the fraction of his min-cost that his connection costs in this solution. The first set, B_0, will contain the z customers for which the above calculated value was maximal. The other customers are arbitrarily divided into y sets of r customers, $B_1, ..., B_y$.

Consider a set B_i, $1 \leq i \leq y$. For each customer $a \in B_i$, $c(a) \leq \sum_{b \in B_i} d(s_b, t)$ (since this is the cost of assigning r customers, including customer a, to facility t in OPT). Summing this over all the customers in B_i, we get $\sum_{a \in B_i} c(a) \leq r \cdot \sum_{a \in B_i} d(s_a, t)$. This is true for every $1 \leq i \leq y$, which means that the cost of assigning the customers of $\bigcup_{i=1}^{y} B_i$ in OPT is at least a $(1/r)$-fraction of the sum of their min-costs.

Now consider a customer $a \in B_0$. If we replace one of the customers of B_1 by customer a, then the previous argument still holds for this modified set of r customers. So the total cost of assigning the customers in this modified set to t is at least a $(1/r)$-fraction of the sum of their min-costs. From the way B_0 has been selected, it follows that $d(s_a, t)/c(a) \geq 1/r$ (otherwise this ratio must have been smaller than $1/r$ for all the customers in this set, and thus also for the sums). Since this is true for any customer in B_0, it is true for the whole B_0, *i.e.*, the cost of assigning these customers to t is at least a $(1/r)$-fraction of the sum of their min-costs.

All the above is true for any facility t opened by an optimal solution, which means that the cost of an optimal solution is at least a $(1/r)$-fraction of the sum of min-costs, as required.

Theorem 2. *Algorithm* **Best-or-Rest** *finds a $2r$-approximate solution for* unweighted min-sum r-gathering, *and can be implemented to run in $O(n(m + r) + (n + mr) \log n)$ time for this problem.*

Proof. The approximation-ratio follows from combining the last two lemmas. It is easy to see that the running-time is the same as in Theorem 1, since we can similarly implement the first stage of the algorithm.

4   Assigning to a Nearest Open Facility

In this section we consider the *min-max r-gathering* problem with the additional constraint that each customer should be assigned to the nearest open facility (or to one of the nearest open facilities in case of a tie). We start by presenting a 9-approximate algorithm which satisfies this constraint.

In the following we say that a customer *prefers* a facility if there is no other open-facility nearer to his input location. We use the term *unsatisfied* for a customer who is not assigned to a nearest open facility. We use algorithm **Move-to-Solid**, described in Figure 2, for finding an approximate solution.

Algorithm Move-to-Solid

1. Run algorithm **Best-or-Rest**. If there are no unsatisfied customers, we are done. Otherwise, re-assign customers according to the following stages (initially no customer is considered *re-assigned*).
2. For each customer who has not been re-assigned yet, check which of the existing open facilities he prefers (in case of a tie choose the facility with smallest index). If a facility is preferred by at least r such customers, we say that it became *solid*.
3. Move to each solid facility all the customers who prefer it that have not been re-assigned yet. All the customers in solid facilities are now considered re-assigned.
4. If there are non-solid facilities which contain less than r customers now, re-assign their remaining customers to the facilities they most prefer out of the solid ones (and close these empty facilities).
5. If there are any non-solid facilities left, return to (2).
6. If there are unsatisfied customers, move them to the facilities they prefer out of the remaining (solid) facilities.

Fig. 2. A 9-approximation algorithm for *min-max r-gathering*, in which each customer is assigned to a nearest open facility

Theorem 3. Algorithm Move-to-Solid *finds a 9-approximate solution for* min-max r-gathering, *in which each customer is assigned to a nearest open facility. It requires* $O(n^3/r + m(n + r \log n))$ *time.*

Proof. We first observe that the algorithm runs in the stated polynomial time. We call an execution of stages (2)-(5) *an iteration*. Clearly, there are at most n/r open facilities after stage (1), so there can be at most n/r iterations in which facilities become *solid*. Note that since there are at least r customers in each facility after stage (1), there must be at least one *solid* facility. If at a certain iteration no facility becomes solid, it means that at least one customer assigned to a non-solid facility preferred one of the solid facilities at that iteration, and was therefore re-assigned to it (the customers in non-solid facilities have not been re-assigned yet, and if they all prefer non-solid facilities in (2) then at least one of these facilities must be preferred by at least r such customers). Since customers re-assigned to solid facilities are not re-assigned again until stage (6), there can be at most n such iterations. Thus the number of iterations is smaller than $n+n/r$. Clearly, each iteration requires $O(n^2/r)$ time (this is what stage (2) may require at the worst case). Stage (1) requires $O(n(m + r) + (n + mr) \log n)$ time according to Theorem 1, and stage (6) can clearly be implemented in $O(n^2/r)$ time. Summing these bounds yields the time bound stated in the theorem (as $r \leq n$).

We next explain why the algorithm indeed finds a solution for the problem. Since each solid facility has at least r customers who preferred it over all the other remaining facilities, at least r customers are left at each of the open facilities at the end (note that facilities are only canceled and not created, so a cheaper assignment option cannot appear later). Since each customer is assigned at stage (6) to a facility that he most prefers out of the remaining open facilities, each is assigned to a nearest open facility (by definition). We thus turn to considering the cost.

We proved that algorithm **Best-or-Rest** finds a 3-approximate solution. We denote its cost by C. We now prove that the re-assignments of **Move-to-Solid** increase the cost of the solution by a factor of at most 3. Note that the cost of open facilities does not increase (since we only close facilities), so we only need to consider the increase in the customers' connection-costs (distances).

Clearly, moving unsatisfied customers to a facility they prefer can only decrease their connection-cost. A customer's connection-cost can increase only when he is moved from a canceled facility (a facility found at stage (1) which was left with less than r customers) to the solid facility that he most prefers (at stage (4)). Let u be such a canceled facility. If u was canceled, then one of the customers assigned to it at stage (1) must have preferred one of the solid facilities at that iteration, v, and was moved to it. Let customer i be the first such customer.

It is clear that $d(u, v) \leq 2C$, since $d(u, s_i) \leq C$, and $d(s_i, v) \leq d(s_i, u)$ (since customer i preferred v). Thus, moving any customer assigned to u at stage (1) to the solid facility that he most prefers adds at most $2C$ to his connection-cost, which is therefore at most $3C$. After reaching a solid facility, the cost of a customer does not increase again (he is re-assigned again only if he is unsatisfied at the end, which may only decrease his cost). Therefore, the maximum connection-cost of any customer in this solution is at most $3C$, which is at most 9 times the optimum. Thus, the theorem is proven.

We note that the procedure described in the last proof can be used to transform any solution into a solution in which each customer is assigned to a nearest open facility, while increasing the total cost by a factor of at most 3. Thus, by applying it to a 2-approximate solution found by the algorithm of [1] for r-gather clustering, we can obtain a 6-approximate solution for r-gather clustering which satisfies the proximity requirement. In the context of [1], it is a clustering solution in which each object is assigned to a nearest cluster center (which is clearly a plausible quality of a clustering solution).

4.1   Improved Results for $r = 2$

Recall that *min-max r-gathering* is polynomial for $r = 2$ [2]. However, the solution found by [2] does not guarantee anything regarding the proximity requirement. We start by showing that for any $r \geq 2$, there are problem instances of *r-gather clustering*, for which the minimal cost solution that satisfies the proximity requirement costs almost twice the optimum. We then provide algorithm **Nearest-Neighbor**, that indeed finds a 2-approximate solution which satisfies the proximity requirement for *2-gather clustering*.

Algorithm Nearest-Neighbor

1. For each customer i, find the customer j closest to him (his nearest-neighbor), and let $c(i) = d(s_i, s_j)$. (In case of a tie, pick the customer with smallest index).
2. Consider the customers' $c(i)$ values in non-increasing order, and do the following for each such value x:
   (a) Build a graph $G = (V, E)$, where V contains a vertex for each customer i who satisfies $c(i) = x$ that was not assigned yet. For every $u, v \in V$, $(u, v) \in E$ iff $d(u, v) = x$.
   (b) Remove isolated vertices from G. Repeatedly remove edges whose both endpoints have a degree > 1 as long as there are such edges, *i.e.*, until the graph becomes a set of vertex-disjoint stars.
   (c) Open facilities in the star centers, and assign the customers in the remaining vertices of G to their star's center (in case of a single edge, arbitrarily pick one of its endpoints to be the center)
   (d) For each customer in V which was not assigned so far, open a facility at the input location of his nearest-neighbor, and assign that customer and his nearest neighbor to that facility.

Fig. 3. Finding a 2-approximate solution for *2-gather clustering*, in which each customer is assigned to a nearest open facility

Claim. For every $r > 1$ and $\epsilon > 0$, there are instances of *r-gather clustering* such that the minimal cost solution which satisfies the proximity requirement costs at least $(2 - \epsilon)$ times the optimum.

Proof. Omitted from this version due to space limitations.

For the approximation we use algorithm **Nearest-Neighbor**, described in Figure 3.

Theorem 4. *Algorithm* **Nearest-Neighbor** *finds a 2-approximate solution for 2-gather clustering, in which each customer is assigned to a nearest open facility. It requires $O(n^2)$ time.*

Proof. We start by showing that the algorithm finds a solution for the problem, which costs at most the maximum of the customers' $c(i)$ values. The cost of assigning a customer i at stage 2(c) is clearly at most $c(i)$, since the assignment described uses at most one edge of E for each customer. Each open facility is assigned at least two customers at this stage (those who are at the same star).

At stage 2(d), an unassigned customer i is assigned to the input location s_j of his *nearest neighbor* j. We observe that if customer j is the nearest neighbor of customer i then $c(j) \leq c(i)$ (since customer i is at a distance of $c(i)$ from customer j). If $c(j) < c(i)$, then clearly customer j was not assigned yet, and it is assigned to the same location s_j by the algorithm (with zero cost). So this is a valid assignment, and the cost of assigning customer i is exactly $c(i)$.

If $c(j) = c(i)$, then customer j must have been previously assigned to his own location s_j, when another customer, k (satisfying $c(k) > c(j)$), has been assigned to it (otherwise s_i would not have been isolated in G, and customer i would have already been assigned at stage 2(c)). So this is again a valid assignment, which costs $c(i)$. Thus all the customers are assigned, and each facility contains at least 2 customers.

All this is true for each of the $c(i)$ values and for each of the customers. Therefore, the total cost of the assignment is at most the maximum of the customers' $c(i)$ values. Clearly, the optimal solution costs at least half of this (the customers might be able to meet at the middle of a shortest path between them).

Finally, we explain why each customer is indeed assigned by the algorithm to a nearest open facility. Facilities are only opened by the algorithm in locations of customers, and each customer is either assigned to his own location or to the location of one of his nearest neighbors (in which case there is no facility in his own location). As the algorithm progresses, there can only be less assignment options (since some of the customers are already assigned to locations of other customers). Therefore, at the end there can be no nearer open facility for any of the customers. It is easy to see that each stage of the algorithm requires a total of at most $O(n^2)$ time.

5 Hardness Results

We match the approximation-ratio for *min-max r-gathering* with the following hardness result.

Theorem 5. *For any $r \geq 3$, it is NP-hard to approximate min-max r-gathering within less than a factor of 3, even if there are no facility costs.*

Proof. We prove the theorem by a reduction from the *Exact-k-Cover* problem (also called *Exact-Cover by k-Sets*), which is known to be strongly NP-hard for any $k \geq 3$ [7,8]. The input consists of a set of elements $S = \{x_1, ... x_{kn}\}$, and m subsets of this set of elements, $S_1, ..., S_m$, where $|S_i| = k$ for every $1 \leq i \leq m$. The question is whether there exists a collection of n subsets $S_{i_1}, ..., S_{i_n}$, such that each element is included in exactly one of them. Our reduction first proves that *min-max r-gathering* is NP-hard, and we later see that this implies that it is NP-hard to approximate within less than a factor of 3.

We construct the following input for *min-max r-gathering*. The set of customer locations is $S = \{s_1, ... s_{kn}\}$, *i.e.*, there is one customer for each element x_i, $1 \leq i \leq kn$. There is one potential facility location $f_i \in F$ for each subset S_i ($1 \leq i \leq m$), with $p(f_i) = 0$. For every $x_i \in S_j$, $d(s_i, f_j) = 1$. The other distances are those implied by this definition (*i.e.*, the distances in the graph $G = (S \bigcup F, E)$, where $(u, v) \in E$ iff $d(u, v) = 1$ and the weight of each edge is 1). We set $r = k$. We now prove that the cost of an optimal solution for this problem is 1 iff the answer to the *Exact-k-Cover* problem is "yes".

Assume the answer to the *Exact-k-cover* problem is "yes". Opening facilities in the locations corresponding to the cover subsets $S_{i_1}, ..., S_{i_n}$, and assigning each customer to the facility corresponding to the subset which covers his corresponding element, provides a solution in which each facility is assigned r customers and the cost is 1 for each customer. Thus, the optimal cost is indeed 1.

On the other hand, if the optimal cost is 1, then the answer to the *Exact-k-Cover* problem is "yes". A solution with a cost of 1 can only exist if each customer is assigned to a facility which corresponds to a subset containing his corresponding element. Thus, there are exactly r such customers assigned to each open facility in that solution, since each facility has only r customers at a distance of 1. Therefore there must be n such facilities, since all the customers are assigned. These facilities correspond to n subsets, each of them containing r different elements. Thus these subsets form an *Exact-k-Cover*. So both sides of the reduction are proven. Since *Exact-k-Cover* is NP-hard for any $k \geq 3$, it proves that our problem is NP-hard for any $r \geq 3$.

Clearly, the cost is at least 3 iff the answer is "no", since there is no potential facility location at distance 2 from a customer. Thus, the theorem is proven.

The problem remains hard to approximate even for the following special case.

Theorem 6. *For any $r \geq 3$, the special case of* min-max r-gathering *in which $S = F$ and there are no facility costs, is NP-hard to approximate within less than a factor of 2.*

Proof. Proving NP-hardness for the special case where $S=F$ requires a change in the reduction described in the previous proof. Instead of having only one location f_i corresponding to each subset S_i, r locations correspond to each subset S_i: $f_{i_1}, ..., f_{i_r}$. For each $x_j \in S_i$ we define $d(s_j, f_{i_1}) = 1$. Also, for every $1 \leq j < r$, we define $d(f_{i_j}, f_{i_r}) = 1$. Again, the other distances are those implied by those we defined. Each location both contains a customer and is a potential location of a facility ($S = F$).

It is not difficult to see that a solution has cost 1 iff the customers corresponding to each subset S_i are assigned to f_{i_r}, and customers who correspond to elements are assigned to neighboring locations of type f_{i_1} (*i.e.*, the new locations and customers have no influence on them). Otherwise the cost is at least 2. Therefore the reduction holds due to the same arguments, and the problem cannot be approximated within less than a factor of 2, assuming $P \neq NP$.

Since *r-gather clustering* is a generalization of the problem mentioned in the last theorem, this hardness result also holds for *r-gather clustering*, thus matching the approximation-ratio obtained by[1]. Previously this was known for *r-gather clustering* only for $r \geq 7$ [1].

Corollary 1. *For any $r \geq 3$, it is NP-hard to approximate the* r-gather cluster-ing *problem within less than a factor of 2.*

We next prove the hardness of a related problem described in the Introduction, *min-max minimum-movement facility-location*, which was introduced by [5].

They observed that this problem is approximable within a factor of 2, and we prove a matching lower bound on the approximability, thus resolving an open question they presented [5].

Theorem 7. *It is NP-hard to approximate the* min-max minimum-movement facility-location *problem within less than a factor of 2.*

Proof. (**sketch**) The reduction is from SAT. We build an unweighted graph with the following vertices: A *"server"* for each variable, a *"client"* for each clause, and an empty vertex for each literal, connected to the clauses which contain it and to its variable (a facility may be opened at any vertex). The connection-costs are defined according to the distances in this graph. Thus, there is a satisfying assignment to the formula iff there is a solution of cost 1 to the minimum-movement facility-location problem (facilities are opened in vertices corresponding to true literals).

6   Concluding Remarks and Open Problems

We considered the *min-max* version of the *r-gathering problem*, and provided constant-approximation algorithms and hardness-of-approximation results for several variants, some of which are tight. Some of our results improve previous results for special cases or related problems, including an improved approximation for *min-sum r-gathering* without facility costs and improved results for *r-gather clustering* and *min-max minimum-movement facility-location*.

Obvious remaining open problems are providing improved approximation algorithms or hardness results for *min-max r-gathering* with the proximity requirement and for *min-sum r-gathering*. Other problems which remain for future research are the generalizations in which each customer may have a different demand and each facility must serve a total demand of at least r, while the connection-costs are the product of distance and demand (previously considered by [12,13] for the *min-sum* version).

References

1. Aggarwal, G., Feder, T., Kenthapadi, K., Khuller, S., Panigrahy, R., Thomas, D., Zhu, A.: Achieving anonymity via clustering. In: Proceedings of the 25th ACM SIGMOD-SIGACT-SIGART Symposium on Principles of Database Systems, pp. 153–162 (2006)
2. Anshelevich, E., Karagiozova, A.: Terminal backup, 3D matching, and covering cubic graphs. In: Proceedings of the 39th Annual ACM Symposium on Theory of Computing, pp. 391–400 (2007)
3. Arya, V., Garg, N., Khandekar, R., Meyerson, A., Munagala, K., Pandit, V.: Local search heuristics for k-median and facility location problems. SIAM J. Comput. 33(3), 544–562 (2004)
4. Byrka, J.: An optimal bifactor approximation algorithm for the metric uncapacitated facility location problem. In: Proceedings of the 10th International Workshop on Approximation Algorithms for Combinatorial Optimization Problems (APPROX), pp. 29–43 (2007)

5. Demaine, E.D., Hajiaghayi, M., Mahini, H., Sayedi-Roshkhar, A.S., Oveisgharan, S., Zadimoghaddam, M.: Minimizing movement. In: Proceedings of the 18th Annual ACM-SIAM Symposium on Discrete Algorithms, pp. 731–740 (2007)

6. Drezner, Z.: Facility Location. Springer, Heidelberg (1995)

7. Ergun, F., Kumar, R., Rubinfeld, R.: Fast approximate PCPs. In: Proceedings of the 31st Annual ACM Symposium on Theory of Computing, pp. 41–50 (1999)

8. Garey, M.R., Johnson, D.S.: Computers and Intractability – A Guide to the Theory of NP-Completeness, p. 221. Freeman publishing, San Francisco (1979)

9. Gonzalez, T.F.: Clustering to minimize the maximum intercluster distance. Theoretical Comput. Sci. 38, 293–306 (1985)

10. Gudmundsson, J., van Kreveld, M.J., Narasimhan, G.: Region-restricted clustering for geographic data mining. In: Proceedings of the 14th Annual European Symposium on Algorithms, pp. 399–410 (2006)

11. Guha, S., Khuller, S.: Greedy strikes back: Improved facility location algorithms. J. Algorithms 31(1), 228–248 (1999)

12. Guha, S., Meyerson, A., Munagala, K.: Hierarchical placement and network design problems. In: Proceedings of the 41st Annual Symposium on Foundations of Computer Science, pp. 603–612 (2000)

13. Karger, D.R., Minkoff, M.: Building steiner trees with incomplete global knowledge. In: Proceedings of the 41st Annual Symposium on Foundations of Computer Science, pp. 613–623 (2000)

14. Mahdian, M., Ye, Y., Zhang, J.: Approximation algorithm for the soft-capacitated facility location problem. In: Proceedings of the 6th International Workshop on Approximation Algorithms for Combinatorial Optimization Problems (APPROX), pp. 129–140 (2003)

15. Schneier, B.: Applied Cryptography, pp. 71–73. John Wiley and Sons, Chichester (1996)

16. Sviridenko, M.: An improved approximation algorithm for the metric uncapacitated facility location problem. In: Cook, W.J., Schulz, A.S. (eds.) IPCO 2002. LNCS, vol. 2337, pp. 240–257. Springer, Heidelberg (2002)

17. Truta, T.M., Bindu, V.: Privacy protection: p-sensitive k-anonymity property. In: Proceedings of the Workshop on Privacy Data Management, in conjunction with the 22nd IEEE International Conference of Data Engineering (ICDE), pp. 94–103 (2006)

18. Vygen, J.: Approximation algorithms for facility location problems (2005), citeseer.ist.psu.edu/vygen05approximation.html

19. Zhang, J., Chen, B., Ye, Y.: Multi-exchange local search algorithm for the capacitated facility location problem. In: Bienstock, D., Nemhauser, G.L. (eds.) IPCO 2004. LNCS, vol. 3064, pp. 219–233. Springer, Heidelberg (2004)

On the Max Coloring Problem

Leah Epstein[1] and Asaf Levin[2]

[1] Department of Mathematics, University of Haifa, 31905 Haifa, Israel
lea@math.haifa.ac.il
[2] Department of Statistics, The Hebrew University, Jerusalem, Israel
levinas@mscc.huji.ac.il

Abstract. We consider max coloring on hereditary graph classes. The problem is defined as follows. Given a graph $G = (V, E)$ and positive node weights $w : V \to [1, \infty)$, the goal is to find a proper node coloring of G whose color classes $C_1, C_2, \ldots, C_k$ minimize $\sum_{i=1}^{k} \max_{v \in C_i} w(v)$. We design a general framework which allows to convert approximation algorithms for standard node coloring into algorithms for max coloring. The approximation ratio increases by a multiplicative factor of at most e for deterministic offline algorithms and for randomized online algorithms, and by a multiplicative factor of at most 4 for deterministic online algorithms. We consider two specific hereditary classes which are interval graphs and perfect graphs.

For interval graphs, we study the problem in several online environments. In the List Model, intervals arrive one by one, in some order. In the Time Model, intervals arrive one by one, sorted by their left endpoint. For the List Model we design a deterministic 12-competitive algorithm, a randomized $3e$-competitive algorithm, and prove a lower bound of 4 on the (deterministic or randomized) competitive ratio. For the Time Model, we use simplified versions of the algorithm and the lower bound of the List Model, to achieve a deterministic 4-competitive algorithm, a randomized e-competitive algorithm, and lower bounds of $\phi \approx 1.618$ on the deterministic competitive ratio and $\frac{4}{3}$ on the randomized competitive ratio. The former lower bounds hold even for unit intervals. For unit intervals in the List Model, we obtain a deterministic 8-competitive algorithm, a randomized $2e$-competitive algorithm and lower bounds of 2 on the deterministic competitive ratio and $\frac{11}{6} \approx 1.8333$ on the randomized competitive ratio.

Finally, we employ our framework to obtain an offline e-approximation algorithm for max coloring of perfect graphs, improving and simplifying a recent result of Pemmaraju and Raman.

1 Introduction

The (offline) max coloring problem is defined as follows: Given a graph $G = (V, E)$ and positive node weights $w : V \to [1, \infty)$, the goal is to find a proper node coloring of G (i.e., each pair of adjacent nodes are assigned distinct colors) whose color classes $C_1, C_2, \ldots, C_k$ minimize $\sum_{i=1}^{k} \max_{v \in C_i} w(v)$.

C. Kaklamanis and M. Skutella (Eds.): WAOA 2007, LNCS 4927, pp. 142–155, 2008.
© Springer-Verlag Berlin Heidelberg 2008

An interval graph has the property that its nodes can be presented as intervals on the real line so that two nodes share an edge if and only if their respective intervals intersect. Motivated by a design of dedicated memory managers problem, Pemmaraju, Raman and Varadarajan introduced the max coloring problem [19]. They designed a 2-approximation algorithm for the max coloring problem on interval graphs. Further, they showed that the First-Fit algorithm, which colors nodes in the first available color in a order in which they are given, when the intervals are considered in a monotone non-increasing order of their weights, is a 10-approximation algorithm for the max coloring problem on interval graphs. In that paper it is mentioned that the problem is actually interesting in the online environment, but it is not studied in that context.

In the online max coloring problem the nodes arrive one by one, and each time a node v arrives the set of edges connecting v to the earlier nodes is revealed. In this paper we consider the online max coloring problem where G is an interval graph. In this case we assume the graph is given via its intervals representation. The intervals are presented to the algorithm one by one clairvoyantly, that is, all information regarding the interval is revealed upon arrival. That is, we assume that each time an interval arrives its two endpoints are revealed. Each interval is to be colored before the next one is presented and this color assignment can not be changed afterwards. We are interested in two online versions of the problem. In the List Model, the intervals are given in an arbitrary order. In the Time Model, the intervals arrive sorted by their left endpoints. The study of the Time Model is motivated by the application of the design of memory managers in which each interval corresponds to memory requests that arrives along time (so the requests are ordered according to their left endpoints).

For an algorithm $\mathcal{A}$, we denote its cost by $\mathcal{A}$ as well. The cost of an optimal offline algorithm that knows the complete sequence of intervals is denoted by OPT. Since the problem is scalable, we consider the absolute competitive ratio and the absolute approximation ratio criteria. For an online algorithm we use the term competitive ratio whereas for an offline algorithm we use the term approximation ratio. The competitive ratio of $\mathcal{A}$ is the infimum $\mathcal{R}$ such that for any input, $\mathcal{A} \leq \mathcal{R} \cdot \text{OPT}$. If $\mathcal{A}$ is randomized, the last inequality is replaced by $E(\mathcal{A}) \leq \mathcal{R} \cdot \text{OPT}$. If the competitive ratio of an online algorithm is at most $\mathcal{R}$ we say that it is $\mathcal{R}$-competitive. If an algorithm has an unbounded competitive ratio, we say that it is not competitive. The approximation ratio of a polynomial time offline algorithm is defined similarly to be the infimum $\mathcal{R}$ such that for any input, $\mathcal{A} \leq \mathcal{R} \cdot \text{OPT}$. If the approximation ratio of a polynomial time offline algorithm is at most $\mathcal{R}$ we say that it is a $\mathcal{R}$-approximation.

In [17] Pemmaraju, Raman and Varadarajan designed an approximation algorithm with an approximation ratio of $O(\log n)$ for the (offline) max coloring of chordal graphs. They also analyzed empirically several heuristics.

In [18] Pemmaraju and Raman presented a 4-approximation algorithm for the (offline) max coloring of perfect graphs. Since every chordal graph is also a perfect graph, this result improves the earlier $O(\log n)$-approximation algorithm of [17] for chordal graphs. We recall that perfect graphs are such that the graph can be colored using ω colors, where ω is the size of the largest clique in the graph. Note that ω is a clear lower bound on the chromatic number of the graph. An algorithm that finds such a coloring is implied using the ellipsoid algorithm [7] (see also chapter 67 in [20]). Further results on the max coloring problem are provided in [8,3,15,5].

Coloring interval graphs has been intensively studied, Kierstead and Trotter [13] constructed an online algorithm which uses at most $3\omega - 2$ colors where ω is the maximum clique size of the interval graph. They also presented a matching lower bound of $3\omega - 2$ on the number of colors in a coloring of an arbitrary online algorithm. Note that the chromatic number of interval graphs equals to the size of a maximum clique, which is equivalent in the case of interval graphs to the largest number of intervals that intersect any point (see [10,6]). This means that the optimal offline algorithm can color every interval graph with ω colors. This can be actually done by applying First-Fit to the intervals sorted by their left end points. Therefore, a 1-competitive algorithm exists for this problem in the Time Model. Many papers studied the performance of First-Fit for this problem [11,12,19,2]. The last paper shows that the performance of First-Fit is strictly worse than the one of the algorithm of [13].

Interval coloring received much attention recently. In [19], a simple reduction from offline max coloring to online interval coloring was shown. The upper bounds in this paper were shown by exploiting the algorithm of [13] (which becomes a 2-approximation instead of the 3-competitive algorithm, since a part of the computation can be done offline), and First-Fit (this paper first improved the known bound on First-Fit and then used it). The reduction simply applies the online algorithm to the set of intervals, sorted by non-increasing order of weight. Adamy and Erlebach [1] introduced the interval coloring with bandwidth problem. In this problem each interval has a bandwidth requirement in $(0, 1]$. The intervals are to be colored so that at each point, the sum of bandwidths of intervals colored by a certain color does not exceed 1. This problem was studied also in [16,4].

Our results: We first present the positive results of this paper. I.e., we present a randomized online algorithm that uses as a sub-routine a node coloring algorithm. This sub-routine is applied to color graphs that are subgraphs of the original graph. We then show how to choose the parameters of our algorithm to obtain a deterministic online algorithm though with inferior competitive ratio. Note that though we reduce the max coloring problem to an interval coloring problem, which is also done in [19]. However our reduction does not require pre-sorting of the intervals, and therefore our algorithms for interval graphs are online. Using known results for online minimum coloring of interval graphs we obtain the following results. For the List Model we design a deterministic 12-competitive algorithm, a randomized $3e$-competitive algorithm, and prove a lower bound of 4 on the deterministic or randomized competitive ratio. For the Time Model, we use simplified versions of the algorithm and lower bound of the List Model, to achieve a deterministic 4-competitive algorithm, a randomized e-competitive algorithm, a lower bound of $\phi \approx 1.618$ on the deterministic competitive ratio, and a lower bound of $\frac{4}{3}$ on the randomized competitive ratio. The lower bound holds even for unit intervals. For unit intervals and the List Model, we obtain a deterministic 8-competitive algorithm, a randomized $2e$-competitive algorithm and improved lower bounds of 2 and $\frac{11}{6} \approx 1.8333$ on the deterministic and randomized competitive ratios, respectively. Our upper bounds for online algorithms are based on using a general reduction which we introduce in this paper, that allows to convert a r-competitive algorithm for standard coloring into a $4r$-competitive ($e \cdot r$-competitive) deterministic (randomized) algorithm for max coloring. Finally, we use our randomized algorithm with a derandomization procedure to obtain an offline (deterministic) e-approximation algorithm for max coloring of perfect graphs. We present the algorithms in Section 2, and the lower bounds in Section 3.

2   Algorithms

Before we define our algorithms, we would like to discuss the performance of First-Fit. This is clearly a natural algorithm for coloring. As shown in a sequence of papers [11,12,19], applying First-Fit to interval graphs for the standard coloring problem results in a constant competitive algorithm, though First-Fit is worse than the algorithm of Kierstead and Trotter [13,2]. However, we can show that First-Fit is not competitive for the max coloring problem.

Proposition 1. *First-Fit is not competitive even in the Time Model and unit intervals.*

Proof. Let M be a large constant fixed later. We introduce the input in blocks, all intervals are of length 2. Block i ($i \geq 0$) consists of i copies of the interval $[4i, 4i + 2]$, with weight 1 each, and one interval $[4i + 1, 4i + 3]$ of weight M. Clearly, the i intervals are colored using colors $1, \ldots, i$, since they arrive first, and do not overlap with any previous intervals. The next interval which has larger weight is colored with color $i + 1$. Therefore, the cost of the algorithm after block $j - 1$ is $M \cdot j$. An optimal offline algorithm would use one color for all intervals with larger weight, and $j - 1$ colors for all other intervals. This results in the cost $M + (j - 1)$. Taking $M = j^2$ we get a competitive ratio of $\frac{j^3}{j^2 + j - 1}$. When j grows to infinity, this competitive ratio becomes arbitrarily bad.   □

We design a framework for converting a deterministic C-competitive algorithm for online coloring of a given class of graphs into a randomized $e \cdot C$-competitive algorithm for max coloring on the same class of graphs. Our framework applies to hereditary class of graphs (i.e., if a graph belongs to this class, then every induced subgraph belongs to this class). We apply the scheme using deterministic algorithms only. This results in deterministic algorithms using a deterministic reduction scheme and in randomized algorithms using a randomized reduction scheme. Clearly, the randomized scheme can be used for converting a randomized algorithm to a randomized one.

Our algorithm has a positive integer parameter k and another (real value) parameter $\alpha > 1$. Our algorithm chooses uniformly at random an integer value

$0 \leq \ell < k$. Upon arrival of a new node we round down its weight as follows. We find the largest integer value t such that $\alpha^{kt+\ell}$ is no larger than the weight of the new node. The rounded weight of the node becomes $\alpha^{kt+\ell}$. Let OPT be the total weight (i.e., cost) of an optimal offline solution for the original sequence. For a given color, the weight of this color is defined to be the largest weight of any node which is colored by OPT with this color. Let OPT_i be the number of colors that OPT uses which have weight in the interval $[\alpha^{i-1}, \alpha^i)$. Denote by p the largest integer, such that OPT opens colors with weight in the interval $[\alpha^{k(p-1)}, \alpha^{kp})$. Note that p is unknown to the algorithm and is used only for the analysis.

Lemma 1. OPT *satisfies* $\text{OPT} \geq \sum_{i=1}^{kp} \alpha^{i-1} \cdot \text{OPT}_i$.

The input is partitioned into subsequences (also called classes), such that each one of them is colored independently, using its own set of colors. The subsequence S_0 consists of all nodes with weight that is smaller than α^ℓ. The subsequence S_i for $i \geq 1$ contains all nodes whose weight is in the interval $[\alpha^{\ell+(i-1)k}, \alpha^{\ell+ik})$.

Once we are coloring such a class S_i, all weights are considered as identical weights and the problem is reduced to the classical online coloring problem. We use a C-competitive algorithm to color such a class.

Lemma 2. *The number of colors that are used to color S_i is at most $C \cdot \sum_{j=\ell+(i-1)k+1}^{kp} \text{OPT}_j$ for $i \geq 1$ and at most $C \cdot \sum_{j=1}^{kp} \text{OPT}_j$ for $i = 0$.*

Proof. For $i \geq 1$, OPT can use the colors with weight at least $\alpha^{\ell+(i-1)k}$ to color the nodes of S_i. Therefore, there are at most $\sum_{j=\ell+(i-1)k+1}^{kp} \text{OPT}_j$ colors that are used by OPT to color S_i. Since we use a C-competitive algorithm to color S_i, the claim follows. For $i = 0$, OPT uses at most $\text{OPT} = \sum_{j=1}^{kp} \text{OPT}_j$ colors to color S_0, and the claim follows similarly.   □

It remains to analyze the (randomized) algorithm.

Lemma 3. *Assuming the existence of a C-competitive algorithm for online coloring, the randomized online algorithm has a competitive ratio of at most $\frac{C \cdot \alpha^{k+1}}{k(\alpha-1)}$.*

Proof. Since each color that our algorithm uses to color S_i has a weight of at most $\alpha^{\ell+ik}$, by Lemma 2, we conclude that for a given value of ℓ the cost of the solution returned by the algorithm is at most $C \cdot \sum_{i=1}^{p} \alpha^{\ell+ik} \sum_{j=\ell+(i-1)k+1}^{kp} \text{OPT}_j + C \cdot \alpha^\ell \cdot \sum_{j=1}^{kp} \text{OPT}_j$. We now consider the expected cost of the returned solution (the expectation is over the randomized value of ℓ). Since ℓ is chosen uniformly in the set $\{0, 1, \ldots, k-1\}$, the expected cost is at most the following:

$$\frac{\sum_{\ell=0}^{k-1} \left(C \cdot \sum_{i=1}^{p} \alpha^{\ell+ik} \sum_{j=\ell+(i-1)k+1}^{kp} \text{OPT}_j + C \cdot \alpha^\ell \cdot \sum_{j=1}^{kp} \text{OPT}_j \right)}{k} =$$

$$\frac{C \sum_{j=1}^{kp} \text{OPT}_j \cdot \sum_{t=0}^{j+k-1} \alpha^t}{k} \leq \frac{C \sum_{j=1}^{kp} \text{OPT}_j \cdot \alpha^{j+k-1} \cdot \sum_{t=0}^{\infty} \left(\frac{1}{\alpha} \right)^t}{k} =$$

$$\frac{C \sum_{j=1}^{kp} \text{OPT}_j \cdot \alpha^{j+k} \cdot \frac{1}{\alpha-1}}{k}$$

where the first equation holds by changing the order of summation, and the inequality holds since $\text{OPT}_j \geq 0$ for all j. Recall that by Lemma 1, $\text{OPT} \geq \sum_{j=1}^{kp} \alpha^{j-1} \cdot \text{OPT}_j$. We next note that the coefficients of OPT_j in the lower bound of OPT is at most $\frac{C \cdot \alpha^{k+1}}{k(\alpha-1)}$ times the coefficient of OPT_j in the upper bound on the expected cost of the solution returned by the algorithm, and thus the claim follows. □

Theorem 1. *Assuming the existence of a C-competitive algorithm for online coloring in one of the models, and a given hereditary graph class, there is a (randomized) online algorithm for max coloring (in the same model) with competitive ratio at most $e \cdot C$ for the same graph class.*

Proof. By setting $k = \infty$ in the algorithm, we carry the algorithm where $\alpha^k = e$. Instead of choosing ℓ to be a uniformly random integer, we set $\ell = u \cdot k$ where u is randomly chosen real number in the interval $[0,1]$, and in this case $\alpha^\ell = e^u$. Then, by Lemma 3, where $\alpha = 1 + \frac{1}{k}$ and k approaches ∞, the claim follows. □

Theorem 2. *Assuming the existence of a C-competitive algorithm for online coloring in one of the models, and a given hereditary graph class, there is a deterministic online algorithm for max coloring (in the same model) with competitive ratio at most $4 \cdot C$ for the same graph class.*

Proof. By Lemma 3, and setting $\alpha = 2$ and $k = 1$. We note that for $k = 1$ our algorithm is deterministic as ℓ has a unique possible value of 0. □

For max coloring of interval graphs we can use the following results: For the List Model, we use the 3-competitive algorithm of Kierstead and Trotter [13]. For the List Model with unit intervals, we use the 2-competitive algorithm of Epstein and Levy [4]. For the Time Model, we color each class optimally using First-Fit [10]. Therefore, we establish the following:

Corollary 1. *For online max coloring of interval graphs there is a randomized algorithm whose competitive ratio is $3e$ in the List Model, $2e$ in the List Model with unit intervals and e in the Time Model.*

For online max coloring of interval graphs there is a deterministic algorithm whose competitive ratio is 12 in the List Model, 8 in the List Model with unit intervals, and 4 in the Time Model.

We next note that for an offline algorithm, we can use a derandomization procedure to transform the (online) randomized algorithm into a deterministic approximation algorithm. To obtain the derandomization note that for each node v, v belongs to at most two adjacent classes S_i and S_{i+1} for the different values of ℓ. Therefore, there are at most n threshold values $\mathcal{S}$ that can be found in advance. We choose $\alpha = 1 + \frac{1}{k}$ and k is a huge integer number, and then we have to calculate only n solutions (the ones that correspond to the threshold values $\mathcal{S}$), and pick the best solution. Therefore, we establish the following theorem.

Theorem 3. *Given a hereditary class of graphs that has a ρ-approximation algorithm for the (offline) minimum node coloring problem, then there is a $(e \cdot \rho)$-approximation algorithm for the offline max coloring problem.*

Note that for perfect graphs (which are known to be hereditary class of graphs) there exists such an optimal algorithm for the minimum node coloring (see [7]). Using a classification as above, and running a node coloring algorithm for each class, we obtain an e-approximation algorithm for perfect graphs improving the 4-approximation algorithm of [18].

Corollary 2. *There is a deterministic e-approximation algorithm for (offline) max coloring of perfect graphs.*

3 Lower Bounds

We start with a lower bound for the Time Model.

Theorem 4. *The competitive ratio of any deterministic online max coloring algorithm of interval graphs in the Time Model is at least $\phi \approx 1.618$, which holds even if the input is restricted to unit intervals. For randomized algorithms, the competitive ratio is at least $\frac{4}{3}$.*

Proof. The input consists of a large enough number of blocks N, unless it stops earlier. Let $a < 1$ be a parameter (fixed to be $a = \frac{\sqrt{5}-1}{2} = \phi - 1 \approx 0.618$ in the deterministic case, and $a = \frac{1}{2}$ in the randomized case). A block is a clique, where block i (for some $i \geq 1$) consists of $i - 1$ intervals of weight a (regular intervals) and one interval of weight 1 (the expensive interval). An incomplete block has only the regular intervals. The sequence is either processed till the end, or stops in a situation where some number $i \geq 1$ of blocks is complete, and the last block $i + 1$ is incomplete.

All the regular intervals of a block i are identical copies of the interval $[4i, 4i+2]$, and the expensive interval of the same block is the interval $[4i + 1, 4i + 3]$. Clearly, intervals of different blocks do not intersect. All intervals in one block should receive distinct colors, but any pair of intervals from different blocks can receive the same color. We now compute the optimal offline cost for t blocks. If all blocks are complete, there are N blocks, and each has one expensive interval. We color all expensive intervals using one color, and at most one regular interval per block with each one of $N - 1$ additional colors. This gives a total cost of $1 + a(N - 1)$. If there are $i \geq 1$ complete blocks, and one incomplete block, there are at most i intervals in each block. Therefore OPT needs only i colors, where one of these colors is used for all expensive intervals (in the incomplete block, it is used for a regular interval). We get the cost OPT $= 1 + a(i - 1)$.

Next, we consider the behavior of the algorithm. In the *deterministic* case, we make sure that the algorithm uses exactly i colors immediately after i complete blocks have arrived (if they indeed arrive) . Note that the regular intervals in each block arrive first. If the algorithm uses at least one new color, we stop the sequence. Otherwise, it uses a single new color for the expensive block. Consider first the case that N complete blocks arrive. This means that the algorithm used a new color for each expensive interval. Its cost is therefore N. Otherwise, let $i + 1$ be the index of the incomplete block. The algorithm used i distinct colors for the expensive intervals of blocks $1, \ldots, i$. It uses a new color for one regular interval of block $i + 1$. Therefore its cost is $i + a$.

Consider the competitive ratio if all blocks are complete. The ratio is $\frac{N}{aN+1-a}$. For large enough N, the ratio tends to $\frac{1}{a} = \frac{\sqrt{5}+1}{2} = \phi \approx 1.618$. If the sequence stopped at an incomplete block $i + 1$, the ratio is $\frac{i+a}{ia+1-a} = \phi$ (for any value of i). The result for randomized algorithm is shown in the full version of the paper. □

The lower bound for the List Model is based on blocks as well, however blocks are not simple cliques, and their construction is similar to the construction of the lower bound of 3 in [13].

Theorem 5. *The competitive ratio of any deterministic or randomized online max coloring algorithm of interval graphs in the List Model is at least* 4.

Proof. We start with a proof of the deterministic lower bound and later show how to extend it for randomized algorithms. To prove the theorem, we use gadgets. We describe them now, and afterwards we show how to obtain them.

Let $b < 1$ be a constant (later chosen to be $\frac{1}{2}$). A full (k, i) gadget is a construction of intervals, all contained in the interval $(i - 1, i)$ (so that intervals of different gadgets do not intersect), where the minimum number of colors needed to color them is at most ki. Out of these colors, at most $k(i - 1)$ are of weight b. If there are exactly $k(i - 1)$ colors using weight b, there are at most k additional colors which use weight 1. The numbers k and i are known in advance to an online algorithm. Any online algorithm is forced to use exactly $(3k - 2)i$ colors for this gadget. (Note that it is possible to actually force it to use a slightly larger amount of colors which is $3ki - 4$ for $i > 1$, but the amount we use is large enough for our purposes and makes the analysis more convenient.) A partial (k, i) gadget is the same as a full (k, i) gadget, only it does not contain any intervals of weight 1. It will be created in the same way as a full gadget, but the construction will stop before any intervals of weight 1 are presented. To color a partial gadget, the minimum number of colors needed is at most $k(i - 1)$ (all of weight b), and any online algorithms can be forced to use $(3k - 2)(i - 1)$ colors (Again, this is a convenient amount for our calculations, we can actually force it to use $3k(i - 1) - 2$ colors.) We call this set of colors "the colors used in the gadget". If a set of $(3k - 1)i$ colors is used at some point during the construction, we stop the construction immediately. In the randomized version of the lower bound, we do not stop the construction, but we charge the algorithm for $(3k-1)i$ of the colors it uses in a full (k, i) gadget, and treat the rest of the colors as new in the future. We describe the case in which we construct only a partial gadget later.

It is left to describe how to obtain a gadget. We use a construction which is very similar to the lower bound of 3 in [13]. A difference with [13], already used in [4] is the assumption that some information on the optimal cost (which is either $k(i - 1)$ or ki in our case) is known in advance.

The construction of a gadget consists of ki phases, where in the first $k(i - 1)$ phases, all intervals are of weight b, whereas the last k phases consist of intervals of weight 1. If the gadget is partial, the last k phases are not introduced. If it is full, all the phases are given. Let $U = (3k - 2)i > 3$ be the number of colors we would like to force in this gadget. Let $S = U^{3ki}$ be the initial number of intervals presented in the first phase. As mentioned above, all intervals presented are contained in $(i - 1, i)$. It is possible to introduce these intervals of length $\frac{1}{2S}$ starting from point $i - 1 + \frac{1}{4S}$ with distances of $\frac{1}{2S}$ between them. In each phase, the number of intervals which can be used for the next phase decreases by a factor of at most U^3 (actually at most $\frac{4(U)^3}{6}$).

After a phase is defined, we shrink some parts of the line into single points. Given a point p, that is a result of shrinking an interval $[a, b]$. Every interval presented in the past which is contained in $[a, b]$ is also shrunk into p and therefore the point p inherits a list of colors that such intervals received. These colors cannot be assigned to any interval that contains the point p. The shrinking is done only for simplification purposes. In practice it means that for a given point p that is the result of shrinking, every future interval either contains this point or not, i.e., it either contains all intervals that were shrunk into this point, or has no overlap with any of them.

If an algorithm uses more than U colors, we can stop the construction. Therefore we assume that the algorithm is initially given a palette of U colors. As soon as all these colors are used, the proof is complete. This is just one stopping condition, we may stop the sequence earlier as well, after the partial gadget has been constructed.

Since the algorithm is using at most U colors, this means that there exists a set of $\frac{S}{U}$ intervals that share the exact same color c. We shrink all intervals into single points. Later phases result in additional points.

We now define phase j (for $j \geq 2$). The phases are constructed in a way that in the beginning of phase j there is a set of at least $U^{3(ki-j+1)}$ points that contain a given subset of the U colors. These points are called points of interest.

There exist some other points containing other subsets of colors. All these points are called void points. At this time, we partition the points of interest into consecutive sets of four. At most three points of interest that do not participate in the partition become void points.

We next define additional intervals, increasing the size of the largest cardinality clique (with respect to the number of intervals, i.e., ignoring weights) by exactly one. Given a set of four points listed from left to right a_1, a_2, a_3, a_4, let b_1 be the leftmost void point on the right hand side of a_1, between a_1 and a_2. If no such point exists, then let $b_1 = \frac{a_1 + a_2}{2}$, i.e., the point which is halfway between a_1 and a_2. Similarly, let d be the rightmost void point between a_3 and a_4, and if no such point exists then $d = \frac{a_3 + a_4}{2}$. Let f be a point between a_2 and a_3 that is not a void point. We introduce the intervals $I_1 = [a_1, \frac{a_1 + b_1}{2}]$ and $I_2 = [\frac{d + a_4}{2}, a_4]$.

If they both receive the same color, we introduce the intervals $I_3 = [\frac{a_1 + b_1}{2}, f]$ and $I_4 = [f, \frac{d + a_4}{2}]$. The interval I_3 intersects with a_2, and with I_1. The second interval I_4 intersects I_3, a_3 and I_2, therefore two new colors must be used. In total, three new colors were used.

If I_1, I_2 receive distinct colors, we introduce the interval $I_5 = [\frac{a_1+b_1}{2}, \frac{d+a_4}{2}]$. Interval I_5 intersects with I_1, I_2, a_2, a_3, and thus gets a new color. In total, three new colors were used.

We shrink every such interval $[a_1, a_4]$ into a single point. Each of the new shrunk points received three new colors.

Note that we do not use more than U colors, and each new shrunk point receives three new colors. Four intervals are introduced only if the first two received the same color. There are less than $\frac{U^3}{6}$ options to choose from the set of three new colors. We can choose at least $6 \cdot U^{3(ki-j)}$ points having the same set of used colors. The points containing these exact sets of colors become the points of interest of the next phase, and the others become void points of the next phase. Points that are void points of previous phases and are not contained in shrunk intervals remain void points. Note that the points where the new intervals intersect are points with no previous intervals, and therefore the clique size increases by exactly 1.

After the first $k(i-1)$ phases, we start presenting intervals of weight 1 instead of b. The first phase of intervals of larger weight is different from all other phases, as we would like the set of all intervals of weight 1 to be k colorable. Thus, the first such phase we introduce has a clique size of exactly 1. Therefore, we introduce single intervals $[a_1, a_4]$ instead of the construction above, in this phase only. In this phase the algorithm uses a single new color.

Recall that multiple gadgets are used, but they are built in a way that no two intervals from different gadgets can intersect. Specifically, we can replace the single intervals from the proof of Theorem 4 by gadgets. A complete block i is replaced by a full (k, i) gadget for a large value of k. An incomplete block is replaced by a partial (k, i) gadget. However, the decision which defines a block as complete or incomplete is different. We again use at most N blocks.

Recall that in a full gadget i, the online algorithm is forced to use exactly $(3k-2)i$ colors. After the construction of a partial gadget, it is possible to count the total number of colors that were ever used, $m_i \leq (3k-2)i$. Since the total number of colors in block i must be $(3k-2)i$, we define $n_i = (3k-2)i - m_i$ to be the number of new colors needed to make block i a full gadget.

We define the construction as follows. When block $i+1$ is presented, the partial gadget is presented first. If the algorithm used at least $(3k-2)i + 2k$ colors by that time, the sequence terminates. Otherwise, the gadget is presented in full. The analysis of this construction is left for the full version. □

Finally, the lower bound for unit intervals can be improved in the List Model.

Theorem 6. *The competitive ratio of any deterministic online max coloring algorithm of unit interval graphs in the List Model is at least 2. For randomized algorithms, the competitive ratio is at least $\frac{11}{6} \approx 1.8333$.*

Proof. We first define a base block of size i. This is a set of intervals, where the largest clique size is $2i$, all weights in this block are the same. There is no overlap between intervals of this block and intervals of other blocks. We consider the deterministic case first.

We force the algorithm to use at least $3i$ colors for this block, no matter what their weight is, or which intervals or what weights appeared previously in the sequence. Another feature is that there exists a range of length 1, $[x, x + 1]$, where the largest clique size is $2i$, and all $3i$ colors are represented in this range, i.e., given the $3i$ colors the algorithm was forced to use in the block, every one of them is used to color some interval that overlaps $[x, x + 1]$ at least in one point.

The construction is similar to the lower bound of $\frac{3}{2}$, shown in [4] for online coloring of unit intervals.

The construction of the block is partitioned into three phases. In the initial phase we provide i identical requests for an interval $[4i, 4i + 1]$. The online algorithm has to color these intervals with exactly i colors, denote those colors by $c_1,...,c_i$ and the set of those colors by C.

In the next phase we present at most $2i$ intersecting intervals. These intervals are presented one by one in a way that all intervals colored by some color, c, where $c \in C$ are slightly shifted to the right with respect to any interval that is colored by a color $\bar{c}$, where $\bar{c} \notin C$. We present intervals until exactly i of them are colored by colors that are not in C. We now show how this goal is achieved.

Let $I_1 = [a, a + 1]$ be the rightmost interval colored by $\bar{c} \notin C$ and let $I_2 = [d, d+1]$ be the leftmost interval colored by $c \in C$ among intervals introduced so far in the current phase (not in the initial phase or in another block). If there is no interval colored $\bar{c}$ we say that I_1 is empty and if there is no interval colored c we say that I_2 is empty. Let $\varepsilon = \frac{1}{64i}$. A new interval, I, is presented as follows.
1. If both I_1 and I_2 are empty (this holds only when we introduce the first interval) then $I = [4i + \frac{3}{2}, 4i + \frac{5}{2}]$. **2.** If only I_1 is empty, $I = [d - \varepsilon, d + 1 - \varepsilon]$. **3.** If only I_2 is empty, $I = [a + \varepsilon, a + 1 + \varepsilon]$. **4.** If I_1 and I_2 are not empty then, $I = [\frac{d+a}{2}, \frac{d+a}{2} + 1]$, i.e. the interval is halfway between I_1 and I_2 with unit length, intersecting all previous intervals presented in this step.

Note that none of the intervals in this phase intersect intervals of the initial phase. Moreover, the left endpoints of all the intervals in the phase are located within a distance of less than 1 from the right endpoints of the intervals of the initial phase. Also note that the algorithm stops after introducing at most $2i$ intervals, at that time, if it is reached, there are exactly i intervals with a color that is not in C, since $|C| = i$.

Assume now that $[y_i + 1, y_i + 2]$ is the rightmost interval with color $\bar{c} \notin C$ (from the construction we have $4i + \frac{1}{4} < y_i < 4i + \frac{3}{4}$) after all intervals from phase 2 were presented. We present i requests for the interval $[y_i, y_i + 1]$. This interval intersects all the intervals with color not in C from the previous phase. They also intersect all the intervals from the initial phase.

To complete the analysis, note that the intervals presented in the last phase all intersect with intervals of exactly $2i$ different colors. There are i colors in C and i colors not in C from the second phase. This gives a coloring of $3i$ colors while the largest clique has cardinality $2i$. The range $[y_i, y_i + 1]$ intersects intervals of all $3i$ colors as needed.

Let $0 < \alpha < 1$. We now define the lower bound sequence. The sequence consists of at most N extended blocks, where an extended block contains a base block i of size i where all weights are α, and two additional requests for the interval $[y_i, y_i + 1]$, of weight 1 (thus the cardinality of largest clique becomes $2i + 2$). This forces the algorithm to use two additional colors. We call a block without the extension a base block, and the pair of intervals of unit weight is called the extension. The algorithm is forced to use a total of $3i + 2$ colors for the extended block.

The very first block, which is called extended block 0 has a different structure, but is still called a block of size 0. This is simply a base block of size 1, where all weights are 1. The block is started with requests for $[0, 1]$ (instead of $[4, 5]$), and the first interval of the second phase is $[\frac{3}{2}, \frac{5}{2}]$. In this extended block, the algorithm must use 3 colors.

The sequence consists of at most $N + 1$ blocks, each time a new block is built, its size is larger by 1 than the previous block. The first block is an extended block 0, and each other block i is of size i.

In a block of size i ($i \geq 1$), we first construct the base block, and check the set of colors which was used for this block. Before the block is presented, the algorithm was forced to use at least $3i - 1$ colors (the only exception is for $i = 1$ where the algorithm was already forced to use 3 colors). Consider the previous set of $3i - 1$ colors that the algorithm uses. (If $i > 1$ and there are already more than $3i - 1$ colors, we only make the algorithm pay for three new colors in each extended block $1 < j < i$, which overlap the range $[y_j, y_j + 1]$, and consider the set of colors that is built inductively in this way.) In block 0 the number of colors is exactly three, and if block 1 is extended then the number of colors it is charged for is exactly 5. If the new base block uses at least three new colors compared to the previous set of colors, we stop the construction (in block 1, since in block 0 three colors are used, we stop the construction after the base block if it contains two new colors). Otherwise we add the extension, and build the next base block (unless $i = N$).

We compute the optimal cost of the sequence up to base block i. The largest clique has size $2i$. Every block has two intervals of weight 1 and all others of weight α. Clearly, the very first block can be colored using two colors of weight 1. Therefore, to color the sequence, two colors of weight 1 and $2i - 2$ colors of weight α are needed. This gives a cost of $2 + 2(i - 1)\alpha$. If the sequence terminates at phase N with an extended block, then the cost is the same as if the base block $N + 1$ were presented, i.e. $2 + 2N\alpha$.

After block 0 is presented, the algorithm uses three colors of weight 1. If the sequence terminates after base block $i \geq 1$, three additional colors of weight α were used in this phase, and therefore the cost increases by at least 3α. Otherwise, still three new colors are used in this phase, and therefore the cost must increase by at least the cost of these colors. At least one of them is of weight 1, thus the cost increases by at least $1 + 2\alpha$. To make these calculations correct also for block 1, we charge block 0 only by $3 - \alpha$. Each base block is charged by α for a new color it must have, and either by an additional 2α if it has three new colors, or by an additional $\alpha + 1$ if it has at most two new colors.

Therefore, if the sequence terminates after base block i, the cost of the algorithm is at least $3 - \alpha + (2\alpha + 1)(i - 1) + 3\alpha = 2 + (2\alpha + 1)i$, whereas the optimal cost is $2 + 2(i - 1)\alpha$. The cost of the algorithm if the sequence is completed is $3 - \alpha + (2\alpha + 1)N$, whereas the optimal cost is $2 + 2N\alpha$. We get the ratio $\frac{3+2\alpha+(2\alpha+1)(i-1)}{2+2(i-1)\alpha}$ in the first case, and $\frac{3-\alpha+(2\alpha+1)N}{2+2N\alpha}$ in the second case. We choose a value of α such that $\frac{2\alpha+3}{2} = \frac{1+2\alpha}{2\alpha}$. The value $\alpha = \frac{1}{2} = 0.5$ satisfies this requirement. The ratio in the first case is $\frac{4+2(i-1)}{i+1} = 2$. The ratio in the second case tends to the same value for large enough values of N. The claim follows.

We provide the extension of this proof for randomized algorithms (which is similar to the proof for deterministic algorithms) in the full version of the paper.

$\square$

4   Concluding Remarks

We presented a framework for converting a deterministic C-competitive algorithm for online coloring of a given hereditary class of graphs into a deterministic $4C$-competitive algorithm for max coloring on the same class of graphs, and a randomized $e \cdot C$-competitive algorithm. For example, consider bipartite graphs. Lovász, Saks and Trotter [14] showed a deterministic online algorithm which colors such a graph on n nodes (which is 2 colorable) using $O(\log n)$ colors. Note that Gyárfás and Lehel [9] proved a deterministic lower bound of $\Omega(\log n)$ on the online coloring of bipartite graphs (this holds already for trees). This immediately implies a deterministic $O(\log n)$-competitive algorithm for max coloring of bipartite graphs. Note that the deterministic lower bound of $\Omega(\log n)$ holds for max coloring since node coloring is a special case of max coloring (using a common weight 1 for all nodes). The best offline approximation unless $P = NP$ for bipartite graphs has an approximation ratio of $\frac{8}{7}$ [18,3,15].

References

1. Adamy, U., Erlebach, T.: Online coloring of intervals with bandwidth. In: Solis-Oba, R., Jansen, K. (eds.) WAOA 2003. LNCS, vol. 2909, pp. 1–12. Springer, Heidelberg (2004)
2. Chrobak, M., Ślusarek, M.: On some packing problems relating to dynamical storage allocation. RAIRO Journal on Information Theory and Applications 22, 487–499 (1988)
3. Demange, M., de Werra, D., Monnot, J., Paschos, V.T.: Time slot scheduling of compatible jobs. Journal of Scheduling 10(2), 111–127 (2007)
4. Epstein, L., Levy, M.: Online interval coloring and variants. In: Caires, L., Italiano, G.F., Monteiro, L., Palamidessi, C., Yung, M. (eds.) ICALP 2005. LNCS, vol. 3580, pp. 602–613. Springer, Heidelberg (2005)
5. Escoffier, B., Monnot, J., Paschos, V.T.: Weighted coloring: Further complexity and approximability results. Information Processing Letters 97(3), 98–103 (2006)
6. Golumbic, M.C.: Algorithmic Graph Theory and Perfect Graphs. Academic Press, London (1980)

7. Grötschel, M., Lovász, L., Schrijver, A.: Geometric algorithms and combinatorial optimization. Springer, Heidelberg (1993)
8. Guan, D.J., Zhu, X.: A coloring problem for weighted graphs. Information Processing Letters 61(2), 77–81 (1997)
9. Gyárfás, A., Lehel, J.: On-line and first-fit colorings of graphs. Journal of Graph Theory 12, 217–227 (1988)
10. Jensen, T.R., Toft, B.: Graph coloring problems. Wiley, Chichester (1995)
11. Kierstead, H.A.: The linearity of first-fit coloring of interval graphs. SIAM Journal on Discrete Mathematics 1(4), 526–530 (1988)
12. Kierstead, H.A., Qin, J.: Coloring interval graphs with First-Fit. SIAM Journal on Discrete Mathematics 8, 47–57 (1995)
13. Kierstead, H.A., Trotter, W.T.: An extremal problem in recursive combinatorics. Congressus Numerantium 33, 143–153 (1981)
14. Lovász, L., Saks, M., Trotter, W.T.: An on-line graph coloring algorithm with sublinear performance ratio. Discrete Math. 75, 319–325 (1989)
15. Monnot, J., Paschos, V.T., de Werra, D., Demange, M., Escoffier, B.: Weighted coloring on planar, bipartite and split graphs: Complexity and improved approximation. In: Fleischer, R., Trippen, G. (eds.) ISAAC 2004. LNCS, vol. 3341, pp. 896–907. Springer, Heidelberg (2004)
16. Narayanaswamy, N.S.: Dynamic storage allocation and online colouring interval graphs. In: Chwa, K.-Y., Munro, J.I.J. (eds.) COCOON 2004. LNCS, vol. 3106, pp. 329–338. Springer, Heidelberg (2004)
17. Pemmaraju, S.V., Penumatcha, S., Raman, R.: Approximating interval coloring and max-coloring in chordal graphs. In: Ribeiro, C.C., Martins, S.L. (eds.) WEA 2004. LNCS, vol. 3059, pp. 399–416. Springer, Heidelberg (2004)
18. Pemmaraju, S.V., Raman, R.: Approximation algorithms for the max-coloring problem. In: Caires, L., Italiano, G.F., Monteiro, L., Palamidessi, C., Yung, M. (eds.) ICALP 2005. LNCS, vol. 3580, pp. 1064–1075. Springer, Heidelberg (2005)
19. Pemmaraju, S.V., Raman, R., Varadarajan, K.R.: Buffer minimization using max-coloring. In: SODA 2004. Proc. of 15th Annual ACM-SIAM Symposium on Discrete Algorithms, pp. 562–571 (2004)
20. Schrijver, A.: Combinatorial Optimization Polyhedra and Efficiency. Springer, Heidelberg (2003)

Full and Local Information in Distributed Decision Making

Panagiota N. Panagopoulou[1,2] and Paul G. Spirakis[1,2]

[1] Computer Engineering and Informatics Department, Patras University
[2] Research Academic Computer Technology Institute, Greece
panagopp@cti.gr, spirakis@cti.gr

Abstract. We consider the following distributed optimization problem: three agents $i = 1, 2, 3$ are each presented with a load drawn independently from the same known prior distribution. Then each agent decides on which of two available bins to put her load. Each bin has capacity α, and the objective is to find a distributed protocol that minimizes the probability that an overflow occurs (or, equivalently, maximizes the *winning probability*).

In this work, we focus on the cases of *full information* and *local information*, depending on whether each agent knows the loads of both other agents or not. Furthermore, we distinguish between the cases where the agents are allowed to follow different decision rules (*eponymous model*) or not (*anonymous model*). We assume no communication among agents.

First, we present optimal protocols for the full information case, for both the anonymous and the eponymous model.

For the local information, anonymous case, we show that the winning probability is upper bounded by 0.622 in the case where the input loads are drawn from the uniform distribution.

Motivated by [3], we present a general method for computing the optimal single-threshold protocol for *any* continuous distribution, and we apply this method to the case of the exponential distribution.

Finally, we show how to compute, in exponential time, an optimal protocol for the local information, eponymous model for the case where the input loads are drawn from a discrete-valued, bounded distribution.

1 Introduction

In a *distributed optimization problem* there are n agents, each of whom is presented with a private *input*. Then each agent decides on an *output*, and her decision depends on her private input as well as on any information she has about the inputs presented to all or a subset of the other agents. All agents have the same objective, which is to maximize a common function, but they are not allowed to cooperate in order to reach their objective.

Naturally, the more information the agents have about the inputs of the other agents, the better decisions they can make. On the other hand, sharing such information among the agents induces a *communication cost* to the solution of the problem. A natural problem arising then is to evaluate this trade-off between

C. Kaklamanis and M. Skutella (Eds.): WAOA 2007, LNCS 4927, pp. 156–169, 2008.
© Springer-Verlag Berlin Heidelberg 2008

the quality of the solution and the computational resources needed to achieve it, i.e. to understand the *value of information*.

In this work, we focus on a *load balancing* problem, where there are *three* agents, each presented with a *load* with size drawn from the same known prior distribution on $[0, \alpha]$. Each agent must then decide on which of *two* available bins (bin 0 and bin 1) to put her load. Each bin has capacity α, and the objective is to put the loads on the bins so that the probability that no overflow occurs (referred to as the *winning probability*) is maximized. We assume no cooperation among the agents.

Related work. The distributed optimization problem studied in this work was originally introduced by Papadimitriou and Yannakakis [3] in an effort to understand the crucial economic value of information [1] as a computational resource in a distributed system (e.g. in the context of Computational Complexity [4]). In order to understand how the optimum solution achieved by the agents varies as a function of the amount of information available to them, Papadimitriou and Yannakakis [3] considered each possible communication pattern and discovered the corresponding optimal decision protocol to be unexpectedly sophisticated. For the special case where no communication is allowed, i.e. when each agent i is aware of only her own load x_i, it was conjectured that the simple decision rule $\mathcal{Q}$: "if $x_i \leq (1 - \frac{1}{\sqrt{7}})\alpha$ then put x_i on bin 0 else put x_i on bin 1" is optimal; however no proof has been found until now verifying or rejecting this conjecture. Georgiades et al. [2] studied the extension of the load balancing problem to the case of n agents. Their work was focused on the special cases of oblivious decision rules, for which agents do not "look at" their inputs, and non-oblivious decision rules, for which they do. In either case, optimality conditions were derived in the form of combinatorial polynomial equations.

Contribution. In this work, we re-examine the load balancing distributed optimization problem for the case of three agents. First of all, we distinguish the case where the agents know their ids (eponymous model) and the case where the agents do not know their ids (anonymous model). If an agent is aware of her id, then her decision rule need not be identical to that of any other agent. We show that knowing one's id allows for better, with respect to the winning probability, optimal protocols.

Furthermore, we introduce *randomization* on decision rules, i.e. we allow for an agent to put her load on bin 0 with some probability $p > 0$ and on bin 1 with probability $1 - p > 0$. We show that, in some cases, randomization indeed helps in maximizing the winning probability.

We give optimal protocols for the full information setting, i.e. when each agents knows not only her own load, but also the loads of the other agents; we do so both for the anonymous and the eponymous model.

Next we focus on the case where only local information is provided, i.e. when each agent is aware of only her own load. We show that the winning probability is upper bounded by 0.622 in the case where the input loads are drawn from the uniform distribution. Furthermore, we present a general method for computing the threshold t^* for which the winning probability of the parameterized (single-threshold) protocol $\mathcal{Q}_t$: "if $x_i \leq t$ then put x_i on bin 0 else put x_i on bin 1" is maximized for *any* continuous distribution. We apply this method to the case of the exponential distribution.

We conclude by studying the eponymous case of the local information setting, for which we give an exponential-time algorithm that computes an optimal, deterministic protocol for the case where the input loads are drawn from a discrete-valued, bounded distribution.

2 Framework

2.1 Setting

Three non-cooperating *agents* with ids 1, 2 and 3 are each presented with a *load* $x_i \in [0, \alpha]$ ($i \in \{1, 2, 3\}$) for some $\alpha \in \mathbb{R}_+$. The loads x_1, x_2 and x_3 are independent, identically distributed random variables drawn from a known distribution F on $[0, \alpha]$. Denote Δ the set of all distributions on $[0, \alpha]$.

Each agent i must then decide on which one of two available bins (bin 0 and bin 1) to put her load. Each bin has *capacity* α. Let $p_i(x_i) \in \{0, 1\}$ denote the decision of agent i when presented with the load x_i. The agents *win* if the total load on each bin does not exceed its capacity α, i.e. if $\sum_{i:p_i(x_i)=0} x_i \leq \alpha$ and $\sum_{i:p_i(x_i)=1} x_i \leq \alpha$.

Note that the agents are not allowed to communicate, so the objective is to find a distributed protocol (i.e. a procedure that defines for each possible x_i a bin $p_i(x_i) \in \{0, 1\}$, for all agents $i = 1, 2, 3$) that *maximizes* the probability of winning, defined in the following subsection.

2.2 The Winning Probability

Suppose that the input loads are independently drawn from the same distribution $F \in \Delta$. The performance of a protocol $\mathcal{M}$ for a distribution $F \in \Delta$ of the input loads is measured by means of its probability of winning $P_w(\mathcal{M}; F)$:

$$P_w(\mathcal{M}; F) = \int_{\mathbf{x} \in [0, \alpha]^3} \Pr\{\mathbf{x}\} \Pr\{\mathcal{M} \text{ wins } | \mathbf{x}\}$$

where the integration is over all possible ordered triples of input loads $\mathbf{x} = \langle x_1, x_2, x_3 \rangle \in [0, \alpha]^3$.

2.3 Models of Information

As already mentioned, the agents are not allowed to cooperate in order to maximize their probability of winning. So each agent's decision should only depend on the information provided to her regarding the triple of the input loads. Thus if each agent knows only her own load x_i, then her decision must depend only on x_i. On the other hand, if she is also aware of the loads x_j, x_k presented to both other agents, then her decision can depend on the triple of input loads $\mathbf{x} = \{x_i, x_j, x_k\}$.

Furthermore, if we assume that the agents do not know their ids, then any two agents i, j presented with the same input must make their decisions according to exactly the same rule (i.e. in this setting the agents are indistinguishable). In contrast, if an agent knows her id then she can base her decision on it, so in this case any two agents presented with the same input could decide on which bin to put it according to different rules.

Hence we distinguish the following four models of information:

Local information, anonymous. Each agent knows only her own load (and not her id).
Local information, eponymous. Each agent knows her own load and her own id.
Full information, anonymous. Each agent knows her load as well as the loads of both other agents, but not their ids.
Full information, eponymous. Each agent knows her load and id, as well as the loads and ids of both other agents.

Let $\mathcal{L}^{\mathcal{A}}, \mathcal{L}^{\mathcal{E}}, \mathcal{F}^{\mathcal{A}}, \mathcal{F}^{\mathcal{E}}$ denote the families of protocols for the local information anonymous, local information eponymous, full information anonymous and full information eponymous model respectively.

For any $\mathcal{T} \in \{\mathcal{L}^{\mathcal{A}}, \mathcal{L}^{\mathcal{E}}, \mathcal{F}^{\mathcal{A}}, \mathcal{F}^{\mathcal{E}}\}$, a protocol $\mathcal{M} \in \mathcal{T}$ is *deterministic* if it decides on the same bin (0 or 1) whenever executed with the same input. Clearly, if $\mathcal{M}$ is a deterministic protocol then, for any fixed $\mathbf{x}$, $\Pr\{\mathcal{M} \text{ wins} \mid \mathbf{x}\}$ is either 0 or 1.

On the other hand, $\mathcal{M}$ is *randomized* if there exists an input y and an agent i for which it decides $p_i(y) = 0$ with some probability $p > 0$ (that depends on $\mathcal{M}$, i and y) and $p_i(y) = 1$ with probability $1 - p > 0$. Observe that any randomized protocol in $\mathcal{L}^{\mathcal{A}}$ or $\mathcal{L}^{\mathcal{E}}$ is a probability distribution over the protocols in $\mathcal{L}^{\mathcal{E}}$. Similarly, any randomized protocol in $\mathcal{F}^{\mathcal{A}}$ or $\mathcal{F}^{\mathcal{E}}$ is a probability distribution over the protocols in $\mathcal{F}^{\mathcal{E}}$.

2.4   A Conjecture on the Local Information Model

In [3] it is conjectured that protocol $\mathcal{Q}$, described in Fig. 1, is optimal for the local information model, when the distribution of the input loads is the uniform distribution on $[0, \alpha]$. In fact it was shown that, in this case, $\mathcal{Q}$ is indeed optimal among all (*single-threshold*) protocols that involve at most one "switch" of bins per agent. i.e. among all protocols of the form "if $x_i \leq t$ then put x_i on bin 0 else put x_i on bin 1".

Protocol $\mathcal{Q}$
Input: The agent's load x_i
Output: A bin $p(x_i) \in \{0, 1\}$
1. if $x_i \leq \left(1 - \frac{1}{\sqrt{7}}\right) \alpha$ then $p(x_i) = 0$
2. else $p(x_i) = 1$

Fig. 1. Protocol $\mathcal{Q}$ for the local information, anonymous model

3 Full Information, Anonymous Model

In this section we focus on the case where each agent is aware of her own load as well as of the loads of both other agents, but she does not know her or any of the other agents' id. We show how to construct a randomized protocol that maximizes the probability of winning for this full information, anonymous case.

Protocol $\mathcal{R}^{\mathcal{A}}$
Input: The agent's load x_i and the loads x_j, x_k of both other agents
Output: A bin $p(x_i) \in \{0, 1\}$
1.   if $x_i = \max\{x_i, x_j, x_k\} \neq \max\{x_j, x_k\}$ then $p(x_i) = 0$
2.   else if $x_i = \max\{x_i, x_j, x_k\} = \max\{x_j, x_k\}$ then
       $p(x_i) = 0$ with probability $1/2$
       $p(x_i) = 1$ with probability $1/2$
3.   else $p(x_i) = 1$

Fig. 2. Protocol $\mathcal{R}^{\mathcal{A}}$ for the full information, anonymous model

For this full information and anonymous model, we should seek for a protocol that does not depend on the agents' ids; in other words, all agents must decide on the same bin when introduced with the same *unordered* triple of input loads $\mathbf{x} = \{x_1, x_2, x_3\}$. Consider the simple randomized protocol $\mathcal{R}^{\mathcal{A}}$ described in Fig. 2.

Theorem 1. $\mathcal{R}^{\mathcal{A}}$ *is an optimal protocol for the full information, anonymous model.*

Proof. It suffices to prove that whenever protocol $\mathcal{R}^{\mathcal{A}}$ fails, then any full information, anonymous protocol fails as well. Consider a triple of input loads $\{x_i, x_j, x_k\}$. Without loss of generality assume that $x_i = \max\{x_i, x_j, x_k\}$. Assume that $x_i \neq \max\{x_j, x_k\}$. Then loads x_k and x_j are put on bin 1, while load x_i is put on bin 0. Since $\mathcal{R}^{\mathcal{A}}$ fails, it holds that $x_k + x_j > \alpha$. But then $x_i + x_k > x_j + x_k > \alpha$ and $x_i + x_j > x_k + x_j > \alpha$ hence any protocol would fail as well. Assume now that $x_i = \max\{x_j, x_k\}$. Without loss of generality assume that $x_i = x_j = \max\{x_i, x_j\}$. If $x_k < x_i$, then x_i is put on bin 0 with probability $1/2$, x_j is put on bin 0 with probability $1/2$ and x_k is put on bin 1. An optimal protocol would maximize the probability that the two maxima loads x_i and x_j are put on separate bins. Since we are in the anonymous model and $x_i = x_j$, any protocol puts x_i and x_j on bin 0 with the same probability. Let p be the probability that the optimal protocol puts x_i on bin 0. Then the probability that x_i and x_j are put on separate bins is $p(1 - p) + (1 - p)p = -2p^2 + 2p$. Observe that this probability is maximized for $p = 1/2$, thus in this case $\mathcal{R}^{\mathcal{A}}$ behaves optimally. Finally, if $x_i = x_j = x_k$ then $\mathcal{R}^{\mathcal{A}}$ puts each load on bin 0 with probability $1/2$. An optimal protocol would maximize the probability that not all loads are put on the same bin. This probability is $1 - p^3 - (1 - p)^3 = -3p^2 + 3p$. Again, this is maximized for $p = 1/2$. Thus $\mathcal{R}^{\mathcal{A}}$ is optimal. $\qquad\square$

The proof of the above theorem immediately implies:

Corollary 1. *If the input loads are drawn from a discrete valued distribution, then an optimal protocol for the full information, anonymous model can not be deterministic[1].*

4   Full Information, Eponymous Model

In this section we focus on the case where each agent is aware of her own load and id as well as of the loads and ids of both other agents. We show how to construct a protocol that maximizes the probability of winning for this full information, eponymous case. Suppose that each agent executes protocol $\mathcal{R}^{\mathcal{E}}$ described in Fig. 3. Then:

Protocol $\mathcal{R}^{\mathcal{E}}$
  Input: The agent's load x_i, the agent's id i,
         and the loads x_j, x_k and ids j, k of both other agents
  Output: A bin $p_i(x_i) \in \{0, 1\}$
  1.   if $x_i = \max\{x_i, x_j, x_k\} \neq \max\{x_j, x_k\}$ then $p_i(x_i) = 0$
  2.   else if $x_i = \max\{x_i, x_j, x_k\} = x_t$ for some $t \in \{j, k\}$
        and $i < t$ then $p_i(x_i) = 0$
  3.   else $p_i(x_i) = 1$

Fig. 3. Protocol $\mathcal{R}^{\mathcal{E}}$ for the full information, eponymous model

Theorem 2. $\mathcal{R}^{\mathcal{E}}$ *is an optimal protocol for the full information, eponymous model.*

Proof. It suffices to prove that whenever protocol $\mathcal{R}^{\mathcal{E}}$ fails, then any (deterministic or probabilistic) protocol fails as well. Suppose that $\mathcal{R}^{\mathcal{E}}$ fails for a triple of input loads $\langle x_i, x_j, x_k \rangle$. Without loss of generality assume that agent i is the agent with the minimum index among all agents of maximum load. Note that, under all circumstances, protocol $\mathcal{R}^{\mathcal{E}}$ will put load x_i on bin 0 and loads x_j and x_k on bin 1. Thus $\mathcal{R}^{\mathcal{E}}$ will fail if and only if $x_j + x_k > \alpha$. In this case however, since $x_i \geq x_j$ and $x_i \geq x_k$, it holds that $x_i + x_j > \alpha$ and $x_i + x_k > \alpha$ as well, hence any protocol would also fail.   □

5   Local Information, Anonymous Model

In this section we focus on the case where each agent knows her own load but not her id (thus the agents are indistinguishable).

---

[1] The reason why this conclusion does not hold for continuous distributions is that, in such cases, the probability that two input loads are equal is zero.

5.1 Bounds on the Winning Probability for the Uniform Distribution

Theorem 3. *Assume that the three loads are independent random variables, uniformly distributed on $[0,1]$, and the capacity of each bin is $\alpha = 1$. Then for any local information, anonymous protocol $\mathcal{P}$,*

$$\frac{1}{6} \leq \Pr\{\mathcal{P} \text{ wins}\} < 0.622 \ .$$

Proof. Consider an arbitrary protocol $\mathcal{P}$. Denote x, y and z the loads of the three agents. Assume that the sum of the three loads is less or equal to 1. Then $\mathcal{P}$ wins, so the joint probability that $\mathcal{P}$ wins and the sum of the three loads is not greater than 1 is

$$\Pr\{\mathcal{P} \text{ wins}, x+y+z \leq 1\} = \Pr\{x+y+z \leq 1\}$$
$$= \int_0^1 \int_0^{1-x} \int_0^{1-x-y} 1 \, dz \, dy \, dx$$
$$= \frac{1}{6} \ .$$

Thus

$$\Pr\{\mathcal{P} \text{ wins}\} \geq \Pr\{\mathcal{P} \text{ wins}, x+y+z \leq 1\} \geq \frac{1}{6} \ .$$

Assume now that $x+y > 1$ and $y+z > 1$ and $z+x > 1$. Then $\mathcal{P}$ fails, so

$$\Pr\{\mathcal{P} \text{ fails}, x+y > 1, y+z > 1, z+x > 1\}$$
$$= \Pr\{x+y > 1, y+z > 1, z+x > 1\} = 6 \int_{\frac{1}{2}}^1 \int_{\frac{1}{2}}^x \int_{1-y}^y 1 \, dz \, dy \, dx = \frac{1}{4} \ .$$

Now define $S_0, S_1 \subseteq \left[\frac{1}{2}, 1\right]$ as

$$S_0 = \left\{ w \in \left[\frac{1}{2}, 1\right] : \mathcal{P} \text{ puts } w \text{ on bin } 0 \right\} \ ,$$
$$S_1 = \left\{ w \in \left[\frac{1}{2}, 1\right] : \mathcal{P} \text{ puts } w \text{ on bin } 1 \right\} \ .$$

Denote by s_0 and s_1 the total length of S_0 and S_1 respectively, and let $s = s_0$. Then $s_1 = 1/2 - s$. Similarly, define $T_0, T_1 \subseteq \left[\frac{1}{4}, \frac{1}{2}\right)$ as

$$T_0 = \left\{ w \in \left[\frac{1}{4}, \frac{1}{2}\right) : \mathcal{P} \text{ puts } w \text{ on bin } 0 \right\} \ ,$$
$$T_1 = \left\{ w \in \left[\frac{1}{4}, \frac{1}{2}\right) : \mathcal{P} \text{ puts } w \text{ on bin } 1 \right\} \ .$$

Denote by t_0 and t_1 the total length of T_0 and T_1 respectively, and let $t = t_0$. Then $t_1 = 1/4 - t$.

We call a triple $(x, y, z) \in [0, 1]^3$ of input loads *feasible* if $x + y \leq 1$ or $y + z \leq 1$ or $z + x \leq 1$. Observe that

$$\Pr\{\mathcal{P} \text{ fails}, x + y + z > 1, (x, y, z) \text{ feasible}\}$$
$$\geq \Pr\left\{\mathcal{P} \text{ fails}, (x, y, z) \text{ feasible, 2 loads} \in \left[\frac{1}{2}, 1\right]\right\}$$
$$+ \Pr\left\{\mathcal{P} \text{ fails, 1 load} \in \left[\frac{1}{2}, 1\right] \text{ and 2 loads} \in \left[\frac{1}{4}, \frac{1}{2}\right)\right\}.$$

Now,

$$\Pr\left\{\mathcal{P} \text{ fails}, (x, y, z) \text{ feasible, 2 loads} \in \left[\frac{1}{2}, 1\right]\right\}$$
$$\geq 3 \int_{S_0} \int_{S_0} \int_0^{\max\{1-x, 1-y\}} 1 \, dz \, dy \, dx + 3 \int_{S_1} \int_{S_1} \int_0^{\max\{1-x, 1-y\}} 1 \, dz \, dy \, dx$$
$$= 3 \int_{S_0} \int_{S_0} \int_0^{1-\min\{x, y\}} 1 \, dz \, dy \, dx + 3 \int_{S_1} \int_{S_1} \int_0^{1-\min\{x, y\}} 1 \, dz \, dy \, dx$$
$$= 3 \int_{S_0} \int_{S_0} (1 - \min\{x, y\}) \, dy \, dx + 3 \int_{S_1} \int_{S_1} (1 - \min\{x, y\}) \, dy \, dx$$
$$= 3s_0^2 + 3s_1^2 - 3 \int_{1/2}^1 \int_{S_0} \min\{x, y\} + 3 \int_{S_1} \int_{S_0} \min\{x, y\} \, dy \, dx$$
$$- 3 \int_{S_1} \int_{1/2}^1 \min\{x, y\} \, dy \, dx + 3 \int_{S_1} \int_{S_0} \min\{x, y\} \, dy \, dx$$
$$= 3s^2 + 3\left(\frac{1}{2} - s\right)^2 - 3 \int_{1/2}^1 \int_{1/2}^1 \min\{x, y\} \, dy \, dx + 6 \int_{S_1} \int_{S_0} \min\{x, y\} \, dy \, dx$$
$$= 6s^2 - 3s + \frac{1}{4} + 6 \int_{S_0} \int_{S_1} \min\{x, y\} \, dy \, dx$$
$$\geq 6s^2 - 3s + \frac{1}{4} + 6 \int_{1/2}^{1/2+s} \int_{1/2+s}^1 x \, dy \, dx$$
$$= -3s^3 + \frac{9}{2}s^2 - \frac{3}{2}s + \frac{1}{4}.$$

Moreover,

$$\Pr\left\{\mathcal{P} \text{ fails, 1 load} \in \left[\frac{1}{2}, 1\right] \text{ and 2 loads} \in \left[\frac{1}{4}, \frac{1}{2}\right)\right\}$$
$$\geq 3s_0 t_0^2 + 3s_1 t_1^2$$
$$= 3st^2 + 3\left(\frac{1}{2} - s\right)\left(\frac{1}{4} - t\right)^2$$
$$= \frac{3}{2}t^2 + \frac{3}{2}st - \frac{3}{4}t - \frac{3}{16}s + \frac{3}{32}.$$

Thus

$$\Pr\{\mathcal{P} \text{ fails}, x + y + z > 1, (x, y, z) \text{ feasible}\}$$

$$\geq -3s^3 + \frac{9}{2}s^2 - \frac{3}{2}s + \frac{1}{4} + \frac{3}{2}t^2 + \frac{3}{2}st - \frac{3}{4}t - \frac{3}{16}s + \frac{3}{32}$$

$$= -3s^3 + \frac{9}{2}s^2 + \frac{3}{2}t^2 + \frac{3}{2}st - \frac{27}{16}s - \frac{3}{4}t + \frac{11}{32} .$$

In order to lower bound the joint probability

$$\Pr\{\mathcal{P} \text{ fails}, x + y + z > 1, (x, y, z) \text{ feasible}\}$$

it suffices to minimize

$$f(s, t) = -3s^3 + \frac{9}{2}s^2 + \frac{3}{2}t^2 + \frac{3}{2}st - \frac{27}{16}s - \frac{3}{4}t + \frac{11}{32}$$

subject to $s \in [0, 1/2], t \in [0, 1/4]$. It can be proved that the minimum is achieved at $t = \frac{1+\sqrt{37}}{48}$, $s = \frac{11-\sqrt{37}}{48}$ and equals $\frac{521-37\sqrt{37}}{2304}$. Thus, for any protocol $\mathcal{P}$,

$$\Pr\{\mathcal{P} \text{ fails}\} = \Pr\{\mathcal{P} \text{ fails}, x + y + z > 1, (x, y, z) \text{ feasible}\}$$
$$+ \Pr\{\mathcal{P} \text{ fails}, (x, y, z) \text{ not feasible}\}$$
$$\geq \frac{521 - 37\sqrt{37}}{2304} + \frac{1}{4}$$
$$> 0.378 ,$$

and therefore

$$\Pr\{\mathcal{P} \text{ wins}\} < 1 - 0.378 = 0.622 . \qquad \square$$

5.2 Optimal Single-Threshold Protocols for Continuous Distributions

We deal with the problem of computing the optimal single-threshold protocol for the general case where the three input loads are independent, identically distributed random variables drawn from the same arbitrary continuous distribution D on $[0, \infty)$. Let $f(x)$ be the probability density function of an input load x, and let $F(x) = \int_0^x f(w)\,dw$. Let x, y, z be three random variables corresponding to the three input loads. The joint probability density function of these three random variables is $f(x, y, z) = f(x)f(y)f(z)$, since x, y and z are mutually independent. Assume again that each bin has capacity α.

We seek for a $t \in [0, \alpha]$ that maximizes the probability of winning of the single-threshold protocol $\mathcal{Q}_t$: "if $x \leq t$ then put x on bin 0, else put x on bin 1". Note that there are 4 possible outcomes: (a) two loads are put on bin 0 and the other on bin 1, (b) one load is put on bin 0 and the other two on bin 1, (c) all three loads are put on bin 0 and (d) all three loads are put on bin 1. Thus the winning probability can be expressed as the following sum of joint probabilities:

$$P_w(\mathcal{Q}_t; D) = \Pr\{\text{(a) occurs}, \mathcal{Q}_t \text{ wins}\} + \Pr\{\text{(b) occurs}, \mathcal{Q}_t \text{ wins}\}$$
$$+ \Pr\{\text{(c) occurs}, \mathcal{Q}_t \text{ wins}\} + \Pr\{\text{(d) occurs}, \mathcal{Q}_t \text{ wins}\} .$$

We consider the following cases.

Case 1: $0 \le t < \frac{\alpha}{3}$. In this case, both outcome (a) occurs and the agents win if and only if the load on bin 1 lies in $(t, \alpha]$ and the other two loads lie in $[0, t]$. The former occurs with probability $F(\alpha) - F(t)$, while the latter occurs with probability $\int_0^t \int_0^t f(x) f(y) \, dy \, dx$, since their sum can be at most $2t < \alpha$. Since there are 3 ways for outcome (a) to occur (depending on which of the three loads is put on bin 1),

$$\Pr\{(a) \text{ occurs}, \mathcal{Q}_t \text{ wins}\} = 3 \cdot (F(\alpha) - F(t)) \cdot \int_0^t \int_0^t f(x) f(y) \, dy \, dx \ .$$

Outcome (b) occurs and the agents win if and only if one load lies in $[0, t]$ and the other two lie in $(t, \alpha]$, while their sum does not exceed α. Thus

$$\Pr\{(b) \text{ occurs}, \mathcal{Q}_t \text{ wins}\} = 3 \cdot (F(t) - F(0)) \cdot \int_t^{\alpha - t} \int_t^{\alpha - x} f(x) f(y) \, dy \, dx \ ,$$

where the integral denotes the probability that two random variables exceed t while their sum is no more than α. Outcome (c) occurs and the agents win if and only if all three loads lie in $[0, t]$, since their sum can not exceed $3t < \alpha$. Thus

$$\Pr\{(c) \text{ occurs}, \mathcal{Q}_t \text{ wins}\} = \int_0^t \int_0^t \int_0^t f(x) f(y) f(z) \, dz \, dy \, dx \ .$$

Outcome (d) occurs and the agents win if and only if all three loads lie in $(t, \alpha]$ while their sum does not exceed α. Thus

$$\Pr\{(d) \text{ occurs}, \mathcal{Q}_t \text{ wins}\} = \int_t^{\alpha - 2t} \int_t^{\alpha - x - t} \int_t^{\alpha - x - y} f(x) f(y) f(z) \, dz \, dy \, dx \ .$$

Thus the (total) winning probability of $\mathcal{Q}_t$ when $0 \le t < \frac{\alpha}{3}$ is

$$P^1 = 3 \cdot (F(\alpha) - F(t)) \cdot \int_0^t \int_0^t f(x) f(y) \, dy \, dx$$

$$+ 3 \cdot (F(t) - F(0)) \cdot \int_t^{\alpha - t} \int_t^{\alpha - x} f(x) f(y) \, dy \, dx$$

$$+ \int_0^t \int_0^t \int_0^t f(x) f(y) f(z) \, dz \, dy \, dx$$

$$+ \int_t^{\alpha - 2t} \int_t^{\alpha - x - t} \int_t^{\alpha - x - y} f(x) f(y) f(z) \, dz \, dy \, dx \ .$$

Case 2: $\frac{\alpha}{3} \le t < \frac{\alpha}{2}$. Similar reasoning as above yields

$$P^2 = 3 \cdot (F(\alpha) - F(t)) \cdot \int_0^t \int_0^t f(x)f(y)\,dy\,dx$$

$$+ 3 \cdot (F(t) - F(0)) \cdot \int_t^{\alpha-t} \int_t^{\alpha-x} f(x)f(y)\,dy\,dx$$

$$+ \int_0^t \int_0^t \int_0^t f(x)f(y)f(z)\,dz\,dy\,dx$$

$$- \int_{\alpha-2t}^t \int_{\alpha-x-t}^t \int_{\alpha-x-y}^t f(x)f(y)f(z)\,dz\,dy\,dx \ .$$

Case 3: $\frac{\alpha}{2} \le t < \alpha$. Using again similar arguments as in case 1, the winning probability in this case is

$$P^3 = 3 \cdot (F(t) - F(\alpha)) \cdot \left(\int_0^t \int_0^t f(x)f(y)\,dy\,dx - \int_{\alpha-t}^t \int_{\alpha-x}^t f(x)f(y)\,dy\,dx \right)$$

$$+ \int_{\alpha-t}^t \int_0^{\alpha-x} \int_0^{\alpha-x-y} f(x)f(y)f(z)\,dz\,dy\,dx$$

$$+ \int_0^{\alpha-t} \int_0^{\alpha-x-t} \int_0^t f(x)f(y)f(z)\,dz\,dy\,dx$$

$$+ \int_0^{\alpha-t} \int_{\alpha-x-t}^t \int_0^{\alpha-x-y} f(x)f(y)f(z)\,dz\,dy\,dx \ .$$

Thus

$$P_w(\mathcal{Q}_t; D) = \begin{cases} P^1 & \text{if } 0 \le t < \frac{\alpha}{3} \\ P^2 & \text{if } \frac{\alpha}{3} \le t < \frac{\alpha}{2} \\ P^3 & \text{if } \frac{\alpha}{2} \le t \le \alpha \end{cases} \ ,$$

and the optimal threshold is $t^* = \arg\max_t \{P_w(\mathcal{Q}_t; D)\}$.

The Exponential Distribution. We apply the previous analysis to the case where the input loads are independently drawn from the exponential distribution $\mathcal{E}$. The probability density function of $\mathcal{E}$ is $g(x) = \frac{1}{\lambda}e^{-\frac{x}{\lambda}}$ for all $x \in [0, \infty)$, for some parameter $\lambda > 0$. Then P^1, P^2 and P^3 are as follows:

$$P^1 = P^2 = -3e^{-\frac{\alpha+2t}{\lambda}} + \left(9 + \frac{3\alpha - 6t}{\lambda}\right)e^{-\frac{\alpha+t}{\lambda}}$$

$$+ \left(\frac{6\alpha t - 9t^2 - \alpha^2}{2\lambda^2} + \frac{9t - 4\alpha}{\lambda} - 7\right)e^{-\frac{\alpha}{\lambda}} + 1$$

$$P^3 = -6e^{-\frac{2t}{\lambda}} + \left(\frac{3\alpha - 6t}{\lambda} + 9\right)e^{-\frac{\alpha+t}{\lambda}} + \left(\frac{6t - 3\alpha}{\lambda} - 3\right)e^{-\frac{2\alpha}{\lambda}}$$

$$+ \left(\frac{2\alpha^2 - 6\alpha t + 3t^2}{2\lambda^2} + \frac{2\alpha - 3t}{\lambda} - 1\right)e^{-\frac{\alpha}{\lambda}} + 1 \ .$$

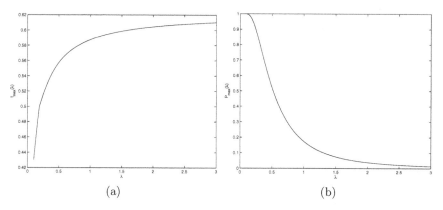

Fig. 4. (a) Optimal threshold as a function of λ (b) Optimal probability of winning as a function of λ

Figure 4(a) shows the optimal threshold for the exponential distribution as a function of its parameter λ, for $\alpha = 1$. Observe that, for large values of λ, the optimal threshold approaches the optimal threshold $1 - \frac{1}{\sqrt{7}}$ for the uniform distribution, since in this case $g(x \mid x \le \alpha)$ approaches the uniform distribution. Figure 4(b) shows the winning probability (corresponding to the optimal threshold) as a function of λ. As expected, larger values of λ (and thus larger input loads) give lower winning probabilities.

6 Local Information, Eponymous Model

Suppose now that each agent knows her own load as well as her id. In this case any two agents are allowed to decide on a different bin when presented with the same load. We focus on the following, wide class of input distributions $\Phi^r \subseteq \Delta$:

$$\Phi^r = \{F \in \Delta \ : \ \exists n \in \mathbb{N} \text{ and } \beta_1 = 0 < \beta_2 < \ldots < \beta_{n-1} < \beta_n = \alpha$$

$$\text{such that } \Pr\{\beta_i\} \in \mathbb{Q} \quad \forall i \in \{1, \ldots, n\} \quad \text{and} \quad \sum_{i=1}^{n} \Pr\{\beta_i\} = 1\}.$$

Theorem 4. *The optimal deterministic protocol for the local information, anonymous model when the input loads are independent, identically distributed random variables drawn from some $F \in \Phi^r$, can be constructed in time $\Theta\left(n^3 2^{3n}\right)$, i.e. in exponential time.*

Proof. Observe that any deterministic protocol $\mathcal{P}$ in $\mathcal{L}^\mathcal{E}$ can be fully described by a $3 \times n$ matrix $B^\mathcal{P} \in \{0,1\}^{3 \times n}$, such that $B^\mathcal{P}(i,k)$ is the bin that agent i puts her load when introduced with load β_k. Hence there are 2^{3n} possible deterministic protocols $\mathcal{P}_t$, $t = 1, \ldots, 2^{3n}$, and each $\mathcal{P}_t$ is represented by a matrix B^t.

Given any possible protocol $\mathcal{P}_t$, the total load on bin 0 and the total load on bin 1 for the triple of input loads $\langle \beta_{k_1}, \beta_{k_2}, \beta_{k_3} \rangle$ are, respectively,

$$\Lambda_0^t(k_1, k_2, k_3) = \sum_{i=1}^{3} \left(1 - B^t(i, k_i)\right) \cdot \beta_{k_i}$$

and

$$\Lambda_1^t(k_1, k_2, k_3) = \sum_{i=1}^{3} B^t(i, k_i) \cdot \beta_{k_i} .$$

Let $\lambda^t(k_1, k_2, k_3)$ be a variable indicating whether $\mathcal{P}^t$ wins when the triple of input loads is $\langle \beta_{k_1}, \beta_{k_2}, \beta_{k_3} \rangle$, i.e.

$$\lambda^t(k_1, k_2, k_3) = \begin{cases} 1 \text{ if } \Lambda_0^t(k_1, k_2, k_3) \leq \alpha \text{ and } \Lambda_1^t(k_1, k_2, k_3) \leq \alpha \\ 0 \text{ else} \end{cases} .$$

Now, for each protocol $\mathcal{P}_t$, we can compute (in time $\Theta(n^3)$) its winning probability as follows:

$$P_w(\mathcal{P}_t; F) = \sum_{k_1=1}^{n} \sum_{k_2=1}^{n} \sum_{k_3=1}^{n} \Pr\{\beta_{k_1}\} \Pr\{\beta_{k_2}\} \Pr\{\beta_{k_3}\} \cdot \lambda^t(k_1, k_2, k_3) .$$

Hence, by exhaustive search, we can find the protocol which gives the maximum winning probability in time $\Theta\left(n^3 2^{3n}\right)$.   □

7   Concluding Remarks and Directions for Future Research

We re-examined here the distributed decision making problem in the context of a, seemingly simple, special load-balancing setting, which is nevertheless quite complex.

It is clear that the distinction between anonymous and eponymous models is essential in order to derive optimal protocols, since, as observed, if we assume that the agents are aware of their ids then we can achieve higher winning probability. Moreover, the fact that an optimal protocol for the full information, anonymous case can not be deterministic shows that randomization indeed helps in maximizing the winning probability. It is interesting to examine whether this holds for the local information case as well.

For the local information, anonymous case, we showed how to compute the optimal single-threshold protocol for any known continuous distribution on $[0, \infty)$. However, just as in the special case of the uniform distribution, it remains open whether such single-threshold protocols are globally optimal. For the eponymous case, it would be interesting to derive more efficient, optimal protocols for discrete-valued input distributions.

The extension of all the above to more agents and bins is also a matter of future research.

References

1. Arrow, K.: The Economics of Information. Harvard University Press (1984)
2. Georgiades, S., Mavronicolas, M., Spirakis, P.: Optimal, Distributed Decision-Making: The Case of No Communication. In: Proceedings of the 12th International Symposium on Fundamentals of Computation Theory, pp. 293–303 (1999)
3. Papadimitriou, C.H., Yannakakis, M.: On the Value of Information in Distributed Decision-Making. In: Proceedings of the 10th Annual ACM Symposium on Principles of Distributed Computing, pp. 61–64 (1991)
4. Yao, A.C.: Some Complexity Questions Related to Distributive Computing. In: STOC 1979. Proceedings of the 11th ACM Symposium on Theory of Computing, pp. 209–213 (1979)

The Minimum Substring Cover Problem

Danny Hermelin[1,*], Dror Rawitz[2,*],
Romeo Rizzi[3,**], and Stéphane Vialette[4,**]

[1] Department of Computer Science,
University of Haifa, Haifa 31905, Israel
`danny@cri.haifa.ac.il`
[2] School of Electrical Engineering,
Tel-Aviv University, Tel-Aviv 69978, Israel
`rawitz@eng.tau.ac.il`
[3] Dipartimento di Matematica ed Informatica (DIMI),
Università di Udine, I-33100 Udine, Italy
`Romeo.Rizzi@dimi.uniud.it`
[4] Laboratoire de Recherche en Informatique (LRI),
Université Paris-Sud, 91405 Orsay, France
`vialette@lri.fr`

Abstract. In this paper we consider the problem of covering a set of strings S with a set C of substrings in S, where C is said to cover S if every string in S can be written as a concatenation of the substrings in C. We discuss applications for the problem that arise in the context of computational biology and formal language theory. We then proceed to show that this problem is at least as hard as the MINIMUM SET COVER problem. In the main part of the paper, we focus on devising approximation algorithms for the problem using two generic paradigms – the local-ratio technique and linear programming rounding.

1 Introduction

In a *covering problem* we are faced with the following situation: We are given two (not necessarily disjoint) sets of elements, the *base elements* and the *covering elements*, and the goal is to find a minimum (weight) subset of covering elements that "covers" all the base elements. The exact notion of covering differs from problem to problem, yet this abstract setting is common to many classical combinatorial problems in various application areas. Two famous examples are MINIMUM SET COVER – where the covering elements are subsets of the base elements and the notion of covering corresponds to set inclusion – and MINIMUM VERTEX COVER – where the setting is graph-theoretic and the notion of covering corresponds to incidence between vertices and edges. Ever since the early days of combinatorial optimization, research on covering problems such as the two examples above proved extremely fruitful in laying down fundamental techniques and ideas. The early work of Johnson [1] and Lovász [2] on MINIMUM SET

---

* Partially supported by the Caesarea Rothschild Institute.
** Supported by the Italian-French PAI Galileo Project 08484VH.

C. Kaklamanis and M. Skutella (Eds.): WAOA 2007, LNCS 4927, pp. 170–183, 2008.

COVER pioneered the greedy analysis approach, while Chvátal [3] gave the first analysis based on linear programming (LP) while tackling the same problem. The first LP-rounding algorithm by Hochbaum [4] was also designed for MIN-IMUM SET COVER, while Bar-Yehuda and Even gave the first Primal-Dual [5] and Local-Ratio [6] algorithms for MINIMUM VERTEX COVER.

In this paper we introduce a new covering problem which resides in the realm of strings. A string c is a *substring* of a string s, if c can be obtained by deleting any number of consecutive letters from both ends of s. In our covering problem, the base elements are strings and the covering elements are their substrings. The notion of covering corresponds to string-factorization, or to the generation of strings by substring concatenation. More formally, for a given set of strings S, let $\mathcal{C}(S)$ denote the set of all substrings of strings in S. We define a *cover* of S to be a subset $C \subseteq \mathcal{C}(S)$ such that any string $s \in S$ can be written as a concatenation of strings in C. If each string in S can be written as a concatenation of at most ℓ strings in C, we say that C is an ℓ-*cover* of S. Given a weight function $w : \mathcal{C}(S) \to \mathbb{Q}^+$, we are interested in computing an ℓ-cover of S with minimum possible weight:

MINIMUM SUBSTRING COVER

Instance: A set of strings S, a weight function $w : \mathcal{C}(S) \to \mathbb{Q}^+$, and an integer $\ell \geq 2$.

Solution: An ℓ-cover C of S. That is, a set of strings $C \subseteq \mathcal{C}(S)$, where for each $s \in S$ there exist $c_1, \ldots, c_p \in C$, $p \leq \ell$, with $s = c_1 \cdots c_p$.

Measure: Total weight of the cover, *i.e.* $w(C) = \sum_{c \in C} w(c)$.

Example 1. Consider the set of strings $S = \{\text{'a'}, \text{'aab'}, \text{'aba'}\}$. Then $\mathcal{C}(S) = \{\text{'a'}, \text{'b'}, \text{'aa'}, \text{'ab'}, \text{'ba'}, \text{'aab'}, \text{'aba'}\}$, and $C_1 = \{\text{'a'}, \text{'b'}\}$ and $C_2 = \{\text{'a'}, \text{'ab'}\}$ are covers of S. The cover C_1 is a 3-cover of S, while C_2 is a 2-cover.

Throughout the paper, we use n to denote the number of strings in S, and m to denote the maximum length of any string in S, *i.e.* $n = |S|$ and $m = \max\{|s| : s \in S\}$.

Note that in case $\ell \geq m$, there is no actual bound on the concatenation length of the required cover, and this case is denoted by $\ell = \infty$. An ∞-cover is referred to simply as a cover. Another interesting special case is when $\ell = 2$. In this case, we are required to cover S with a set of prefixes and suffixes in S, where a *prefix* (resp. *suffix*) of a string s is a substring of s which is obtained by removing consecutive letters only from the end (resp. beginning) of s. As we will see, these two extremal cases both give a certain amount of combinatorial leverage, and therefore deserve particular consideration. We also wish to point out that our use of general weight functions $w : \mathcal{C}(S) \to \mathbb{Q}^+$ allows for more robustness in modeling different scenarios. For instance, when w is the *unitary function*, *i.e.* $w(c) = 1$ for every $c \in \mathcal{C}(S)$, this corresponds to the situation where we want to minimize the size of a cover of S. When $w(c) = |c|$, *i.e.* the weight of every substring is its length (w is the *length-weighted function*), this corresponds to the case where we want to minimize the total length of the cover. Often some sort of middle ground between these two situations might also be desirable.

Example 2. Consider the two covers C_1 and C_2 of the set of strings S in Example 1. If w is the unitary function, then $w(C_1) = w(C_2) = 2$. However, if w is the length-weighted function, we have $w(C_1) = 2 < w(C_2) = 3$.

Our initial inspiration for studying MINIMUM SUBSTRING COVER came from a paper by Bodlaender *et al.* [7], who described an application for this problem in the context of protein folding (The authors of [7] actually referred to our problem as the DICTIONARY GENERATION problem, and considered its unweighted variant under the parameterized complexity framework.) Protein folding is the problem of determining the folding structure of proteins using their amino-acid sequential description. This problem is extremely important, since most of the functionally of a protein is determined by its folding structure, and because current biological methods for extracting the sequential description of a given protein exceed by far the methods for extracting the folding structure of the protein. In [7], it is argued that since all known approaches for protein folding are **NP**-hard, a possible heuristic for this problem is to break the protein sequence into small segments, small enough for allowing efficient folding computation. This heuristic is justified by the fact that many proteins seem to be composed of relatively small regions which fold independently of other regions. The theory of *exon shuffling* proposes that all proteins are concatenations of such regions, where the regions are drawn from a common ancestral dictionary [8,9].

MINIMUM SUBSTRING COVER can also model interesting computational issues which arise in formal language theory, and in particular, in the area of combinatorics of words. Our notion of cover actually corresponds to the notion of *combinatorial rank*, an important parameter of a set of words (*cf.* [10]). Neraud [11] studied the problem of determining whether a given set of words is *elementary*, where a set of strings is said to be elementary if it does not have a cover of size strictly less than its own. Neraud describes a direct application of this notion to the famous D0L-sequence equivalence problem (*cf.* [12]) via so-called elementary morphisms [13]. He also argues that this notion appears frequently in numerous sub-areas such as test sets, code theory, representation of formal languages, and the theory of equations in free monoids. His main result is in showing that deciding whether a given set of words is elementary is **coNP**-complete, which implies that MINIMUM SUBSTRING COVER is **NP**-hard.

Apart from the work of Bodlaender *et al.* [7] and Neraud [11], there has also been some recent work on problems closely related to MINIMUM SUBSTRING COVER, especially for the case of $\ell = 2$. The MINIMUM SET COVER WITH PAIRS problem introduced by Hassin and Segev in [14], is a variant of MINIMUM SET COVER where base elements are now covered by pairs of sets, and the goal is to cover all base elements using a minimum weight collection of sets. Hassin and Segev gave an $\mathcal{O}(\sqrt{n \lg n})$ approximation algorithm for the unweighted version of this problem, along with a few other algorithms for special cases of this problem. Another closely related problem is the HAPLOTYPE INFERENCE BY MAXIMUM PARSIMONY, an important problem in computational biology. Huang *et al.* [15] gave an algorithm for this problem, which translates to an $\mathcal{O}(m^2 \lg n)$ algorithm for MINIMUM SUBSTRING COVER with $\ell = 2$. Hajiaghayi *et al.* [16] introduced the MINIMUM MULTICOLORED SUBGRAPH problem within the same context, and gave an algorithm which in our terms obtains a performance ratio of $\mathcal{O}(\lg n \cdot \sqrt{m})$ with high probability. We discuss this algorithm and how to extend it to MINIMUM SUBSTRING COVER with general values of ℓ in Section 4.

The rest of this paper is organized as follows. In Section 2 we present some lower bounds on the approximation factors of polynomial-time algorithms for MINIMUM SUBSTRING COVER. We show that, in general, the problem is **NP**-hard to approximate within a factor of $c \ln n$ for some $c > 0$, and within $\lfloor m/2 \rfloor - 1 - \varepsilon$ and all $\varepsilon > 0$. We also show that the problem remains **APX**-hard even when m is constant, and the given weight function is either the unitary or the length-weighted function. Following this, in Section 3, we apply the local-ratio technique [17,6] to obtain three approximation algorithms with performance ratios $\binom{m+1}{2} - 1$, $m - 1$, and m, where the last two are specializations of the first to the cases of $\ell = 2$ and $\ell = \infty$ (the latter only applies for restricted types of weight functions). Finally, we present in Section 4 an algorithm based on rounding the linear programming relaxation of the problem, which achieves a performance ratio of $\mathcal{O}(\lg^{1/\ell} n \cdot m^{(\ell-1)^2/\ell})$ with high probability. This algorithm is an extension of an algorithm of Hajiaghayi *et al.* [16], with a slightly tighter analysis.

2   Approximation Lower Bounds

We begin our discussion by presenting some lower bounds on the performance ratios of polynomial-time approximation algorithms for MINIMUM SUBSTRING COVER. We show that in general, MINIMUM SUBSTRING COVER is **NP**-hard to approximate within factors of $c \ln n$ and $\lfloor m/2 \rfloor - 1 - \varepsilon$, for some $c > 0$ and all $\varepsilon > 0$ (recall that $n = |S|$ and $m = \max_{s \in S} |s|$). We also show that the problem is **APX**-hard even when all strings in S have length at most 4, and the given weight function $w : \mathcal{C}(S) \rightarrow \mathbb{Q}^+$ is either the unitary or the length-weighted function.

To prove our approximation lower bounds, we present an L-reduction [18] from the MINIMUM HYPERGRAPH VERTEX COVER problem, which is no more than the MINIMUM SET COVER problem when the roles of the covering and base elements are reversed. In MINIMUM HYPERGRAPH VERTEX COVER, we are given a vertex-weighted hypergraph $H = (V(H), E(H))$, $w_H : V(H) \rightarrow \mathbb{Q}^+$, and the goal is to find a minimum weight vertex cover of H. That is, a subset of vertices $V \subseteq V(H)$ of minimum weight, such that $V \cap e \neq \emptyset$ for each hyperedge $e \in E(H)$. It is known that the problem is **NP**-hard to approximate within a factor of $c \ln |E(H)|$ for some constant c [19], and also **NP**-hard to approximate within $\max_{e \in E(H)} |e| - 1 - \varepsilon$ for any $\varepsilon > 0$ (assuming $\max_{e \in E(H)} |e| > 2$) [20].

Let (H, w_H) be a given instance of MINIMUM HYPERGRAPH VERTEX COVER. From (H, w_H), we construct an instance (S, w, ℓ) for MINIMUM SUBSTRING COVER as follows. Let $e_{\max}$ denote the largest edge of H. The set of strings S is defined over an alphabet Σ which consists of two unique letters 'v', 'V' $\in \Sigma$ for each vertex $v \in V(H)$, and an additional special unique letter '\$' $\in \Sigma$

which we use for padding. We refer to the substring 'vV' as the *encoding* of the vertex $v \in V(H)$. For each edge $e \in E(H)$, we construct a string s_e by concatenating (in any arbitrary order) the encodings of all vertices $v \in e$. In addition, we concatenate $2(|e_{max}| - |e|)$ '$' letters to the end of s_e. The set of strings S is defined by $S = \{s_e : e \in E(H)\}$. Note that $n = |S| = |E(H)|$, and that $m = \max_{s \in S} |S| = 2|e_{max}|$.

Example 3. Suppose $V(H) = \{a, b, c, d\}$ and $E(H) = \{\{a, b\}, \{b, d\}, \{a, c, d\}\}$. The set of strings S is then constructed as $S = \{\text{'aAbB$$'}, \text{'bBdD$$'}, \text{'aAcCdD'}\}$.

Next, we define the weight function $w : \mathcal{C}(S) \to \mathbb{Q}^+$ by

$$w(c) = \begin{cases} 0 & : c \in \Sigma, \\ w_H(v) & : c \text{ is the encoding of } v \in V(H), \\ \infty & : \text{otherwise.} \end{cases}$$

Finally, to complete the construction, we set $\ell = m - 1$.

Lemma 1. *H has a vertex cover with total weight k iff S has an ℓ-cover with total weight k.*

Proof. Suppose $V \subseteq V(H)$ is a vertex cover of H with $w_H(V) = \sum_{v \in V} w_H(v) = k$, and consider the set of substrings $C = \Sigma \cup \{\text{'vV'} : v \in V\}$. Clearly, $w(C) = k$. Furthermore, C is an ℓ-cover of S, since C can cover any string $s_e \in S$ using $\ell - 1$ letters and a single encoding of a vertex $v \in V \cap e$.

Conversely, suppose S has an ℓ-cover C with $w(C) = k$. Write $C = C_1 \cup C_2$, where $C_1 = C \cap \Sigma$. Then $w(C_2) = k$ and C_2 consists only of substrings which are encodings of vertices in H. This is because no ℓ-cover can cover any string in S using only letters, and all non-encoding substrings of length at least 2 in $\mathcal{C}(S)$ have infinite weight. Let $V \subseteq V(H)$ be the vertices in H corresponding to the encodings in C_2. Then $w_H(V) = w(C_2) = k$. Moreover, since C uses at least one vertex-encoding in C_2 to cover any string $s_e \in S$, it follows by our construction that $V \cap e \neq \emptyset$ for all $e \in E(H)$.   □

The lemma above implies that any α-approximation algorithm for MINIMUM SUBSTRING COVER would give an α-approximation algorithm for MINIMUM HYPERGRAPH VERTEX COVER. Hence, due to the results of [19] and [20], we can conclude that it is **NP**-hard to approximate MINIMUM SUBSTRING COVER within $c \ln n$ for some constant c, and within $\lfloor m/2 \rfloor - 1 - \varepsilon$ for any $\varepsilon > 0$. However, the construction in the lemma relies on a somewhat unnatural weight function, and on the fact that the strings in S are allowed to be fairly long. Nevertheless, we can show that a special case of this construction can be used to relax both these conditions at the cost of reducing the lower bounds to only a constant.

Consider our construction for the case where $H = G = (V(G), E(G))$ is a graph rather than a hypergraph. This special case, better known as the MINIMUM VERTEX COVER problem, is known to be **NP**-hard to approximate within some constant, even in the unweighted case, and even if each vertex in G is incident to at most three edges in $E(G)$ [21,18]. Note that in this case, any vertex cover of

G must be of size at least $|V(G)|/4$. Consider the set of strings $\{s_e : e \in E(G)\}$ constructed as defined above. This set consists of four letter strings, each of which is a concatenation of two encodings of vertices of G (no '$' letters). Now define the input set of strings S by $S = \Sigma \cup \{s_e : e \in E(G)\}$. We can prove the following relationship between the size of a 3-cover of S and the size of a vertex cover in G.

Lemma 2. *G has a vertex cover of size k iff S has a 3-cover of size $2|V(G)|+k$.*

Proof. Suppose G has a vertex cover $V \subseteq V(G)$ of size k. Then the set of substrings $C = \Sigma \cup \{'vV' : v \in V\}$ is a 3-cover of S, and furthermore, $|C| = 2|V(G)| + k$.

Conversely, suppose S has a 3-cover C. Since $\Sigma \subseteq S$, C must include every letter in Σ. Hence, C is of size $2|V(G)| + k$, for some k. Let us say that C is *normalized* if it consists solely of letters and vertex-encodings. If C is normalized, we can write $C = \Sigma \cup C_1$, where C_1 is the set of $|C| - |\Sigma| = k$ vertex encodings in C, and by a similar argument used for Lemma 1, we can show that the set of k vertices $V \subseteq V(G)$ corresponding to the vertex-encodings in C_1 is a vertex cover of G. Otherwise, if C is not normalized, we can always normalize C at no cost to its total size. Indeed, note that any string $s_e \in S$ can be covered using a vertex encoding of a vertex incident to e and two additional letters in Σ. Furthermore, notice that any non-encoding substring $c \notin \Sigma$ can only be used to cover a single word in S. Hence, if C covers some string $s_e \in S$ using a non-encoding substring $c \notin \Sigma$, we can replace c with a vertex encoding of some vertex incident to e without violating the fact that C is a cover and with no increase to its total size. Doing this for all non-encoding substrings $c \in C \setminus \Sigma$, we obtain a normalized cover of S whose size is at most $|C|$. □

Using similar arguments, we can also prove that:

Lemma 3. *G has a vertex cover of size k iff S has an 3-cover of total length $2|V(G)| + 2k$.*

Using the last two lemmas and the fact that any vertex cover of G must be of size at least $|V(G)|/4$ it is not hard to see that the above construction constitutes an L-reduction from MINIMUM VERTEX COVER on graphs with bounded degree to both unweighted MINIMUM SUBSTRING COVER and length-weighted MINIMUM SUBSTRING COVER. It follows that there is some constant c for which MINIMUM SUBSTRING COVER for constant length strings and unitary/length-weighted weight functions is **NP**-hard to approximate. Combining this with the implications of Lemma 1, we obtain the main result of this section:

Theorem 1. MINIMUM SUBSTRING COVER *is* **NP**-*hard to approximate*

- *within $c \ln n$ for some constant c, and within $\lfloor m/2 \rfloor - 1 - \varepsilon$ for any $\varepsilon > 0$.*
- *within some constant c, when m and ℓ are constant, and w is either the unitary or the length-weighted function.*

3   Local-Ratio Algorithms

In the previous section we gave some negative results for the MINIMUM SUB-STRING COVER problem. In this section we show how to apply the local-ratio technique [17,6] to obtain positive results in the form of approximations algorithms with performance ratios depending on the length of the longest word in S. In particular, if m is the maximum length of any word in S, we show how to find in polynomial time an $(\binom{m+1}{2} - 1)$-approximate ℓ-cover for S for general values of ℓ. For $\ell = 2$, we show how to obtain $(m - 1)$-approximate covers, and for $\ell = \infty$, we show how to compute m-approximate covers. (The latter case applies only for a restricted type of weight functions.) We begin by giving a brief overview of the local-ratio technique.

The local-ratio technique [17] is based on the Local-Ratio Lemma [6], which in our terms is stated as follows:

Lemma 4 (Local-Ratio). *Let C be a cover for S, and let w_1 and w_2 be weight functions for $C(S)$. If C is an α-approximate, both with respect to w_1 and with respect to w_2, then C is also α-approximate with respect to $w_1 + w_2$.*

A local-ratio α-approximation algorithm is typically recursive and works as follows. Given a problem instance with a weight function w, we find a non-negative weight function $w_1 \leq w$ such that (1) every solution of a certain type is α-approximate with respect to w_1, and (2) there exists some element e in our input for which $w(e) = w_1(e)$. We subtract w_1 from w and remove some zero weight element from the problem instance. Then, we recursively solve the new problem instance, while assuring that the solution returned can be fixed so that it becomes of the above mentioned type. If fixing the solution does not increase its w_1 weight, nor its $w - w_1$ weight, the Local-Ratio Lemma guarantees that this solution is α-approximate with respect to our original weight function w. The base of the recursion occurs when the problem instance has degenerated into a trivial instance.

Figure 1 gives an approximation algorithm for MINIMUM SUBSTRING COVER which is based on the local-ratio technique. We call this algorithm LR. We first show that algorithm LR computes $(\binom{m+1}{2} - 1)$-approximate ℓ-covers for general values of ℓ. Following this, we show that some fine tuning of the algorithm allows us to achieve approximation ratios of $m - 1$ and m, for the special cases of $\ell = 2$ and $\ell = \infty$ respectively.

The general outline of algorithm LR is as follows: First, the algorithm adds all substrings $c \in C(S)$ with zero weight to an initial partial-solution C, since these do not have effect on the total weight of the optimal solution. Then, if C is not already a cover of S, LR selects a string $s \in S$ not covered by C, and examines all substrings C_s of s not already in C. It then subtracts $\varepsilon = \min\{w(c) : c \in C_s\}$ from the weight of all substrings in C_s, and recurses on the new weight function. The last line of the algorithm ensures that at least one substring of s will not be included in C. Such solutions are shown to be $(\binom{m+1}{2} - 1)$-approximate with respect to w_1, and also with respect to w_2, and therefore due to the Local-Ratio Lemma, are also $(\binom{m+1}{2} - 1)$-approximate with respect to w.

Algorithm LR(S, w, ℓ)

Data : A set of strings S, a weight function $w : \mathcal{C}(S) \to \mathbb{Q}^+$, and an integer
$\ell \geq 2$.
Result : An ℓ-cover C for S.
begin

1. $C \leftarrow \{c \in \mathcal{C}(S) : w(c) = 0\}$.
2. if C is an ℓ-cover of S **then return** C.
3. Let $s \in S$ be a string not ℓ-covered by C of maximum length.
4. $C_s \leftarrow \{c \in \mathcal{C}(S) \setminus C : c$ is a substring of $s\}$.
5. Set $\varepsilon = \min\{w(c) : c \in C_s\}$.
6. Define $w_1(c) = \begin{cases} \varepsilon & c \in C_s, \\ 0 & \text{otherwise.} \end{cases}$
7. Define $w_2 = w - w_1$.
8. $C \leftarrow \text{LR}(S, w_2, \ell)$.
9. if $C \setminus \{s\}$ is an ℓ-cover for S **then** $C \leftarrow C \setminus \{s\}$.
   return C.

end

Fig. 1. A local ratio approximation framework

Note that at each recursive call of the algorithm, at least one substring in $\mathcal{C}(S)$ which has positive weight with respect to w, will have zero weight with respect to w_2. Hence, the algorithm is guaranteed to terminate, and furthermore, it is also guaranteed to terminate after at most polynomial-many recursive calls. It is not difficult to see that each recursive call can be carried out in polynomial-time. The only problematic line could be line 2, but this can be performed efficiently using standard dynamic-programming techniques (details omitted). Finally, observe that by its definition, algorithm LR indeed returns an ℓ-cover of S. In the following lemma we show that this cover is $(\binom{m+1}{2} - 1)$-approximate.

Lemma 5. *Algorithm LR computes an* $(\binom{m+1}{2} - 1)$*-approximate* ℓ*-cover of* S.

Proof. To prove that the cover C returned by algorithm LR is $(\binom{m+1}{2} - 1)$-approximate, we apply induction on the number of recursive calls of the algorithm, and show that at any recursive call, C is $(\binom{m+1}{2} - 1)$-approximate with respect to the given weight function w of that particular call. At the recursive basis, C has zero weight with respect to w so it is indeed $(\binom{m+1}{2} - 1)$-approximate. For the inductive step, consider any recursive call other then the basis, and assume that the cover C returned at Line 8 is $(\binom{m+1}{2} - 1)$-approximate with respect to w_2. Note that C also remains $(\binom{m+1}{2} - 1)$-approximate with respect to w_2 after Line 9.

Let $s \in S$ be the string selected at Line 3. Since s is of length at most m, it has at most $\binom{m+1}{2}$ distinct substrings, and so $|C_s| \leq \binom{m+1}{2}$. Hence, $\sum_{c \in \mathcal{C}(S)} w_1(c) \leq \binom{m+1}{2}\varepsilon$. Furthermore, if C includes s after Line 9, then at least one substring of s is not included in C. This is because, by our selection of s, s can only be used to cover itself among all strings not covered by zero-weight substrings in $\mathcal{C}(S)$. In any case, after Line 9 we have $C_s \not\subseteq C$, and so $|C \cap C_s| \leq \binom{m+1}{2} - 1$.

Hence, $\sum_{c \in C} w_1(c) \leq (\binom{m+1}{2} - 1)\varepsilon$. Furthermore, by our selection of ε, any cover for S has weight at least ε with respect to w_1. It follows that, after Line 9, C is $(\binom{m+1}{2} - 1)$-approximate with respect to w_1 as well as with respect to w_2. According to the Local-Ratio Lemma, the cover returned is $(\binom{m+1}{2} - 1)$-approximate with respect to w, and so the lemma is proved.   □

We next show that with a small modification to algorithm LR, we can achieve an approximation factor of $m - 1$ for the special case of $\ell = 2$. First, when $\ell = 2$, we consider $\mathcal{C}(S)$ to be the set of all prefixes and suffixes of strings in S, rather than the set of all substrings of S. We use algorithm LR with the following modification. We replace Line 4 of the algorithm with:

$$C_s \leftarrow \{c \in \mathcal{C}(S) \setminus C : \exists c' \in C \text{ with } s = cc'\} \cup$$
$$\{c \in \mathcal{C}(S) \setminus C : \exists c' \in \mathcal{C}(S) \text{ with } s = c'c\} \cup \{s\}.$$

That is, for every pair of prefix and suffix of s, C_s either includes the suffix if it is not already in C (*i.e.* does not have zero weight), or it includes the prefix if the suffix is already in C. Note that since $s \in C_s$, $C_s \neq \emptyset$. We denote the modified version of algorithm LR by LR$_2$.

It is clear that algorithm LR$_2$ can be implemented to run in polynomial-time. Furthermore, the analysis of the performance ratio of algorithm LR$_2$ is almost the same as the analysis for algorithm LR. The main difference is in the upper bound of the total w_1 weight of C. First observe that we still have $\sum_{c \in C} w_1(c) \geq \varepsilon$ for any 2-cover C of S, since any cover must still include at least one string of C_s. On the other hand, since $|C_s| \leq m$, we have $\sum_{c \in C_s} w_1(c) \leq m \cdot \varepsilon$. Since after Line 9 we know that $C_s \not\subseteq C$, we have in fact $\sum_{c \in C_s} w_1(c) \leq (m-1) \cdot \varepsilon$.

Lemma 6. *Algorithm LR$_2$ computes an $(m-1)$-approximate 2-cover of S.*

We next consider the case of $\ell = \infty$ (*i.e.* $\ell \geq m$). Given a weight function $w : \mathcal{C}(S) \to \mathbb{Q}^+$, we say that w is *proper* if for any $c, c_1 \in \mathcal{C}(S)$, $w(c_1) \leq w(c)$ whenever c_1 is a prefix or a suffix of c. For example, unitary and length-weighted functions are proper. We show how to modify algorithm LR so that it computes m-approximate covers for proper weight functions. Note that for length-weighted functions the problem is trivial since the solution is always the alphabet of S.

Our modified version of algorithm LR for the case of $\ell = \infty$ is called LR$_\infty$. It is obtained by replacing Line 9 in algorithm LR with the following line:

while $\exists c, c_1 \in C$ with $c = c_1 c_2$ or $c = c_2 c_1$ **do** $C \leftarrow C \setminus \{c\} \cup \{c_2\}$.

Note that this while loop requires polynomial-time because the total length of the substrings in C decreases in every iteration of the while loop. The more important observation is that, since $\ell \geq m$, C remains an ℓ-cover for S after the while loop terminates. Furthermore, after line 9, C is both prefix-free and suffix-free. That is, there are no two strings in C where one is the prefix or suffix of the other. This implies that $|C \cap C_s| \leq m$, and is precisely the property that we use to obtain our m-approximation factor.

Lemma 7. *Algorithm LR_∞ computes an m-approximate cover of S assuming the given weight function $w : C(S) \to \mathbb{Q}^+$ is proper.*

Proof. First observe that if the initial weight function is proper, then all weight functions throughout the entire recursion of the algorithm are proper. This is because whenever the weight of a string decreases, the weight of all its prefixes and suffixes decreases by the same amount. Next note that Line 9 of algorithm LR_∞ does not increase the weight of C with respect to w_2, nor with respect to w_1, since both are proper weight functions. The approximation factor promised by the lemma is therefore obtained due to the observation that $1 \leq |C \cap C_s| \leq m$ after Line 9, and so $\varepsilon \leq \sum_{c \in C} w_1(c) \leq m \cdot \varepsilon$. ☐

Theorem 2. MINIMUM SUBSTRING COVER *is approximable within a factor of:*

- $\binom{m+1}{2} - 1$, *for general values of ℓ.*
- $m - 1$, *for $\ell = 2$.*
- m, *for $\ell = \infty$ and proper weight functions.*

4   Linear Programming Rounding

In [16], Hajiaghayi *et al.* considered the MINIMUM MULTICOLORED SUBGRAPH problem, which is a generalization of MINIMUM SUBSTRING COVER when the given factorization length ℓ is set to 2. In this section, we extend the linear programming rounding algorithm given in [16] to apply for any constant value of ℓ. We also give a tighter analysis. We obtain a $\mathcal{O}(\lg^{1/\ell} n \cdot m^{(\ell-1)^2/\ell})$-approximation algorithm for our problem, which outperforms the algorithm given in the previous section when $\ell < 4$. This algorithm can also be used for solving a generalization of the MINIMUM MULTICOLORED SUBGRAPH problem, namely the MINIMUM MULTICOLORED HYPERGRAPH SUBGRAPH problem. Here ℓ corresponds to the size of the largest hyperedge, and the maximum number of hyperedges colored by any particular color M replaces $\mathcal{O}(m^{\ell-1})$. Hence, the approximation ratio is $\mathcal{O}(\lg^{1/\ell} n \cdot M^{(\ell-1)/\ell})$. The original approximation ratio obtained for the case of $\ell = 2$ by Hajiaghayi *et al.* [16] is $\mathcal{O}(\lg n \cdot \sqrt{M})$.

Given a string s, an ℓ-*factorization* of s is an ordered multiset of substrings $f = (c_1, \ldots, c_p)$ such that $s = c_1 \cdots c_p$ and $p \leq \ell$. Denote by $\mathcal{F}_\ell(s)$ the set of possible ℓ-factorizations of s, and let $\mathcal{F}_\ell(S)$ denote the set of all factorizations of strings in S, i.e. $\mathcal{F}_\ell(S) = \bigcup_{s \in S} \mathcal{F}_\ell(s)$. Now, for every substring $c \in C(S)$, we designate a variable x_c which associated with c, and for every factorization $f \in \mathcal{F}_\ell(S)$, we designate a variable y_f which is associated with f. In these terms, MINIMUM SUBSTRING COVER can be formulated using the following integer linear program:

$$
\begin{array}{lll}
\min & \sum_{c \in C(S)} w(c) x_c & \\
\text{s.t.} & \sum_{f \in \mathcal{F}_\ell(s)} y_f \geq 1 & \forall s \in S \\
& \sum_{c \in f \in \mathcal{F}_\ell(s)} y_f \leq x_c & \forall s \in S, \forall c \text{ substring of } s \qquad \text{(IP)} \\
& x_c, y_f \in \{0, 1\} & \forall c \in C(S), \forall f \in \mathcal{F}_\ell(S)
\end{array}
$$

The variable x_c indicates whether the substring c is in the cover C and the variable y_f indicates whether C covers s by using the factorization f. The first type of constraints make sure that every string is factorized by some factorization. The second type of constraints make sure that if s is covered via the factorization f, then all substrings participating in this factorization are counted in the objective function. A linear programming relaxation of IP is obtained by replacing the integrality constraints by: (i) $x_c \geq 0$ for every $c \in \mathcal{C}(S)$, and (ii) $y_f \geq 0$ for every $f \in \mathcal{F}_\ell(S)$. Notice that the LP-relaxation is solvable in polynomial time since $\max_{s \in S} |\mathcal{F}_\ell(s)| = \mathcal{O}(m^{\ell-1})$, and ℓ is assumed to be constant.

Let $\mu > 1$ be a parameter to be determined later. Given an optimal fractional solution (x^*, y^*) to the LP-relaxation of IP, we construct an integral solution (x, y) for IP by picking every substring c with probability $p(c) = \min\{\mu \cdot x_c^*, 1\}$. That is, $x_c = 1$ with probability $p(c)$, and $x_c = 0$ with probability $1 - p(c)$. If there exists some $f \in \mathcal{F}_\ell(s)$ such that $x_c = 1$ for every $c \in f$ we set $y_f = 1$. We set $y_f = 0$ for any other $f \in \mathcal{F}_\ell(s)$. The resulting set of substrings is denoted by C, namely, $C = \{c : x_c = 1\}$.

The first step is to show that the expected total weight of our solution C is not much more than the total weight of the optimum cover of S. Let us denote the total weight of the optimal cover of S by OPT. We have:

Lemma 8. $\mathbf{E}\big[w(C)\big] \leq \mu \cdot \text{OPT}$.

Proof. $\mathbf{E}\big[w(C)\big] = \mathbf{E}\big[\sum_{c \in \mathcal{C}(S)} w(c)p(c)\big] \leq \sum_{c \in \mathcal{C}(S)} w(c)(\mu \cdot x_c^*) \leq \mu \cdot \text{OPT}$. □

The next step is to show that with a proper selection of μ, the probability that a string $s \in S$ is not covered by C becomes constant.

Lemma 9. *If* $\mu \geq (\ln n + 1)^{1/\ell} \cdot |\mathcal{F}_\ell(s)|^{(\ell-1)/\ell}$ *then for any string* $s \in S$

$$\mathbf{Pr}\big[C \text{ does not cover } s\big] \leq (e \cdot n)^{-1}.$$

Proof. Let $s \in S$ be any arbitrary string. We prove the lemma by suggesting an alternative method for covering s. For this we define the following three families of boolean random variables:

- $\{Z(f, c)\}_{c \in f \in \mathcal{F}(s)}$, where $\mathbf{Pr}\big[Z(f, c) = 1\big] = \min\{\mu \cdot y_f^*, 1\} = p(f)$.
- $\{X(c)\}_{c \in \mathcal{C}(\{s\})}$, where $X(c) = \bigvee_{c \in f \in \mathcal{F}(s)} Z(f, c)$.
- $\{Y(f)\}_{f \in \mathcal{F}_\ell(s)}$, where $Y(f) = \bigwedge_{c \in f} Z(f, c)$.

Note that all variables are independent within each family.

Our alternative method for covering s is done according to the variables $X(c)$. That is, we consider $C_s = \{c : X(c) = 1\}$ as our candidate set of substrings for covering s. We first show that the probability that C_s does not cover s is as least as high as the probability that C does not cover s. We do so by showing that $\mathbf{Pr}\big[X(c) = 1\big] \leq p(c)$ for every substring c of s. Indeed, if $p(c) = 1$ this is trivial. Also, if $p(f) = 1$ for some f with $c \in f$, then $p(c) = 1$. Otherwise, this follows by union bound and the feasibility of (x^*, y^*) with respect to the LP relaxation of IP:

$$\mathbf{Pr}\big[X(c) = 1\big] = \mathbf{Pr}\Big[\bigvee_{c \in f \in \mathcal{F}(s)} Z(f, c) = 1\Big]$$

$$\leq \sum_{c \in f \in \mathcal{F}(s)} \mathbf{Pr}\big[Z(f, c) = 1\big]$$

$$= \sum_{c \in f \in \mathcal{F}(s)} \mu \cdot y_f^*$$

$$\leq \mu \cdot x_c^*$$

$$= p(c) .$$

We next show that C_s covers s with high probability. First, observe that if C_s does not cover s, then for any $f \in \mathcal{F}_\ell(s)$ there exists $c \in f$ such that $X(c) = 0$. From the definition of $X(c)$ it follows that $Z(f, c) = 0$ as well, and this means that $Y(f) = 0$ for every $f \in \mathcal{F}_\ell(s)$. Hence, $\mathbf{Pr}\big[C_s \text{ does not cover } s\big] \leq \mathbf{Pr}\big[\forall f \in \mathcal{F}_\ell(s) : Y(f) = 0\big]$.

Now observe that, for any $f \in \mathcal{F}_\ell(s)$, if $p(f) = 1$ then $Y(f) = 1$. Hence, for the rest of the proof we assume that $p(f) < 1$. We have,

$$\mathbf{Pr}\big[\forall f \in \mathcal{F}_\ell(s) : Y(f) = 0\big] = \prod_{f \in \mathcal{F}_\ell(s)} \mathbf{Pr}\big[Y(f) = 0\big]$$

$$\leq \prod_{f \in \mathcal{F}_\ell(s)} (1 - p(f)^\ell)$$

$$\leq \prod_{f \in \mathcal{F}_\ell(s)} e^{-p(f)^\ell}$$

$$= e^{-\sum_{f \in \mathcal{F}_\ell(s)} p(f)^\ell} ,$$

where the second inequality is due to the fact that $1 - x \leq e^{-x}$ for $x \in [0, 1]$. Since $p(f) = \mu \cdot y_f^* < 1$ for all $f \in \mathcal{F}_\ell(s)$, and since (x^*, y^*) is a feasible solution of the LP relaxation of IP, we have $\sum_{f \in \mathcal{F}_\ell(s)} p(f) = \mu \cdot \sum_{f \in \mathcal{F}_\ell(s)} y_f^* \geq \mu$ for all $f \in \mathcal{F}_\ell(s)$. Due to this, and the the convexity of the function $f(x) = x^\ell$ for $x \in [0, 1]$, we get that

$$\sum_{f \in \mathcal{F}_\ell(s)} p(f)^\ell \geq |\mathcal{F}_\ell(s)| \left(\frac{\sum_{f \in \mathcal{F}_\ell(s)} p(f)}{|\mathcal{F}_\ell(s)|}\right)^\ell$$

$$\geq \frac{\mu^\ell}{|\mathcal{F}_\ell(s)|^{\ell-1}}$$

$$\geq \frac{(\ln n + 1) \cdot |\mathcal{F}_\ell(s)|^{\ell-1}}{|\mathcal{F}_\ell(s)|^{\ell-1}}$$

$$= \ln n + 1,$$

and so

$$\mathbf{Pr}\big[C_s \text{ does not cover } s\big] \leq \mathbf{Pr}\big[\forall f \in \mathcal{F}_\ell(s), Y(f) = 0\big] \leq e^{-\ln n - 1} = (e \cdot n)^{-1},$$

and we are done.   □

The previous lemma implies that by setting

$$\mu = (\ln n + 1)^{1/\ell} \cdot \max_{s \in S} |\mathcal{F}_\ell(s)|^{(\ell-1)/\ell} = \mathcal{O}(\lg^{1/\ell} n \cdot m^{(\ell-1)^2/\ell}) \,,$$

we cover any string $s \in S$ with probability at least $(e \cdot n)^{-1}$. By using union bound on Lemma 9 it follows that

$$\mathbf{Pr}\big[\exists s \text{ not covered by } C\big] \leq \frac{n}{e \cdot n} = e^{-1} \,.$$

Hence, we obtain the main result of this section:

Theorem 3. *With high probability,* MINIMUM SUBSTRING COVER *is approximable within a factor of* $\mathcal{O}(\lg^{1/\ell} n \cdot m^{(\ell-1)^2/\ell})$.

Acknowledgment

We thank Danny Segev for helpful remarks.

References

1. Johnson, D.: Approximation algorithms for combinatorial problems. Journal of Computer and System Sciences 9, 256–278 (1974)
2. Lovász, L.: On the ratio of optimal integeral and fractional solutions. Discrete Mathematics 13, 383–390 (1974)
3. Chvátal, V.: A greedy heuristic for the set-covering problem. Mathematics of Operations Research 4(3), 233–235 (1979)
4. Hochbaum, D.: Approximation algorithms for the set covering and vertex cover problems. SIAM Journal on Computing 11(3), 555–556 (1982)
5. Bar-Yehuda, R., Even, S.: A linear time approximation algorithm for the weighted vertex cover problem. Journal of Algorithms 2, 198–203 (1981)
6. Bar-Yehuda, R., Even, S.: A local-ratio theorem for approximating the weighted vertex cover problem. Annals of Discrete Mathematics 25, 27–46 (1985)
7. Bodlaender, H., Downey, R., Fellows, M., Hallett, M., Wareham, H.: Parameterized complexity analysis in computational biology. Computer Applications in the Biosciences 11(1), 49–57 (1995)
8. Dorit, R., Gilbert, W.: The limited universe of exons. Current Opinions in Structural Biology 1, 973–977 (1991)
9. Patthy, L.: Exons - original building blocks of proteins? BioEssays 13(4), 187–192 (1991)
10. Choffrut, C., Karhumäki, J.: Combinatorics of Words. In: Rozenberg, G., Salomaa, A. (eds.) Handbook of Formal Languages, Springer, Heidelberg (1997)

11. Néraud, J.: Elementariness of a finite set of words is co-NP-complete. Theoretical Informatics and Applications 24(5), 459–470 (1990)
12. Rozenberg, G., Salomaa, A.: The Mathematical Theory of L Systems. Academic Press, London (1980)
13. Ehrenfeucht, A., Rozenberg, G.: Elementary homomorphisms and a solution of the D0L sequence equivalence problem. Theoretical Computer Science 7, 169–183 (1978)
14. Hassin, R., Segev, D.: The set cover with pairs problem. In: Ramanujam, R., Sen, S. (eds.) FSTTCS 2005. LNCS, vol. 3821, pp. 164–176. Springer, Heidelberg (2005)
15. Huang, Y.T., Chao, K.M., Chen, T.: An approximation algorithm for haplotype inference by maximum parsimony. In: Proceedings of the 20'th ACM Symposium on Applied Computing (SAC), pp. 146–150 (2005)
16. Hajiaghayi, M., Jain, K., Lau, L., Mandoiu, I.: Minimum multicolored subgraph problem in multiplex PCR primer set selection and population haplotyping. In: Alexandrov, V.N., van Albada, G.D., Sloot, P.M.A., Dongarra, J.J. (eds.) ICCS 2006. LNCS, vol. 3991, pp. 758–766. Springer, Heidelberg (2006)
17. Bar-Yehuda, R.: One for the price of two: A unified approach for approximating covering problems. Algorithmica 27(2), 131–144 (2000)
18. Papadimitriou, C., Yannakakis, M.: Optimization, approximation, and complexity classes. Journal of Computer and Systems Sciences 43, 425–440 (1991)
19. Raz, R., Safra, S.: A sub-constant error-probability low-degree test, and a sub-constant error-probability PCP characterization of NP. In: Proceedings of the 29th ACM Symposium on the Theory Of Computing (STOC), pp. 475–484 (1997)
20. Dinur, I., Guruswami, V., Khot, S., Regev, O.: A new multilayered PCP and the hardness of hypergraph vertex cover. SIAM Journal on Computing 34(5), 1129–1146 (2005)
21. Alimonti, P., Kann, V.: Hardness of approximating problems on cubic graphs. In: Bongiovanni, G., Bovet, D.P., Di Battista, G. (eds.) CIAC 1997. LNCS, vol. 1203, pp. 288–298. Springer, Heidelberg (1997)

A 5/3-Approximation for Finding Spanning Trees with Many Leaves in Cubic Graphs

José R. Correa[1,*], Cristina G. Fernandes[2,**],
Martín Matamala[3,*], and Yoshiko Wakabayashi[2,**]

[1] School of Business, Universidad Adolfo Ibáñez, Chile
correa@uai.cl
[2] Department of Computer Science, Universidade de São Paulo, Brazil
{cris,yw}@ime.usp.br
[3] Departamento de Ingeniería Matemática, Universidad de Chile, Chile
mmatamal@dim.uchile.cl

Abstract. For a connected graph G, let $L(G)$ denote the maximum number of leaves in a spanning tree in G. The problem of computing $L(G)$ is known to be NP-hard even for cubic graphs. We improve on Loryś and Zwoźniak's result presenting a 5/3-approximation for this problem on cubic graphs. This result is a consequence of new lower and upper bounds for $L(G)$ which are interesting on their own. We also show a lower bound for $L(G)$ that holds for graphs with minimum degree at least 3.

1 Introduction

The MaxLeaf consists of the following problem. Given a connected graph G, find a spanning tree in G with as many leaves as possible. This problem is NP-hard [3] even for cubic graphs [6], and is known to be MAX SNP-complete [2]. Lu and Ravi [9,10] gave the first approximation algorithms for MaxLeaf. Solis-Oba [11] described the currently best approximation algorithm: a greedy 2-approximation.

All graphs considered in this paper are connected, unless otherwise specified. We use n to denote the number of vertices of the graph in question. To the best of our knowledge, Storer [12] was the first to consider MaxLeaf on cubic graphs. He showed that every cubic graph has a spanning tree with at least $\lceil n/4 + 2 \rceil$ leaves. Griggs, Kleitman, and Shastri [4] complemented this result by showing that this bound is tight. As a side note, they also provided a simple polynomial time algorithm (alternative to Storer's) that finds a spanning tree with at least $\lceil n/4 + 2 \rceil$ leaves in a cubic graph. As an illustration, Fig. 1(a) presents a graph that achieves this bound. On the other hand, Linial and Sturtevant [7] proved that Storer's lower bound holds even for graphs with minimum degree three.

* Research partially supported by CONICYT (Chile) through Anillo en Redes ACT08.
** Research partially supported by CNPq (Proc. 490333/04, 307011/03-8, 308138/04-0) and ProNEx - FAPESP/CNPq Proc. No. 2003/09925-5 (Brazil).

C. Kaklamanis and M. Skutella (Eds.): WAOA 2007, LNCS 4927, pp. 184–192, 2008.

Kleitman and West [5] extended the study of Linial and Sturtevant and considered MaxLeaf on graphs with minimum degree at least k, for arbitrary values of k and for small values of k as well.

(a)   (b)

Fig. 1. (a) A cubic graph and a spanning tree with $n/4 + 2$ leaves indicated by the dark edges. (b) A diamond.

For a graph G, we let $L(G)$ denote the maximum number of leaves in a spanning tree of G. As we mentioned, the result of Storer [12] is constructive and can be restated as a proof of a lower bound on $L(G)$ for a cubic graph G. Furthermore, the main result provided by Griggs et al. [4] is a better lower bound on $L(G)$ for the case of 3-connected cubic graphs. It can actually be seen as a constructive proof of the fact that every 3-connected and also every triangle-free cubic graph has a spanning tree with at least $\lceil (n+4)/3 \rceil$ leaves.

A *diamond* is a complete graph on 4 vertices minus an edge, also denoted by $K_4 - e$. We say that a subgraph of a given graph G is a *cubic diamond* if it is a diamond in which all of its vertices have degree 3 in G (see Fig. 1 (b)). In graphs with minimum degree at least 3, we want to distinguish those diamonds that are cubic and those that are not. The 3-*dimensional cube graph* is denoted by Q_3. Specifically, the previous bound by Griggs et al. [4] holds for all cubic graphs that do not contain diamonds. In fact, Griggs et al. observed that their bound is tight for Q_3 and that, for any other cubic graph, the sometimes stronger lower bound of $\lceil (n+5)/3 \rceil$ holds. They also noted that this lower bound is tight for both 3-connected and triangle-free cubic graphs. (See examples in Fig. 2.)

For the purpose of this paper, it is interesting to point out that Griggs et al. result implies a 3/2-approximation for MaxLeaf in 3-connected cubic graphs, since any spanning tree in a cubic graph has at most $n/2+1$ leaves. More recently, there has been some interest in obtaining approximation results for cubic graphs. Indeed, Loryś and Zwoźniak [8] presented a 7/4-approximation for MaxLeaf in cubic graphs. Very recently, Bonsma [1] proved that if G is a connected graph of minimum degree at least 3 with d cubic diamonds, then G has a spanning tree with at least $\lceil (2n - d + 12)/7 \rceil$ leaves.

In this paper, we prove a lower bound on $L(G)$ for a cubic graph G that also takes into account the diamonds present in the graph (but not only their number). Our lower bound is always at least as good as the one for cubic graphs derived from Bonsma's lower bound.

As most previous work, our proof is constructive, so it gives a polynomial algorithm that produces a spanning tree of the given graph with as many leaves as the claimed lower bound. Our algorithm uses the one of Griggs et al. [4] for diamond-free cubic graphs. The better lower bound, together with a related

(a)   (b)

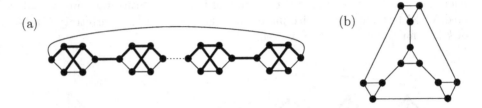

Fig. 2. (a) A triangle-free cubic graph and a spanning tree, indicated by the dark edges, with $n/3 + 2 = \lceil (n + 5)/3 \rceil$ leaves. (b) A 3-connected cubic graph G on $n = 12$ vertices obtained from K_4 by replacing each of its vertices with a triangle. Observe that $L(G) = 6 = \lceil (n + 5)/3 \rceil$.

upper bound, allows us to improve upon the result of Loryś and Zwoźniak [8], obtaining a 5/3-approximation for MAXLEAF in cubic graphs.

This paper is organized as follows. In the next section we derive the new lower bound on $L(G)$, while in Section 3, we prove the new upper bound on $L(G)$. In Section 4, we present the 5/3-approximation with its analysis. Section 5 discusses the extension for graphs with minimum degree at least 3. We conclude with some final remarks in Section 6.

2   A New Lower Bound

The way the diamonds are spread in the graph plays an important role in the new lower bound. It is expressed by a new parameter whose definition follows.

Call *internal* the two vertices in a diamond that have all neighbors within the diamond, and *external* the other two vertices of the diamond (see Fig. 3 (a)). For a cubic graph G, let G^r be the graph obtained from G after the removal of all internal vertices of its diamonds. We denote by c the number of components of G^r. For instance, if G is the graph in Fig. 1(a) with d diamonds, then G^r consists of d disjoint edges and $c = d$ in this case.

The new lower bound is given in the next theorem. It depends on the number n of vertices in the graph and on the parameter c defined above. Recall that Q_3 is the 3-dimensional cube graph.

Theorem 1. *Let $G \neq Q_3$ be a connected cubic graph with d diamonds. Then G has a spanning tree with at least $\max\{lb_1, lb_2\}$ leaves, where $lb_1 = \lceil (n - d + 5)/3 \rceil$ and $lb_2 = 3d - 2c + 2$. Moreover, $\max\{lb_1, lb_2\} \geq \lceil (3n - 2c + 17)/10 \rceil$.*

(a)   (b)

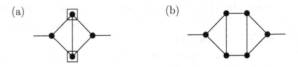

Fig. 3. (a) The squares indicate the internal vertices in a diamond. The other two vertices are the external ones. (b) A double diamond.

Proof. For the first lower bound lb_1 on $L(G)$, let G' be the graph obtained from G after replacing each diamond by the graph in Fig. 3 (b), which we call a *double diamond*. Because of the structure of G', from any spanning tree of G', it is easy to get a spanning tree of G with at most one leaf less per double diamond. The number of vertices in G' is $n' = n + 2d$. Observe that G' is diamond-free. So, from the result of Griggs et al. [4], we conclude that G' has a spanning tree T' with at least $\lceil (n' + 5)/3 \rceil = \lceil (n + 2d + 5)/3 \rceil$ leaves. Thus, from T', we can get a spanning tree T in G with at least $\lceil (n + 2d + 5)/3 \rceil - d = \lceil (n - d + 5)/3 \rceil = lb_1$ leaves. Therefore $L(G) \geq lb_1$.

For the second lower bound lb_2 on $L(G)$, let F be a forest in G consisting of spanning trees in each component of G^r. As G^r has $2d$ vertices of degree one, F has at least $2d$ leaves. Extend F in two phases to obtain a spanning tree in G. In the first phase, add to F edges from $c - 1$ of the diamonds to connect the c components of F and all vertices in these $c - 1$ diamonds. This can be done by losing two leaves and gaining one for each of the $c - 1$ diamonds. In the second phase, add edges from the remaining diamonds to connect its internal vertices to F, losing one leaf and gaining two per diamond. This results in a tree with $2d - (c - 1) + (d - (c - 1)) = 3d - 2c + 2 = lb_2$ leaves. Thus, $L(G) \geq lb_2$.

The maximum of these two lower bounds on $L(G)$ is at least the value they achieve when they are equal. That is, when $(n - d + 5)/3 = 3d - 2c + 2$. From this we deduce that $d = (n + 6c - 1)/10$ and, plugging it back in one of the two lower bounds, we get that $\max\{lb_1, lb_2\} \geq \lceil (3n - 2c + 17)/10 \rceil$. □

There are tight examples for the bound on $L(G)$ given by this theorem. For instance, the graph in Fig. 1(a) is a tight example with $c = n/4$. Indeed, Theorem 1 says that there is a spanning tree in this graph that has at least $\lceil (3n - 2c + 17)/10 \rceil = \lceil n/4 + 17/10 \rceil = n/4 + 2$ leaves. The tree of dark edges in Fig. 1(a) is optimal and has these many leaves. For another tight example, consider the graph indicated in Fig. 4. It consists of d double diamonds connected as a chain and forming a circuit, with one of the edges in each double diamond substituted by a diamond. Call this graph H. The number of vertices in H is $n = 10d$ and in this case $c = 1$. Theorem 1 says that there is a spanning tree in this graph that has at least $\lceil (3n - 2c + 17)/10 \rceil = \lceil (3n + 15)/10 \rceil = 3d + 2$ leaves. The spanning tree in dark edges in Fig. 4 is optimal and has exactly $3d + 2$ leaves.

Based on the example in Fig. 4, one might suspect that any tight example is not 3-connected after we replace each diamond by an edge. Note, however, that

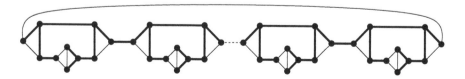

Fig. 4. A tight example for Theorem 1

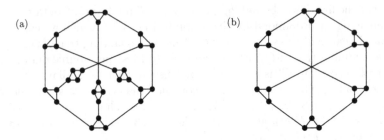

Fig. 5. (a) Another tight example for Theorem 1. (b) The 3-connected graph obtained from the example in (a) after the replacement of each diamond by an edge.

the graph shown in Fig. 5 (a) is a tight example and it remains 3-connected even after we perform these operations, as one can see in Fig. 5 (b).

2.1   Comparison with Bonsma's Lower Bound

Bonsma [1] recently proved that if G is a connected graph with d diamonds and minimum degree at least 3, then $L(G) \geq \lceil (2n - d + 12)/7 \rceil$. It is natural to ask how this result specialized to cubic graphs compares with the lower bound we have given in Theorem 1. To answer this question, let us consider the case $d \neq 0$ (when $d = 0$ the lower bound given by Griggs et al. [4] is as good as the lower bound given by Bonsma, and it is better when $n > 8$).

Let $lb_B = \lceil (2n - d + 12)/7 \rceil$. If $c = d$ then $n = 4d$ and in this case $lb_1 = lb_2 = lb_B$. If $c < d$ then $n > 4d + 1$. Adding $6n - 7d + 35$ on both sides of the last inequality, we obtain $7n - 7d + 35 > 6n - 3d + 36$. Thus, $7(n - d + 5) > 3(2n - d + 12)$, and therefore $lb_1 \geq lb_b$. (If $n \geq 4d + 22$, then $lb_1 > lb_b$.)

We note that the difference between lb_1 and lb_B might be not so negligible. For the tight example shown in Fig. 4, if we take $n = 70p$, where p is a positive integer (that is, G is a necklace with $7p$ double diamonds), we have that $lb_B = 19p + 2$, while $lb_1 = lb_2 = L(G) = 21p + 2$. In this case, lb_B is around 10% smaller than lb_1.

3   New Upper Bound

In this section, we prove a new upper bound on $L(G)$ that involves c. We recall that c is the number of components of G^r, where G^r is the graph obtained from G after the removal of all internal vertices of its diamonds. This upper bound will be useful in the analysis of the proposed approximation, that will be presented in the next section.

Theorem 2. *If G is a connected cubic graph, then any spanning tree of G has at most $\lfloor n/2 - c + 2 \rfloor$ leaves.*

Proof. Let T be an arbitrary spanning tree in G. As G is cubic, T has $(n - d_2 + 2)/2$ leaves, where d_2 is the number of vertices of degree two in T. Indeed, denoting the number of vertices in T of degree i by d_i, for $i = 1, 2, 3$, we have that $n = d_1 + d_2 + d_3$ and $2(n-1) = d_1 + 2d_2 + 3d_3$. From these two equalities, we deduce that $d_1 = (n - d_2 + 2)/2$.

Now observe that, as G^r has c components, edges of at least $c - 1$ diamonds will be used to connect components of G^r in T. Each diamond that is used to connect a component of G^r to another contributes with at least two different vertices of degree two in T. (See Fig. 6.) That is, the number of vertices of degree two in T is at least $2(c-1)$. In symbols, $d_2 \geq 2(c-1)$.

From this and from the previous observation, we deduce that T has at most $\lfloor n/2 - c + 2 \rfloor$ leaves. Hence, $L(G) \leq \lfloor n/2 - c + 2 \rfloor$.   □

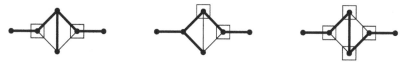

Fig. 6. Possible ways (excluding symmetric cases) to use a diamond to connect components of G^r spanning all vertices. The squared vertices have degree two in the graph of dark edges.

4   The Algorithm

Now we describe an algorithm whose approximation ratio is derived from the lower and upper bounds presented.

Algorithm $A(G)$
Input: a connected cubic graph G
Output: a spanning tree of G with at least $\frac{3}{5}L(G)$ leaves

```
1     d ← number of diamonds in G
2     G′ ← graph obtained from G by substituting each diamond by a double diamond
3     T′ ← GKS(G′)        ▷ T′ is a spanning tree of G′ given by the algorithm of Griggs et al.
4     T₁ ← spanning tree of G obtained from T′ (see proof of Theorem 1)
5     G^r ← graph obtained from G by removing the internal vertices of each diamond
6     F ← forest consisting of a spanning tree in each component of G^r
7     c ← number of components of G^r
8     D ← set of c − 1 diamonds that, if added back to G^r, make it connected
9     for each diamond h in D
10        add to F the three edges of h incident to a same internal vertex
11    for each diamond h not in D
12        add to F the two edges of h incident to a same external vertex
13    let T₂ be the resulting tree
14    let T be the one between T₁ and T₂ with more leaves
15    return T
```

The proof of Theorem 1 gives us immediately an algorithm to construct spanning trees with at least $\max\{lb_1, lb_2\}$ leaves. Just for completeness, we present it in pseudocode. We use GKS to refer to the algorithm of Griggs, Kleitman, and Shastri [4].

Theorem 3. *Algorithm A is a $5/3$-approximation for* MaxLeaf *on cubic graphs.*

Proof. First note that, as GKS is polynomial, A is a polynomial-time algorithm. Indeed, all but lines 3 and 8 can be implemented to run in linear time. For line 8, one can use some disjoint sets data structure and achieve almost linear time. So the most time consuming step is the execution of GKS in line 3.

As for the approximation ratio, let $|A(G)|$ denote the number of leaves in the tree produced by A with G as input. Indeed, A is a $5/3$-approximation, because

$$
\begin{aligned}
\frac{L(G)}{|A(G)|} &\leq \left(\frac{n - 2c + 4}{2}\right)\left(\frac{10}{3n - 2c + 17}\right) \\
&= 5\frac{n - 2c + 4}{3n - 2c + 17} \\
&\leq 5\frac{n - 2c + 4}{3n - 2c - 4c + 12} \\
&= 5\frac{n - 2c + 4}{3(n - 2c + 4)} \\
&= \frac{5}{3}.
\end{aligned}
$$

The first inequality holds by Theorems 1 and 2.   □

5 Constructions and Extension for Minimum Degree 3

Our lower bound shown in Theorem 1 calls attention to the fact that diamonds might not be what makes $L(G)$ smaller, closer to $n/4$. Indeed, we found interesting the following construction that proves this fact. Let H be a diamond-free cubic graph, and let T be an arbitrary spanning tree in G. Let G be the graph obtained from H by substituting every edge not in T by a diamond. Despite the fact that G has many diamonds, there exist spanning trees in G with $n/2 + 1$ leaves, where n is the number of vertices of G, which is as much as it could. (The number of diamonds in G is $n/6 + 1/3$.)

Another general construction that we found interesting is the one already exemplified in Fig. 2 (b). Given a cubic graph H, substitute each vertex of H by a triangle. Let G be the resulting graph. Note that G is (cubic) diamond-free. Then $L(G) = n/3 + 2$. The fact that $L(G) \geq n/3 + 2$ follows immediately from the lower bound of Griggs et al. [4] for cubic diamond-free graphs. On the other hand, let T be an arbitrary spanning tree of G and denote the number of vertices in T of degree i by d_i, for $i = 1, 2, 3$. Then, as already observed, $n = d_1 + d_2 + d_3$ and $2(n - 1) = d_1 + 2d_2 + 3d_3$. From these two equalities, we deduce that $d_1 = d_3 + 2$. But T has at most one degree 3 vertex per triangle. So

$d_3 \leq n/3$ and $L(G) \leq n/3 + 2$. (In fact, a similar construction was described by Griggs et al. [4, p. 671].)

As already mentioned, Bonsma [1] proved that if G is a connected graph of minimum degree at least 3 with d cubic diamonds, then G has a spanning tree with at least $\lceil (2n - d + 12)/7 \rceil$ leaves. We used this bound to obtain a result similar to Theorem 1 for graphs of minimum degree at least 3.

Theorem 4. *Every connected graph G of minimum degree at least three with d cubic diamonds has a spanning tree with at least $\max\{lb_B, lb_2\}$ leaves, where $lb_B = \lceil (2n - d + 12)/7 \rceil$ and $lb_2 = 3d - 2c + 2$. Moreover, $\max\{lb_B, lb_2\} \geq \lceil (3n - c + 19)/11 \rceil$.*

In some cases, the bound lb_2 is better than the bound lb_B of Bonsma [1]. In fact, for the example shown in Fig. 4, if we take $n = 770p$ (that is, a necklace with $77p$ double diamonds) then $lb_B = 209p + 2$ and $lb_2 = 231p$.

Unfortunately, the upper bound for graphs with minimum degree 3 is $n - 1$ (and is tight), and therefore we cannot derive an approximation algorithm better than Solis-Oba's [11] for this case using this lower bound.

6   Final Remarks

Galbiati, Maffioli, and Morzenti [2] proved that MAXLEAF is MAX SNP-complete, but there is no such proof for cubic graphs. We suspect that this case is also MAX SNP-complete. It would be nice to settle this question.

Also, we conjecture that there is a 3/2-approximation algorithm for MAXLEAF on cubic graphs. In fact, in many cases the algorithm described in this paper achieves this ratio.

Acknowledgements

The authors would like to thank the referees for their comments and suggestions.

References

1. Bonsma, P.: Spanning trees with many leaves: new extremal results and an improved FPT algorithm. Technical report, Faculty of EEMCS, University of Twente, The Netherlands (2006)
2. Galbiati, G., Maffioli, F., Morzenti, A.: A short note on the approximability of the maximum leaves spanning tree problem. Information Processing Letters 52(1), 45–49 (1994)
3. Garey, M., Johnson, D.: Computers and Intractability: A Guide to the Theory of NP-Completeness. Freeman, San Francisco (1979)
4. Griggs, J., Kleitman, D., Shastri, A.: Spanning trees with many leaves in cubic graphs. Journal of Graph Theory 13(6), 669–695 (1989)
5. Kleitman, D., West, D.: Spanning trees with many leaves. SIAM Journal on Discrete Mathematics 4(1), 99–106 (1991)

6. Lemke, P.: The maximum-leaf spanning tree problem in cubic graphs is NP-complete. In: IMA Preprint Series 428, Mineapolis (1988)
7. SLinial, N., Sturtevant, D.: Private communication (1987) (see [5])
8. Lory, K., Zwoźniak, G.: Approximation algorithm for maximum leaf spanning tree problem for cubic graphs. In: Möhring, R.H., Raman, R. (eds.) ESA 2002. LNCS, vol. 2461, pp. 686–698. Springer, Heidelberg (2002)
9. Lu, H.I., Ravi, R.: The power of local optimization: Approximation algorithms for maximum-leaf spanning tree. In: Proceedings of 13th Annual Allerton Conference on Communication, Control, and Computing, pp. 533–542 (1992)
10. Lu, H.I., Ravi, R.: Approximating maximum leaf spanning trees in almost linear time. Journal of Algorithms 29(1), 132–141 (1998)
11. Solis-Oba, R.: 2-approximation algorithm for finding a spanning tree with maximum number of leaves. In: Bilardi, G., Pietracaprina, A., Italiano, G.F., Pucci, G. (eds.) ESA 1998. LNCS, vol. 1461, pp. 441–452. Springer, Heidelberg (1998)
12. Storer, J.: Constructing full spanning trees for cubic graphs. Information Processing Letters 13(1), 8–11 (1981)

On the Online Unit Clustering Problem

Leah Epstein[1] and Rob van Stee[2,*]

[1] Department of Mathematics, University of Haifa, 31905 Haifa, Israel
lea@math.haifa.ac.il
[2] Department of Computer Science, University of Karlsruhe, D-76128 Karlsruhe,
Germany
vanstee@ira.uka.de

Abstract. We continue the study of the online unit clustering problem,
introduced by Chan and Zarrabi-Zadeh (*Proc. Workshop on Approximation and Online Algorithms 2006*, LNCS 4368, p.121–131. Springer,
2006). We design a deterministic algorithm with a competitive ratio of
7/4 for the one-dimensional case. This is the first deterministic algorithm
that beats the bound of 2. It also has a better competitive ratio than
the previous randomized algorithm. Moreover, we provide the first non-
trivial deterministic lower bound, improve the randomized lower bound,
and prove the first lower bounds for higher dimensions.

1 Introduction

In clustering problems, a set of points need to be partitioned into groups, also
called clusters, so as to optimize a given objective function. Clustering problems
are fundamental and have many applications, this includes usage of clustering
for computer related purposes, such as information retrieval and data mining,
and various applications in other fields such as medical diagnosis and facility
location.

In the online model, points are presented one by one to the algorithm, and
must be assigned to clusters upon arrival. This assignment cannot be changed
later. We measure the performance of an online algorithm $\mathcal{A}$ by comparing it to
an optimal offline algorithm OPT using the competitive ratio, which is defined
as $\sup_\sigma \mathcal{A}(\sigma)/\text{OPT}(\sigma)$. Here, σ is the input, which is a sequence of points, and
ALG(σ) denotes the cost of an algorithm ALG for this input, which is typically the
number of clusters. For randomized algorithms, we replace $\mathcal{A}(\sigma)$ with $\mathbb{E}(\mathcal{A}(\sigma))$,
and define the competitive ratio as $\sup_\sigma \mathbb{E}(\mathcal{A}(\sigma))/\text{OPT}(\sigma)$. An algorithm with
competitive ratio of at most $\mathcal{R}$ is called $\mathcal{R}$-competitive.

Charikar et al. [2] considered a problem which is called *the online unit covering
problem*. In this problem, a set of n points needs to be covered by balls of
unit radius, and the goal is to minimize the number of balls used. They gave
an upper bound of $O(2^d d \log d)$ and a lower bound of $\Omega(\log d/\log \log \log d)$ on
the competitive ratio of deterministic online algorithms in d dimensions. This

---

* Research supported by the Alexander von Humboldt Foundation.

C. Kaklamanis and M. Skutella (Eds.): WAOA 2007, LNCS 4927, pp. 193–206, 2008.

problem is fully online in the sense that points arrive one by one, each point needs to be assigned to a ball upon arrival, and if it is assigned to a new ball, the exact location of this ball is fixed at this time. The tight bounds on the competitive ratio for $d = 1$ and $d = 2$ are 2 and 4 respectively.

In a recent paper [1], Chan and Zarrabi-Zadeh introduced the unit clustering problem. This problem is still an online problem and is similar to unit covering. However, it is more flexible and does not require that the exact position of the balls is fixed in advance. The algorithm needs to make sure that a set of points which is assigned to one cluster can always be covered by a ball. The goal is still to minimize the total number of balls used. Therefore, the algorithm may terminate with clusters that still have more than one option for their location. In the offline model, this reduces to unit covering. However, in the online model, an algorithm now has the option of moving a cluster after a new point arrives, as long as this cluster still covers all the points that are assigned to it. In [1], the two dimensional problem is considered in the L_∞ norm rather than the L_2 norm. Thus "balls", are actually cubes. For $d = 1$ the two metrics are identical. In this paper, similarly to [1], we consider the L_∞ norm.

Note that online clustering is an online graph coloring problem. If we see the clusters as colors, and the points are seen as vertices, then an edge between two point occurs if they are too far apart to be colored using the same color. The resulting graph for the one dimensional problem is the complement of a unit interval graph (alternatively, the problem can be seen as a clique partition problem in unit interval graphs). See [5] for a survey on online graph coloring. Note that online coloring is a difficult problem that does not admit a constant competitive ratio already for trees [3,6]. There is a small number of classes that admit constant competitive algorithms, one of which is interval graphs [4].

For the one-dimensional case, [1] showed that several naïve algorithms all have a competitive ratio of 2. Some of these algorithms are actually designed to solve already the unit covering problem and thus cannot be expected to overcome this bound (due to [2]). They also showed that any randomized algorithm for unit covering has a competitive ratio of at least 2. To demonstrate the difference between unit covering and unit clustering, they presented a randomized algorithm with a competitive ratio of $15/8 = 1.875$. Finally, they showed a lower bound of $4/3$ on the competitive ratio of any randomized algorithm. The deterministic lower bound that is implied by their work is $3/2 = 1.5$. A multi-dimensional extension of their algorithm, that they design, results in a $15/4 = 3.75$-competitive algorithm for two dimensions, or a $2^d \cdot 15/16$-competitive algorithm for general d. The randomized upper bound for one dimension was improved to $11/6$ by the same authors [8], implying corresponding improvements for higher dimensions.

We improve these results by presenting a relatively simple *deterministic* algorithm which attains a competitive ratio of $7/4 = 1.75$. Using the construction presented by Chan and Zarrabi-Zadeh [1], this implies an upper bound of $2^d \cdot 7/8$ in d dimensions. Moreover, we improve the randomized lower bound to $3/2 = 1.5$ and show a deterministic lower bound of $8/5 = 1.6$. Finally we give a deterministic lower bound of 2 and a randomized lower bound of $11/6 \approx 1.8333$ in two dimensions. The deterministic lower bound holds for the L_2 norm as well. A summary of previous and improved results can be found in Table 1.

Table 1. Summary of new and previous results for one and two dimensions

	Lower bound of [1]	Lower bound (this paper)	Upper bound (this paper)	Upper bound of [8]
$d = 1$ deterministic	1.5	1.6	1.75	2
$d = 1$ randomized	1.3333	1.5	1.75	1.8333
$d = 2$ deterministic	1.5	2	3.5	4
$d = 2$ randomized	1.3333	1.8333	3.5	3.667

We start the paper with additional definitions, afterwards, we present the new algorithm followed by its analysis. Finally, we prove lower bounds, first for one dimension and then for two dimensions. Some proofs are omitted due to space constraints.

2   A Deterministic Algorithm

2.1   Definitions

For a cluster C, denote the leftmost request point contained in it by ℓ_C and the rightmost request point by r_C. A cluster is *single* if there is no cluster which has a common endpoint with it. A cluster is *fixed* if we have defined both its endpoints.

The distance of p to a cluster C is denoted by $d(p, C)$ and is defined as the distance from p to the closest point in C. For a fixed cluster C, this closest point is not necessarily a request point. The distance between two single clusters C and D is defined as the distance between their closest points.

We now define several kinds of pairs of clusters. In these definitions we discuss two clusters, where a cluster C is to the left of cluster D, without overlap. We call a pair of clusters close or far only if there is no cluster between them and there is no fixed cluster 'nearby'. Below, we specify what nearby means in this context.

Definition 1. *A* close pair *consists of two consecutive* single *clusters C and D such that one of the following two properties holds:*

- $d(\ell_C, \ell_D) \leq 1$ *and there is no fixed cluster which overlaps with the interval* $(\ell_D - 1, \ell_D + 1)$
- $d(r_C, r_D) \leq 1$ *and there is no fixed cluster which overlaps with the interval* $(r_C - 1, r_C + 1)$

Definition 2. *A* far pair *consists of two* single *clusters C and D that do not form a close pair and for which $d(r_C, \ell_D) \leq 1$. Moreover, there is no fixed cluster which overlaps with the interval* $(r_C - 1, \ell_D + 1)$.

Note that a cluster which contains a single point cannot be part of a far pair, only of a close pair, since for this cluster its left endpoint and its right endpoint are the same point.

Definition 3. *A fixed pair* consists of two *fixed* clusters C and D that have a common endpoint which is a request point.

Our algorithm avoids close and far pairs to avoid bad examples and in particular, bad examples shown in [1]. Instead, such pairs are turned into fixed pairs using the *attach* operation which we now define.

Our algorithm attaches one cluster to another cluster in one of two ways, left-to-right-attach and right-to-left-attach. Let C, D be a close pair and assume again that D is to the right of C. The algorithm sometimes attaches cluster C to cluster D and sometimes it attaches cluster D to cluster C. In the first case, cluster C is attached to cluster D as follows. Fix the location of D to be the interval $[\ell_D, \ell_D + 1]$ and the location of C to be the interval $[\ell_D - 1, \ell_D]$. This attach operation is only performed if $\ell_C \geq \ell_D - 1$, i.e., in the first case of the definition of a close pair, and called left-to-right-attach, since the rightmost point of the left cluster is fixed to be the leftmost point of the right cluster. If C or D overlaps with an existing cluster as a result of these definitions, we *truncate* it at the point where it starts to overlap. Since there is no cluster between C and D, the overlap can happen only at the right hand side of D or at the left hand side of C.

In the other option, we can attach D to C. To do that, we fix C at $[r_C - 1, r_C]$ and D at $[r_C, r_C + 1]$. This is only done if $r_D \leq r_C + 1$, i.e., in the second case of the definition of a close pair, and called right-to-left-attach. Again, we truncate C or D if this is necessary to avoid overlap.

Thus it can be seen that if a cluster is attached to another, they form a fixed pair. The clusters in a fixed pair always have length 1 unless this would make them overlap with some other cluster. Single clusters are never fixed by our algorithm, and thus their right and left endpoints are request points (possibly, both endpoints are the same request point).

2.2   The Algorithm

The idea of this algorithm is to try and avoid gaps between clusters if requests occur 'close' to one another.

A request inside a cluster is assigned to that cluster. Let a *good cluster for p* be a single cluster C such that p can be assigned to C without creating a new far or close pair. Let a *feasible cluster for p* be a single cluster C such that there is no cluster between C and p and the distance of p to the furthest request point in C is at most 1.

1. If there exists a good cluster C for p, assign p to C.
2. Else, if there exists a feasible cluster C for p such that assigning p to C creates a close pair C and D, assign p to C and attach D to C (perform a right-to-left-attach operation if D is to the right of C and otherwise a left-to-right-attach operation).

3. Else, if there exists a feasible cluster C for p such that assigning p to C creates a far pair C and D, define a new cluster P for p and attach it to D. (P, D form a close pair.)
4. Else, define a new cluster P for p. If there exists a cluster C such that P and C form a close pair, attach P to C.

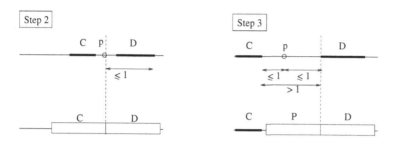

Fig. 1. Creation of a fixed pair in Step 2 and 3

Our algorithm does not allow overlap between clusters (except for endpoints, and even that can only happen if two clusters are attached and fixed, or if a cluster is truncated at the point where another cluster begins). and our algorithm avoids the creation of close and far pairs. A close or far pair can be created if a request point is being assigned to a cluster C and thus making it closer to the closest cluster on the same side of C as the request point. If the point is indeed assigned to C then a single close pair may be created and this pair is fixed right away. Otherwise, if the point is finally assigned to a new cluster, this cluster may form a close pair with each one of two clusters on its both sides. This happens in Step 3 or in Step 4. If it happens in Step 3, it must form a close pair with each one of them, since in this case any fixed cluster is located too far. The algorithm fixes the new cluster with one of the two previously existing clusters.

Therefore, when a close pair or far pair appears, our algorithm immediately fixes at least one half of the pair, possibly leaving the other half unchanged. Thus if there were no close or far pairs before some request, they still do not occur afterwards. Note also that the creation of a new cluster P cannot create a far pair, since the new cluster consists of a single point at this time.

2.3 Analysis

We start by proving several lemmas that clarify the structure of the clusters created by the algorithm. We first consider single clusters.

Lemma 1. *There can be no interval of length 1 which contains two single clusters.*

Proof. Suppose the two single clusters A and B are contained in an interval of length 1. Without loss of generality, denote by A the cluster that is defined earlier by the algorithm. Let b be the first request point in B. We consider the step in which b is assigned to a cluster. Since the point b fits in A (or in the cluster which is closest to b between b and A), it is not assigned by our algorithm in Step 4. In Steps 2 and 3, b is placed in a fixed cluster. In Step 1, b is placed in an existing cluster. But then B has more than one point. In all cases, we find a contradiction. □

Note that this lemma holds even if there are fixed clusters nearby. Specifically, the lemma shows that for two single clusters A and B that both contain only one request point, we have $d(A, B) > 1$.

In the following, we will repeatedly discuss sets of clusters $C_1, C_2, \ldots$ In such cases, denote the leftmost request point contained in C_i by ℓ_i and the rightmost request point by r_i. We now consider a fixed optimal offline algorithm. We call the clusters used by this algorithm "optimal clusters". The clusters used by our algorithm are called "online clusters".

As noted in [1], it is trivial to provide an optimal solution for a given input offline: starting from the left, repeatedly define a cluster of length 1 that has as its left endpoint the leftmost unserved point. It can be seen that in this solution, no two clusters overlap (not even at their endpoints). We will compare our algorithm, which also does not let clusters overlap, to this solution.

Lemma 2. *Consider three consecutive single online clusters, denoted by C_1, C_2 and C_3 from left to right. If there is an optimal cluster X which serves requests from all three clusters, then*

- *there exists a fixed cluster F which overlaps with the interval $(r_1 - 1, \ell_3 + 1)$*
- *there is no single online cluster between F and C_j, where C_j is the cluster among C_1, C_2 and C_3 that is closest to F*
- *there exists a request point in C_j which is served by an optimal cluster Y which does not serve requests from any other single cluster.*

Proof. Suppose there is no such fixed cluster F. The assumption implies that $d(r_1, \ell_3) \leq 1$. Let q be the oldest request point in C_2. If q is newer than r_1 and ℓ_3, C_1 and C_3 formed a close or far pair before q arrived, which our algorithm does not allow. Otherwise, without loss of generality, let r_1 be newer than ℓ_3. Then C_2 and C_3 form a close pair as soon as both q and ℓ_3 have arrived (since $d(\ell_2, \ell_3) \leq d(r_1, \ell_3) \leq 1$), which our algorithm also does not allow.

This proves the existence of the cluster F. Suppose that F is to the left of C_1. (The case where F is to the right of C_3 is symmetric.) By Lemma 1, C_1 and C_2 are not contained in an interval of length 1. This implies that the optimal cluster X which serves r_2 does not serve ℓ_1. By the same Lemma, there is no online single cluster between F and C_1 since $d(F, r_1) < 1$.

Consider the optimal cluster Y which serves ℓ_1. By these observations and the fact that Y does not overlap with X, we have that Y does not cover any point from any single cluster besides C_1 (possibly it covers some points of F). □

This Lemma shows that an optimal cluster X can only serve requests from three consecutive single clusters if these online clusters are the first or last three clusters in a sequence of consecutive single clusters (or the only three, of course).

Definition 4. *A* group *of online clusters is a maximal set of consecutive clusters such that each two successive clusters are 'connected' by an optimal cluster.*

That is, if $C_1, \ldots, C_m$ (numbered from left to right) form a group, there is an optimal cluster which contains both r_i and ℓ_{i+1} for $i = 1, \ldots, m-1$. (These optimal clusters are not necessarily all distinct.) If there is more than one group, for each group we have that the leftmost point of the leftmost online cluster is not to the right of the leftmost point of the leftmost optimal cluster by the way we construct our optimal solution.

Lemma 3. *For $m \geq 3$, at least $m-1$ offline clusters are needed to serve all the request points in m consecutive single clusters that are in one group.*

Proof. Denote these single clusters by $C_1, \ldots, C_m$ from left to right. For $m = 3$, even if there is an optimal cluster X which serves requests from all three single clusters, it cannot cover two of them completely by Lemma 1.

Suppose $m \geq 4$. Clearly, an optimal cluster cannot cover requests from four (or more) different online clusters $C_i, \ldots, C_{i+3}$, because then the two clusters C_{i+1} and C_{i+2} would have to be contained in an interval of length 1, which is impossible by Lemma 1. Lemma 2 shows that if an optimal cluster serves requests from three consecutive single clusters, then these are the clusters C_1, C_2, C_3 or C_{m-2}, C_{m-1}, C_m (or both), since there must be a fixed cluster immediately next to them on one side. Suppose it happens to the first three clusters (the other case is symmetric). Lemma 2 also shows that in this case, there is an optimal cluster Y which serves only requests from C_1. So whether this case occurs or not, the first (and last) three clusters are served by at least two optimal clusters. No other three consecutive clusters can be served by one optimal cluster by Lemma 2. We see that on average, at least one optimal cluster is required to serve requests from each two consecutive single online clusters. Since we have $m-1$ pairs of consecutive single clusters, the lemma is proved.   □

The next few lemmas consider fixed clusters.

Lemma 4. *In each pair of fixed clusters where none of the clusters is truncated, there is at least one optimal cluster which is completely contained inside the pair.*

Proof. This follows immediately from the fact that in any pair of fixed clusters, the shared endpoint of these two clusters is an actual request point. If both fixed clusters are not truncated, each one of them has length 1. The claim follows from the fact that the optimal solution needs to serve the point in the middle.   □

Lemma 5. *Not all requests in a pair of fixed clusters are served by a single optimal cluster.*

Lemma 6. *If a fixed cluster T is truncated, there is an existing fixed cluster F within a distance of less than 1 of the newly fixed cluster T, and exactly one single cluster E between T and F. The clusters T and E have a shared endpoint. Our algorithm does not create additional clusters between E and F.*

Note that when a fixed cluster T is truncated, both its endpoints are request points by Lemma 6. We consider T to be in the same group as both its neighbors, even if there exists an optimal cluster which only serves requests from one of these three clusters.

Lemma 7. *If a fixed cluster T is truncated, at least three optimal clusters are required to serve all request points in T and the two clusters with which T shares an endpoint.*

Proof. T shares one endpoint with another fixed cluster F, and one with a single cluster E by Lemma 6. These three clusters used to be single clusters. Wlog, let the order of them by E, T, F from left to right. Denote the rightmost endpoint of T before it became fixed by t. If the requests in these three clusters are served by only two optimal clusters, one optimal cluster must serve t and all request points in F, since no optimal cluster can serve E and T entirely by Lemma 1. But if this were possible, then T and F would have formed a close pair already before the request p arrived which caused T and F to become fixed. (If they did not form a close pair then, it was because there was a fixed cluster nearby, and in this case the appearance of p would also not have made them close, so T and F would not have become fixed.)

Since our algorithm avoids the creation of close pairs, we have found a contradiction. □

Theorem 1. *Our algorithm has a competitive ratio of $7/4$.*

Proof. By Lemma 5, the request points of a fixed pair are served by at least two optimal clusters. If they are served by three different optimal clusters, and both clusters in the pair are not truncated, we allow the optimal algorithm to move the leftmost of these clusters to the left until it no longer intersects the pair. This may mean that some request point is no longer served by the optimal algorithm, and thus can only make the competitive ratio higher. The request points that remain outside of optimal clusters are only points that the algorithm assigns to fixed clusters. Our further analysis on the optimal clusters is only for points that are inside single clusters. Thus the reduction above is valid.

We consider the groups that exist after this shifting. Note that the endpoint of a fixed pair which is an inner point of its group (i.e., the left endpoint of the pair if this pair is on the right end of its group, and vice versa) is always covered by some optimal cluster.

For each fixed untruncated pair, by Lemma 4 there is an optimal cluster X which does not serve any request from any cluster outside the pair, and an optimal cluster Y which might. We are going to bound the number of optimal clusters needed to serve all the request points that are not in the fixed untruncated pair(s) at the end(s). We then add 2 to the online cost and 1 to the offline cost for each fixed pair (the cluster Y has already been counted).

Truncated clusters By our definition of groups, truncated clusters occur only in the middle of groups. By Lemma 6, if a cluster T is truncated, there is another (older) fixed cluster F within a distance of less than 1 of it, and one single cluster E between them. Also there is a fixed cluster G on the other side of T which shares an endpoint with it. We call three such clusters E, T, G a triplet. The cluster F is either part of the next triplet or one half of a pair of fixed untruncated clusters.

There might be a triplet E, T, G such that G is also truncated. In this case, there is a single cluster E' immediately next to G, followed by the next fixed cluster G' which is older than G and not truncated. In this case we call the set $\{E, T, G, E'\}$ a *quartet*. As above we have that G' is either part of the next triplet or one half of a pair of fixed untruncated clusters.

This leaves only two possibilities for the inside structure of a group (ignoring the possible fixed pairs at the ends):

- $(sTF)^k s^m (FTs)^\ell$, where $k \geq 0, m \geq 0, \ell \geq 0$
- $(sTF)^k (sTTs)(FTs)^\ell$, where $k \geq 0, \ell \geq 0$

In this list, s represents a single cluster, T is a truncated cluster, and F is an untruncated fixed cluster.

Group elements. We see that we have three structural elements inside a group: triplets, quartets and sequences of single clusters. To calculate the number of optimal clusters required to serve the request points in such a group, we upper bound the number of optimal clusters for each element separately, going from left to right, and then add these together. Here we need to take into account that whenever we move from one element to the next, we need to subtract one from the optimal cost, because one optimal cluster gets counted double (once for each element).

By our results so far, we have the following table for the offline cost of each structural element.

Element	Contribution to online cost	Contribution to offline cost
Triplet	3	3 (Lemma 7)
Quartet	4	3 (Lemma 7)
One single cluster	1	1
Two single clusters	2	2 (Lemma 1)
$m \geq 3$ single clusters	m	$m - 1$ (Lemma 3)

We want to show an upper bound of $7/4$. There are only a few cases we need to check. To begin with, we only need to check groups with fixed untruncated pairs at both ends. We are going to add the optimal cost for each element, and subtract one from the optimal cost for each element beyond the first. Note that any triplet beyond the first does not help to show a competitive ratio above $3/2$, and that a sequence of single clusters cannot occur in combination with a quartet.

- No clusters in the group apart from the fixed untruncated pairs. If there is one fixed pair, we have a ratio of 1 by Lemma 5. Else, we have a ratio of at most $3/2$, again by using Lemma 5 (on both pairs), and noting that we are counting at most one optimal cluster double by Lemma 4.

- If the group starts with a quartet, the ratio is at most 8/5 (two fixed pairs, there can be no sequence of single clusters in this group, any triplets decrease the competitive ratio).
- If the group contains at least one triplet followed by a quartet, the ratio is at most 11/7 (two fixed pairs, one triplet, one quartet; no sequence of single clusters possible).
- Else, there is no quartet. If there is also no triplet, the ratio is at most 7/4, given by $m = 3$ single clusters and two fixed pairs.
- If there is a triplet, we find a ratio of 7/5 for $m = 0$ single clusters, 8/5 for $m = 1$, 9/6 for $m = 2$ and at most 10/7 for $m \geq 3$.

A matching lower bound for our algorithm can be shown using the request sequence $0, 1, 3, 4, 6, 7, 2, 5, 2.5, 3.5$. □

3   Lower Bounds

Theorem 2. *No deterministic algorithm can have a competitive ratio below 8/5.*

Point	Cluster	$\mathcal{A}(\sigma)$	OPT(σ)	Explanation
3	A	1	1	
4	A	1	1	Otherwise we get $\mathcal{A}(\sigma) = 2$ and OPT$(\sigma) = 1$.
5	B	2	2	The point does not fit in cluster A.
6	C	3	2	If the point is placed in B, the requests $2, 4.5, 7$ open three new clusters: $\mathcal{A}(\sigma) = 5$, OPT$(\sigma) = 3$.
2	D	4	3	
1	D	4	3	If the point is not placed in D, $\mathcal{A}(\sigma)/$OPT$(\sigma) = \frac{5}{3}$.
0	E	5	4	
2.5	F	6	4	
7	C	6	4	If the point is not placed in C, $\mathcal{A}(\sigma)/$OPT$(\sigma) = \frac{7}{4}$.
4.1	B	6	4	Otherwise we get $\mathcal{A}(\sigma) = 7$.
5.5	G	7	5	The point fits in no other cluster.
8	H	8	5	The point fits in no other cluster.

We use the above instance. In the table, $\mathcal{A}(\sigma)$ is the cost of an online algorithm and OPT(σ) is the current cost of the optimal solution for the instance σ up to now. In each row, "point" is the location of a new point. "Cluster" is the cluster it must belong to, where a new name means that a new cluster must be opened. See figure 2. It can be seen that the construction results in a lower bound of 8/5.

Theorem 3. *No randomized algorithm can have a competitive ratio below 3/2.*

Theorem 4. *No deterministic online algorithm can have a competitive ratio less than 2 in two dimensions.*

Proof. The proof is illustrated in Figure 3. Consider an online algorithm $\mathcal{A}$, and assume by contradiction that it has a competitive ratio of less than 2. First, four points arrive on the corners of a unit square. $\mathcal{A}$ must assign them all to the same

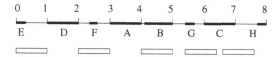

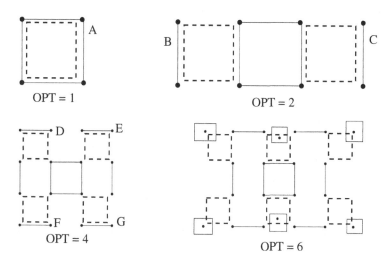

Fig. 2. The general deterministic lower bound: At the top the online clusters (marked in bold), at the bottom the final optimal solution

Fig. 3. The deterministic lower bound in two dimensions: Online clusters are represented by lines and boxes, an optimal solution for each case is represented by dashed boxes

cluster A, otherwise, since there exists a feasible solution consisting of a single cluster, it has a ratio of (at least) 2 and the input stops. It can be seen that A cannot be assigned any further point outside this unit square.

Then, four additional points arrive, two to either side (top right of Figure 3), so that the input now consists of two rows of four points each, one above the other. A must open exactly one cluster for each new pair, since it is possible to cover all existing points with only two clusters. Both of these new clusters, B and C, cannot cover any request above or below them, since the points they contain are already of distance 1 apart from each other (vertically).

In the next phase, eight additional points arrive (bottom left of Figure 3). These are new four points above and below the previous points, so that the 16 points form a square, and the distance between every consecutive pair of points is 1, both vertically and horizontally. Since it is possible to serve all these points with only four clusters, A must open four new clusters for these points; less than four does not cover all the points, and more than four gives a competitive ratio of 2. It can be seen that these clusters D, E, F, G cannot serve any request which is to the left or to the right of them, since each cluster contains two points that are of distance 1 apart (horizontally).

Finally, six additional points arrive. Three of which are in the top row of points, between the two central points, and in distance 1 from the extremal points, and the other three are in the same positions in the bottom row. Algorithm $\mathcal{A}$ is forced to open six new clusters for them, since none of these points fit in an existing cluster: they are to the side of D, E, F, and G, and above or below B and C (bottom right of Figure 3). Clearly, no two new points can be assigned to the same new cluster. Now $\mathcal{A}$ has opened 13 clusters in total while the optimal solution requires only six clusters. This is a contradiction and shows that $\mathcal{A}$ has a competitive ratio of at least 2.   □

Note that this lower bound of 2 for two dimensions implies a lower bound of 2 for any higher dimension as well: we can let all the requests appear in a 2-dimensional subspace.

Theorem 5. *No randomized online algorithm can have a competitive ratio less than 11/6 in two dimensions.*

Proof. We use an adaptation of Yao's principle [7] for proving lower bounds for *randomized algorithms*. It states that a lower bound on the competitive ratio of deterministic algorithms using a fixed distribution on the input, is also a lower bound for randomized algorithms and its value is given by $\frac{E(\mathcal{A}(\sigma))}{\mathrm{OPT}(\sigma)}$.

Let N be a large integer. To simplify presentation, we apply scaling so that the length of a cluster is at most N instead of 1. We give requests only at integer points. There are four phases:

1. $(N+1)^2$ points: $\{N, \ldots, 2N\} \times \{N, \ldots, 2N\}$.
2. Choose an integer i uniformly at random, $0 \le i \le N-1$. In this phase $(N+1)^2$ points appear, so that the set of all points requested so far is now $\{N, \ldots, 2N\} \times \{i, \ldots, 2N+1+i\}$.
3. Choose an integer j uniformly at random, $0 \le j \le N-1$. In this phase $2(N+1)^2$ additional points appear in such a way that the set of points requested so far is $\{j, \ldots, 2N+1+j\} \times \{i, \ldots, 2N+1+i\}$.
4. Choose an integer k uniformly at random, $0 \le k \le N-1$. In this phase $2(N+1)^2$ final points appear in such a way that the set of points requested so far is $\{j, \ldots, 2N+1+j\} \times \{i-k-1, \ldots, 3N+1+i-k\}$.

Thus, the set of request points is first extended vertically, then horizontally, and finally vertically again. We will show that for $N \to \infty$, with high probability, $\mathcal{A}(\sigma) \ge 11$ for any deterministic online algorithm $\mathcal{A}$. It can be seen that the input σ can be covered using only six clusters. These clusters are defined by vertical lines trough the points $(j, 0), (N+j, 0), (N+j+1, 0)$ and $(2N+j+1, 0)$, and horizontal lines through the points $(0, i-k-1), (0, N+i-k-1)$ $(0, N+i-k), (0, 2N+i-k), (0, 2N+i-k+1), (0, 3N+i-k+1)$.

We first focus on the last two phases in our construction. Consider the set of points $S_1 = \{(j,y)|y = i, \ldots, 2N+1+i\}$. Let C_1 be the cluster which contains (j,i). Let $p \le i+N$ be the highest value such that (j,p) is in cluster C_1. Define the set $S_2 = \{(j,y)|y = i-k-1, \ldots, 3N+1+i-k\}$. The following claim is given without proof.

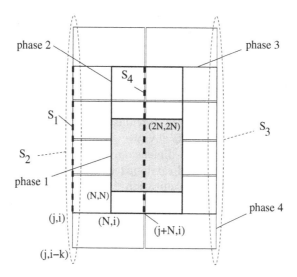

Fig. 4. The randomized lower bound in two dimensions. The values i, j, and k are all chosen uniformly at random from the set $\{1, \ldots, N\}$. The sets $S_1, \ldots, S_5$ are defined in the proof.

Claim. With probability at least $1 - 1/N$, the online algorithm will require four clusters in the fourth phase to cover all the points in the set S_2.

We can apply the same analysis to the point set $S_3 = \{(2N + 1 + j, y) | y = i - k - 1, \ldots, 3N + 1 + i - k\}$, showing that with probability at least $1 - \frac{1}{N}$, four clusters are needed to cover this set as well. Note that there cannot be a cluster which contains points from both S_2 and S_3.

Finally, consider the set of points $S_4 = \{(j + N, y) | y = i, \ldots, 2N + 1 + i\}$. Note that the points in S_4 are requested already in the first two phases of our input sequence. An analysis as in the proof of Theorem 3 shows that with high probability, S_4 requires at least three clusters. We consider the clusters of S_4 and would like to show that with high probability, these are three clusters that are different from the eight clusters that we already found.

To show this, we consider the input after the first two phases. If already at this time, there are at least 11 clusters, we are done. Otherwise, there are at most ten clusters. We say that $N \leq x \leq 2N$ is a border of a cluster X if there exists a point (x, y) that the algorithm assigns to cluster X but no point (x', y') with $x < x' \leq 2N$ and $i \leq y' \leq 2N + 1 + i$ that is assigned to X exists. Consider the clusters that are used by the algorithm to cover S_4. Assume that a cluster C_4 is identical to one of the clusters found for S_2. Then $j + N$ is a border for C_4. Clearly, each cluster has one border. Since j is chosen uniformly at random such that $0 \leq j \leq N - 1$, the probability that $j + N$ is a border of C_4 is at most $\frac{1}{N}$. The probability that among all (at most ten) clusters, at least one has $j + N$ as a border is at most $\frac{10}{N}$. Thus, with probability at least $1 - \frac{10}{N}$, the clusters of S_4 are all different from those of S_2. Clearly, they cannot be the same as these of S_3 (due to the distance).

Thus with high probability we find that the online algorithm requires at least 11 distinct clusters to cover all the requests in the input. Four for S_1, four for S_2, and three for S_4. □

4 Concluding Remarks

This paper significantly improves the previously known bounds. However, many questions still remain open. Specifically, we would like to find out whether the competitive ratio grows with the dimension. Another unresolved issue is the relation between deterministic and randomized algorithms. It is known that for small dimensions ($d = 1, 2$), randomization does not help in the unit covering problem. However, we do not have clear evidence that this is the case for unit clustering as well.

References

1. Chan, T.M., Zarrabi-Zadeh, H.: A randomized algorithm for onine unit clustering. In: Erlebach, T., Kaklamanis, C. (eds.) WAOA 2006. LNCS, vol. 4368, pp. 121–131. Springer, Heidelberg (2007)
2. Charikar, M., Chekuri, C., Feder, T., Motwani, R.: Incremental clustering and dynamic information retrieval. SIAM Journal on Computing 33(6), 1417–1440 (2004)
3. Gyárfás, A., Lehel, J.: On-line and first-fit colorings of graphs. J. Graph Theory 12, 217–227 (1988)
4. Kierstead, H.A., Trotter, W.T.: An extremal problem in recursive combinatorics. Congr. Numer. 33, 143–153 (1981)
5. Kierstead, H.A.: Coloring graphs on-line. In: Fiat, A., Woeginger, G.J. (eds.) Online Algorithms: The State of the Art, pp. 281–305. Springer, Heidelberg (1998)
6. Lovász, L., Saks, M.E., Trotter, W.T.: An on-line graph coloring algorithm with sublinear performance ratio. Discrete Math. 75, 319–325 (1989)
7. Yao, A.C.C.: Probabilistic computations: towards a unified measure of complexity. In: FOCS. Proc. 18th Symp. Foundations of Computer Science, pp. 222–227. IEEE, Los Alamitos (1977)
8. Zarrabi-Zadeh, H., Chan, T.M.: An improved algorithm for online unit clustering. In: COCOON 2007. Proc. 13th Annual International Conference on Computing and Combinatorics. LNCS, vol. 4598, pp. 383–393. Springer, Heidelberg (2007)

Better Bounds for Incremental Medians

Marek Chrobak[1] and Mathilde Hurand[2]

[1] Department of Computer Science, University of California, Riverside, USA
[2] Department d'Informatique (LIX), Ecole Polytechnique, Palaiseau, France

Abstract. In the incremental version of the well-known k-*median problem* the objective is to compute an incremental sequence of facility sets $F_1 \subseteq F_2 \subseteq .... \subseteq F_n$, where each F_k contains at most k facilities. We say that this incremental medians sequence is R-*competitive* if the cost of each F_k is at most R times the optimum cost of k facilities. The smallest such R is called the *competitive ratio* of the sequence $\{F_k\}$. Mettu and Plaxton [6,7] presented a polynomial-time algorithm that computes an incremental sequence with competitive ratio ≈ 30. They also showed a lower bound of 2. The upper bound on the ratio was improved to 8 in [5] and [4]. We improve both bounds in this paper. We first show that no incremental sequence can have competitive ratio better than 2.01 and we give a probabilistic construction of a sequence whose competitive ratio is at most $2 + 4\sqrt{2} \approx 7.656$. We also propose a new approach to the problem that for instances that we refer to as *equable* achieves an optimal competitive ratio of 2.

Keywords: Incremental medians, approximation algorithm, online algorithm, analysis of algorithms.

1 Introduction

The k-*median* problem is one of the most studied facility location problems. We are given two sets: a set $\mathcal{C}$ of *customers* and a set $\mathcal{F}$ of n *facilities*, with a metric function d that specifies the distance d_{xy} between any two points $x, y \in \mathcal{C} \cup \mathcal{F}$. The cost of a facility set $F \subseteq \mathcal{F}$, denoted by $cost(F)$, is defined as the minimum sum, over all customers $c \in \mathcal{C}$, of d_{cF}, where $d_{cF} = \min_{f \in F} d_{cf}$ is the minimum distance from c to F. Given k, the objective is to compute a set of k facilities with minimum cost.

Not surprisingly, the k-median problem is NP-hard. A number of polynomial-time approximation algorithms have been proposed, with the latest one, by Arya *et al.* [1,2] achieving the ratio of $3 + \epsilon$, for any $\epsilon > 0$.

Mettu and Plaxton [6,7] introduced the *incremental medians problem*, where the permitted number k of facilities is not specified in advance. Starting with the empty set, an algorithm receives authorizations for new facilities over time, and after each authorization it is allowed to add another facility to the existing ones. As a result, such an algorithm produces an incremental sequence of facility sets $F_1 \subseteq F_2 \subseteq ... \subseteq F_n$, where $|F_k| \leq k$ for all k. This sequence $\{F_k\}$ is said to

C. Kaklamanis and M. Skutella (Eds.): WAOA 2007, LNCS 4927, pp. 207–217, 2008.

be R-*competitive* if $cost(F_k)$ is at most R times the optimum cost of k facilities, for each k. The smallest such R is called the *competitive ratio* of $\{F_k\}$.

Mettu and Plaxton [6,7] gave a polynomial-time algorithm that computes such an incremental sequence with competitive ratio ≈ 30. This result is quite remarkable, for there is no apparent reason why an incremental sequence $\{F_k\}$ of facility sets, with each $cost(F_k)$ within a constant factor of the the optimum, would even exist – let alone be computed efficiently.

It is thus natural to address the issue of *existence* separately from *computational complexity*, and this is what we focus on in this paper. As shown by Mettu and Plaxton [6,7], no ratio better than 2 is possible, that is, for each $\epsilon > 0$ there is a metric space where each incremental facility sequence has competitive ratio at least $2 - \epsilon$. The upper bound on the ratio was improved to 8 by Lin *et al.* [5] and, independently, by Chrobak *et al.* [4]. In [5], the authors also show that a 16-competitive incremental median sequence can be computed in polynomial time.

Our results. We improve both the lower and upper bounds for incremental medians. For the lower bound, we show that, in general, no competitive ratio better than 2.01 is possible. We also prove, via a probabilistic argument, that each instance has an incremental medians sequence with competitive ratio at most $2 + 4\sqrt{2} \approx 7.656$.

In numerical terms, the improvement of the lower bound is mostly symbolic, as it implies that 2 is not the "right" ratio. For the upper bound, our result shows that the doubling method from [5,4] (see also [3]) is not optimal – even though it gives the optimal ratio of 4 for the closely related "resource augmentation" version of incremental medians [4]. As discussed in Section 6, we believe that our methods can be refined to further improve both the lower and upper bounds.

In addition, we consider a special case of the incremental medians problem where for any fixed value of k, each customer has the same distance to the optimal k-median. We refer to such instances as *equable*. (See Section 5 for a formal definition.) For this case, we show a construction of a 2-competitive incremental medians sequence, matching the lower bound from [6,7]. Our method for this case is very different from previous constructions and we believe that it will be useful in improving the upper bound for general spaces. In fact, this result implies that if there is a constant $\gamma \geq 1$ such that for each fixed k all customers' optimal costs are within factor γ of each other, then our construction achieves ratio at most 2γ – improving our own bound above if $\gamma < 1 + 2\sqrt{2}$.

2   Preliminaries

Let $(\mathcal{F}, \mathcal{C})$ be an instance of the medians problem, where $\mathcal{F}$ is a set of n facilities, $\mathcal{C}$ is the set of customers, and $\mathcal{F} \cup \mathcal{C}$ forms a metric space. By d_{xy} or $d(x, y)$ we denote the distance between points x, y. If Y is a set, we also write $d_{xY} = \min_{y \in Y} d_{xy}$ for the minimum distance from x to Y. For a facility set $F \subseteq \mathcal{F}$, denote by $cost(F)$ the cost of F, that is $\sum_{x \in \mathcal{C}} d_{xF}$.

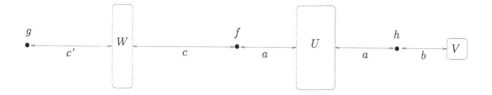

Fig. 1. Metric space used in the lower bound

For a point x and a set Y, denote by $\Gamma_Y(x)$ the point $y \in Y$ that is closest to x, that is $d_{xy} = d_{xY}$ (if this point is not unique, then break the tie arbitrarily.) If X is a set, we also define $\Gamma_Y(X) = \{\Gamma_Y(x) \mid x \in X\}$. Clearly, $|\Gamma_Y(X)| \le |X|$. Note that if F is a facility set and X is a set of customers, then $\Gamma_F(X)$ is exactly the set of facilities in F that serve customers in X if F is the facility set under consideration.

By opt_k we denote the optimum cost of k facilities, that is

$$opt_k \;=\; \min\{cost(F) \mid F \subseteq \mathcal{F} \,\&\, |F| = k\}. \tag{1}$$

By $F_k^* \subseteq \mathcal{F}$ we will denote the optimal set of k facilities, that is, the k-median. (As before, ties are broken arbitrarily.) Thus $cost(F_k^*) = opt_k$.

3   A New Lower Bound

In this section we prove our lower bound of 2.01 on the competitive ratio for incremental medians, improving slightly the previous bound of 2 from [6,7].

Theorem 1. *There is an instance $(\mathcal{C}, \mathcal{F})$ for which no incremental median sequence has competitive ratio smaller than 2.01.*

Proof. The set of customers is $\mathcal{C} = U \cup V \cup W$, where U, V, W are disjoint sets with $|U| + |V| + |W| = n - 3$, where n is a large integer. The set of facilities is $\mathcal{F} = \{f, g, h\} \cup \mathcal{C}$. The distances between customers and facilities are shown in Figure 1. For each set U, V, W, all customers in a set have the same distance to each facility. For example, the distance from f to all $u \in U$ is a, the distance from h to all $v \in V$ is b, etc. Other distances are measured along the shortest paths in the graph from Figure 1. This is also true for two customers from a same set (they are *not* at distance 0 from one-another). For example, if $v, v' \in V$ and $v' \neq v$ then the distance from v to v' is $2b$.

Since for $k = n - 3$ the optimal cost is 0, the first $n - 3$ facilities in any competitive incremental sequence must be chosen from $\mathcal{C}$. In fact, we will only use only three values of k: $k = 1, 2$ and $n - 3$.

To prove that there is no incremental median with ratio better than R, we only need to give some values a, b, c, c', $|U|$, $|V|$ and $|W|$ such that:

$$\min\{cost(v), cost(w)\} \ge R \cdot cost(f), \quad \text{and} \tag{2}$$

$$\min\{cost(u, v), cost(u, w)\} \ge R \cdot cost(g, h). \tag{3}$$

These inequalities imply the lower bound of R, for (2) implies that, for $k = 1$, to beat ratio R we must pick some $u \in U$ as the first facility, and (3) implies that, for $k = 2$, it is not possible to add to u another facility and preserve ratio R.

In order to simplify calculations, we slightly modify the way we compute the costs. If a facility at some point $x \in U \cup V \cup W$ serves a customer $z \neq x$ then the cost of z is the length of the shortest path from z to x via one facility f, g, or h, while the cost of $z = x$ is 0. Our first modification is that we will charge this $z = x$ the cost of such a shortest path as well, that is, c cannot serve itself directly at cost 0. For example, if there is a facility at $x \in U$, then we will charge x the cost of $2a$ to get to this facility. Since this increases the cost by a factor of at most $1 + \Theta(1/n)$, by taking n large enough in the proof below, the argument remains valid for the true cost values.

With this convention in mind, we set $a = 5/4$, $b = 1$, $c = 211/100$, $c' = 141/100$, $|U| = 295\lambda$, $|V| = 25\lambda$, and $|W| = 149\lambda$, for some large integer λ. (Thus $n = 469\lambda + 3$.) Note that $b \leq a \leq c \leq c'$. Then, for $k = 1$ we have

$$cost(f) = |U|a + |V|(b + 2a) + |W|c$$
$$cost(v) = |U|(a + b) + |V|(2b) + |W|(b + 2a + c)$$
$$cost(w) = |U|(a + c) + |V|(b + 2a + c) + |W|(2c')$$

and for $k = 2$ we have

$$cost(g, h) = |U|a + |V|b + |W|c'$$
$$cost(u, v) = |U|(a + b) + |V|(2b) + |W|(a + c)$$
$$cost(u, w) = |U|(2a) + |V|(a + b) + |W|(2c')$$

Then

$$\frac{\min\{cost(v), cost(w)\}}{cost(f)} = \frac{2039}{1014} \geq 2.01, \quad \text{and}$$
$$\frac{\min\{cost(u, v), cost(u, w)\}}{cost(g, h)} = \frac{121393}{60384} \geq 2.01.$$

This implies that inequalities (2), (3) hold with $R = 2.01$, and the lower bound follows.

4   A New Upper Bound

In this section we construct an incremental medians sequence with competitive ratio $R = 2 + 4\sqrt{2}$. First, we show that, given a facility set H we can find subsets $F \subseteq G \subseteq H$ of specified sizes and of appropriately small cost. We then use this result to construct our incremental medians sequence.

4.1   Choosing Two Nested Facility Sets

Let $1 \leq k \leq l \leq m \leq n$. (Recall that $n = |\mathcal{F}|$ is the number of facilities.) Throughout this section we consider three facility sets: H of cardinality m, U of cardinality k, and V of cardinality l. Intuitively, U and V represent optimal $k-$ and $l-$ medians. We use a probabilistic argument to show that there exist two sets F and G, with $|F| = k$, $|G| = l$ and $F \subseteq G \subseteq H$, such that $cost(F)$ and $cost(G)$ are bounded in terms of $cost(U)$, $cost(V)$ and $cost(H)$.

Lemma 1. *Let $1 \leq k \leq l \leq m \leq n$, and let U, V and H be facility sets with $|H| = m$, $|V| = l$ and $|U| = k$. Then there is a set $T \subseteq V$ with $|T| = k$ such that, denoting $\bar{T} = V - T$, we have*

$$cost(\Gamma_H(T)) + cost(\Gamma_H(U \cup \bar{T})) \leq 2 \cdot cost(H) + 4 \cdot cost(V) + 2 \cdot cost(U). \quad (4)$$

Proof. We use a probabilistic argument, by defining a probability distribution on subsets $T \subseteq V$ and proving that inequality (4) holds in expectation.

Define a random mapping $\Phi : U \to \mathcal{C}$, where $\Phi(u)$ is chosen uniformly from the set $\mathcal{C}_u = \{x \in \mathcal{C} \mid \Gamma_U(x) = u\}$. In other words, $\Phi(u)$ is a random customer of u when U is the facility set. Order arbitrarily the elements of V, and for any given Φ define T_Φ as the subset of V that consists of $\Gamma_V(\Phi(U))$ and $k - |\Gamma_V(\Phi(U))|$ smallest elements of V that are not in $\Gamma_V(\Phi(U))$. Thus $|T_\Phi| = k$.

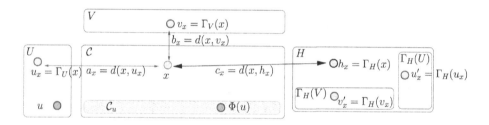

Fig. 2. Notations

For each point x in $\mathcal{C}$, let $u_x = \Gamma_U(x)$, $v_x = \Gamma_V(x)$ and $h_x = \Gamma_H(x)$ be the points serving x respectively in U, V and H. The corresponding distances from x are denoted $a_x = d(x, u_x)$, $b_x = d(x, v_x)$ and $c_x = d(x, h_x)$. Let also $u'_x = \Gamma_H(u_x)$ and $v'_x = \Gamma_H(v_x)$. (See Figure 4.1.)

We now temporarily fix the mapping Φ and a customer $x \in \mathcal{C}$. To simplify notation, we write $T_\Phi = T$ and $u = u_x$. We claim that

$$d(x, \Gamma_H(T)) + d(x, \Gamma_H(U \cup \bar{T})) \leq a_x + 2b_x + c_x + a_{\Phi(u)} + 2b_{\Phi(u)} + c_{\Phi(u)}. \quad (5)$$

To prove the claim, we consider two cases, for $v_x \in T$ and $v_x \in \bar{T}$.

<u>Case 1:</u> $v_x \in \bar{T}$. This case is illustrated in Figure 3.

Since $v'_{\Phi(u)} \in \Gamma_H(T)$, using the definition of $v'_{\Phi(u)}$ and several applications of the triangle inequality, we have $d(x, \Gamma_H(T)) \leq d(x, v'_{\Phi(u)}) \leq a_x + d(u, v_{\Phi(u)}) + d(v_{\Phi(u)}, v'_{\Phi(u)}) \leq a_x + [a_{\Phi(u)} + b_{\Phi(u)}] + d(v_{\Phi(u)}, h_{\Phi(u)}) \leq a_x + a_{\Phi(u)} + 2b_{\Phi(u)} + c_{\Phi(u)}$.

Since $v'_x \in \Gamma_H(U \cup \bar{T})$, using the definition of v'_x and triangle inequality, $d(x, \Gamma_H(U \cup \bar{T})) \leq d(x, v'_x) \leq b_x + d(v_x, v'_x) \leq b_x + d(v_x, h_x) \leq 2b_x + c_x$.

Combining the two bounds, we get

$$d(x, \Gamma_H(T)) + d(x, \Gamma_H(U \cup \bar{T})) \leq a_x + 2b_x + c_x + a_{\Phi(u)} + 2b_{\Phi(u)} + c_{\Phi(u)}.$$

<u>Case 2</u>: $v_x \in T$. This case is illustarted in Figure 4.

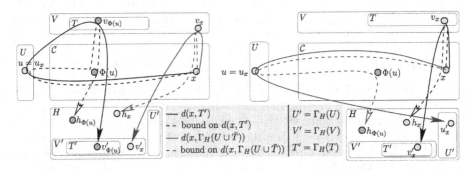

Fig. 3. The proof of (5) when $v_x \in \bar{T}$ **Fig. 4.** The proof of (5) when $v_x \in T$

Since $v'_x \in \Gamma_H(T)$, using the triangle inequality and the definition of v'_x, we have $d(x, \Gamma_H(T)) \leq d(x, v'_x) \leq b_x + d(v_x, v'_x) \leq b_x + d(v_x, h_x) \leq 2b_x + c_x$.

Since $u'_x \in \Gamma_H(U \cup \bar{T})$, using the definition of $u'_x = \Gamma_H(u)$, we have $d(x, \Gamma_H(U \cup \bar{T})) \leq d(x, u'_x) \leq a_x + d(u, u'_x) \leq a_x + d(u, h_{\Phi(u)}) \leq a_x + a_{\Phi(u)} + c_{\Phi(u)}$.

Combining the two bounds we get

$$d(x, \Gamma_H(T)) + d(x, \Gamma_H(U \cup \bar{T})) \leq a_x + 2b_x + c_x + a_{\Phi(u)} + c_{\Phi(u)}$$
$$\leq a_x + 2b_x + c_x + a_{\Phi(u)} + 2b_{\Phi(u)} + c_{\Phi(u)},$$

completing the proof of inequality (5).

From (5), for a fixed Φ we have

$$cost(\Gamma_H(T_\Phi)) + cost(\Gamma_H(U \cup \bar{T}_\Phi))$$
$$\leq \sum_{u \in U} \sum_{x \in C_u} \left[a_x + 2b_x + c_x + a_{\Phi(u)} + 2b_{\Phi(u)} + c_{\Phi(u)} \right]$$
$$\leq cost(H) + 2 \cdot cost(V) + cost(U)$$
$$+ \sum_{u \in U} |C_u| \cdot \left[a_{\Phi(u)} + 2b_{\Phi(u)} + c_{\Phi(u)} \right]. \quad (6)$$

For any facility set Z, we have $cost(Z) = \sum_{u \in U} |C_u| \cdot \text{Exp}_\Phi[d(\Phi(u), Z)]$. Applying it to $Z = U$, V and H, and using the linearity of expectation, inequality (6) yields

$$\text{Exp}_\Phi \left[cost(\Gamma_H(T_\Phi)) + cost(\Gamma_H(U \cup \bar{T}_\Phi)) \right]$$
$$\leq cost(H) + 2 \cdot cost(V) + cost(U)$$
$$+ \sum_{u \in U} |C_u| \cdot \text{Exp}_\Phi \left[a_{\Phi(u)} + 2b_{\Phi(u)} + c_{\Phi(u)} \right]$$
$$= 2 \cdot cost(H) + 4 \cdot cost(V) + 2 \cdot cost(U).$$

This implies that there is a $T = T_\Phi$ that satisfies the lemma.

Theorem 2. *Let* $1 \leq k \leq l \leq m \leq n$. *For any facility sets* H, U *and* V *with* $|U| = k$, $|V| = l$, $|H| = m$, *there exist* $F \subseteq G \subseteq H$ *with* $|F| = k$, $|G| = l$ *such that*

(i) $cost(F) \leq cost(H) + 2 \cdot cost(U)$ *and*
(ii) $cost(G) \leq cost(H) + 4 \cdot cost(V)$.

Proof. Let $U' = \Gamma_H(U)$ and $V' = \Gamma_H(V)$ be the facilities in H that are closest to those in U and V, respectively. Using the triangle inequality, it is not difficult to show (see [5,4], for example) that $cost(U') \leq cost(H) + 2 \cdot cost(U)$ and $cost(V') \leq cost(H) + 2 \cdot cost(V)$.

Let $T \subseteq V$ be the set from Lemma 1. Then either $cost(\Gamma_H(T)) \leq cost(H) + 2 \cdot cost(U)$ or $cost(\Gamma_H(U \cup \bar{T})) \leq cost(H) + 4 \cdot cost(V)$. In the first case, we take $F = \Gamma_H(T)$ and $G = V'$, and in the second case we take $F = U'$ and $G = \Gamma_H(U \cup \bar{T})$. (If $|F| < k$ or $|G| < l$, we can increase their cardinalities by adding a sufficient number of elements of H while preserving the inclusion $F \subseteq G$.) The theorem then follows from Lemma 1 and the bounds on $cost(U')$ and $cost(V')$.

4.2 Competitive Incremental Medians

Recall that n is the number of facilities, F_j^* is the optimal j-median and $opt_j = cost(F_j^*)$, for each $j = 1, 2, ..., n$. Our objective is to construct an incremental medians sequence $F_1 \subseteq F_2 \subseteq ... \subseteq F_n$.

The general approach is similar to that in [5,4]: we construct the sequence backwards, at each step extracting a smaller set of facilities from among those selected earlier. These sets F_j will be constructed only for values of j in a pre-defined sequence $\{\kappa(a)\}$ of indices, for which the optimal costs increase exponentially with a. For the intermediate values of j, we simply let F_j to be $F_{\kappa(a)}$, where a is the smallest index for which $\kappa(a) \leq j$.

The crucial difference between our method and the previous constructions is in how we extract facilities from $F_{\kappa(a)}$ to form $F_{\kappa(a+1)}$. The algorithms in [5] and [4] select $\kappa(a + 1)$ facilities in $F_{\kappa(a)}$ that are closest to those in the optimal set $F_{\kappa(a+1)}^*$. Instead, we use our probabilistic construction from the previous section to simultaneously extract *two* facility sets next in the sequence, namely $F_{\kappa(a+1)}$ and $F_{\kappa(a+2)}$, with Theorem 2 providing an upper bound on their costs.

Construction of incremental medians. Without loss of generality we can assume that $opt_n = 1$, for otherwise we can normalize the instance by dividing all distances by opt_n. (If $opt_n = 0$, instead of n, we can start the process with the largest n' for which $opt_{n'} > 0$.)

We use two parameters $\gamma = 2 + \sqrt{2}/2 \approx 2.71$ and $\lambda = 3\sqrt{2}/2 - 1 \approx 1.16$. We now define a sequence of indices $n = \kappa(0) \geq \kappa(1) \geq ... \geq \kappa(h) = 1$. For $a = 0, 1, ...,$ let

$$\kappa(a) = \begin{cases} \min\{j \mid opt_j \leq \gamma^{a/2}\} & \text{if } a \text{ is even} \\ \min\{j \mid opt_j \leq \lambda\gamma^{(a-1)/2}\} & \text{if } a \text{ is odd} \end{cases}$$

and choose h to be the smallest a for which $\kappa(a) = 1$. For simplicity, we will assume that h is even. Note that we allow some of the elements in the sequence $\{\kappa(a)\}$ to be equal.

We first define facility sets F_j for $j = \kappa(0), \kappa(1), ..., \kappa(h)$. Initially, $F_{\kappa(0)} = \mathcal{F}$, the set of all facilities. Suppose that $F_{\kappa(a)}$ has been already defined for some even $a \geq 0$. In Theorem 2 let $m = \kappa(a)$, $H = F_{\kappa(a)}$, $l = \kappa(a+1)$, $k = \kappa(a+2)$, $V = F^*_{\kappa(a+1)}$ and $U = F^*_{\kappa(a+2)}$. We then choose $F_{\kappa(a+2)} \subseteq F_{\kappa(a+1)} \subseteq F_{\kappa(a)}$ such that

$$cost(F_{\kappa(a+1)}) \leq cost(F_{\kappa(a)}) + 4opt_{\kappa(a+1)}, \quad \text{and} \tag{7}$$

$$cost(F_{\kappa(a+2)}) \leq cost(F_{\kappa(a)}) + 2opt_{\kappa(a+2)}. \tag{8}$$

The existence of such sets is guaranteed by Theorem 2; namely take $F_{\kappa(a+1)} = G$ and $F_{\kappa(a+2)} = F$.

Next, we extend the sequence to other values of j. If $\kappa(a+1) < j < \kappa(a)$, we simply let $F_j = F_{\kappa(a+1)}$. This completes the construction.

Theorem 3. *The incremental sequence $\{F_j\}$ constructed above is R-competitive, where $R = 2 + 4\sqrt{2} \approx 7.656$.*

Proof. For each $j = 1, ..., n$, denote $cost_j = cost(F_j)$. Using the bounds (7), (8), and the definition of the sequence $\{\kappa(a)\}$, each value $cost_{\kappa(a)}$ can be estimated as follows: if a is even, then $cost_{\kappa(a)} \leq 2\sum_{b=1}^{a/2} opt_{2b} \leq 2\sum_{b=1}^{a/2} \gamma^b$, and if a is odd then $cost_{\kappa(a)} \leq 2\sum_{b=1}^{(a-1)/2} \gamma^b + 4\lambda\gamma^{(a-1)/2}$. Summing up the geometric sequences, we thus get

$$cost_{\kappa(a)} \leq \begin{cases} \dfrac{2\gamma^{a/2+1}}{\gamma - 1} & \text{if } a \text{ is even} \\ \dfrac{2\gamma^{(a-1)/2+1}}{\gamma - 1} + 4\lambda\gamma^{(a-1)/2} & \text{if } a \text{ is odd} \end{cases}$$

Fix some number of facilities j, and choose a such that $\kappa(a+1) \leq j < \kappa(a)$. We want to show that $cost_j \leq R \cdot opt_j$. By the construction, $F_j = F_{\kappa(a+1)}$, so $cost_j = cost_{\kappa(a+1)}$. We have two cases.

Suppose first that a is even. By the choice of j and the definition of $\kappa(a)$, we get $opt_j > \gamma^{a/2}$. Since $cost_j = cost_{\kappa(a+1)} \leq 2\gamma^{a/2+1}/(\gamma - 1) + 4\lambda\gamma^{a/2}$, the ratio is

$$\frac{cost_j}{opt_j} \leq \frac{2\gamma}{\gamma - 1} + 4\lambda = R.$$

If a is odd, then by the choice of j and the definition of $\kappa(a)$, we get $opt_j > \lambda\gamma^{(a-1)/2}$. Since $cost_j = cost_{\kappa(a+1)} \leq 2\gamma^{(a+1)/2+1}/(\gamma - 1)$, the ratio is

$$\frac{cost_j}{opt_j} \leq \frac{2\gamma^2}{(\gamma - 1)\lambda} = R,$$

completing the proof.

5 2-Competitive Incremental Medians for Equable Instances

In this section we present a construction of a 2-competitive incremental medians sequence for a special case where, for any fixed value of k, each customer has the same distance to the optimal k-median. More formally, suppose $(\mathcal{F}, \mathcal{C})$ is an instance of the medians problem with $n \leq |\mathcal{C}|$ such that (i) for each $k = 1, 2, ..., n$ there exist an optimal k-median F_k^* such that all distances $d(x, F_k^*)$ are the same, for all $x \in \mathcal{C}$, and that (ii) for $k = n$ we have $d(x, F_n^*) = 0$ for all $x \in \mathcal{C}$ (or, equivalently, $\mathcal{C} \subseteq \mathcal{F}$.) An instance with this property will be called *equable*.

Our method is different from previous constructions of incremental medians, including the one from Section 4. Unlike in these previous approaches, we construct the sequence $F_1, F_2, ..., F_n$ *forward*, maintaining an invariant ensuring that we not only do well at step k, but also that we make good progress towards obtaining a low-cost l-median for all $l > k$.

Throughout this section, $(\mathcal{C}, \mathcal{F})$ denotes an equable instance of the medians problem. For each $k = 1, 2, ..., n$, let F_k^* be the optimal k-median such that $d(x, F_k^*) = \delta_k$ for all $x \in \mathcal{C}$. Thus $opt_k = |\mathcal{C}| \delta_k$ for all k. Without loss of generality, we can assume that $\delta_1 > \delta_2 > ... > \delta_n = 0$.

Incremental spanners. Suppose that for each $k = 1, 2, ..., n$ we have a family $\mathcal{S}_k \subseteq 2^{\mathcal{C}}$ of k sets that forms a partition of $\mathcal{C}$, that is, all sets in $\mathcal{S}_k$ are disjoint and $\bigcup_{A \in \mathcal{S}_k} A = \mathcal{C}$. (Our proof can be modified to work even if the sets in $\mathcal{S}_k$ are not disjoint.) For a set $X \subseteq \mathcal{C}$, define its k-span as

$$Span_k(X) = \bigcup \{A \in \mathcal{S}_i \mid i \geq k \ \& \ A \cap X \neq \emptyset\}.$$

A set $X \subseteq \mathcal{C}$ is called a k-spanner if $Span_k(X) = \mathcal{C}$. Note that if X is a k-spanner then it is also a j-spanner for any $j < k$. A sequence $X_1 \subseteq X_2 \subseteq ... \subseteq X_n$ is called an *incremental spanner* if for each $k = 1, 2, ..., n$, $|X_k| \leq k$ and X_k is a k-spanner.

We now show how to construct an incremental spanner. For $X \subseteq \mathcal{C}$ and any $j = 1, 2, ..., n$, let $setscov_j(X)$ be the collection of sets in $\mathcal{S}_j$ covered by the j-span of X, that is

$$setscov_j(X) = \{A \in \mathcal{S}_j \mid A \subseteq Span_j(X)\}.$$

Note that $|setscov_j(X)| = j$ if and only if X is a j-spanner, because $\mathcal{S}_j$ covers $\mathcal{C}$.

We will construct the sets $\emptyset = X_0 \subseteq X_1 \subseteq ... \subseteq X_n$ so that, for each $k = 0, 1, 2, ..., n$, we will have $|X_k| \leq k$ and the following invariant will hold:

$$|setscov_j(X_k)| \geq k, \quad \text{for all } j = k, k+1, ..., n. \tag{9}$$

Initially, for $k = 0$, we set $X_0 = \emptyset$, and (9) holds trivially. Suppose we have $X_0, X_1, ..., X_{k'}$, for some $k' < n$ and that (9) holds for $k = 0, 1, ..., k'$. This implies, in particular, that $|setscov_{k'}(X_{k'})| \geq k'$, that is, $X_{k'}$ is a k'-spanner. Thus $X_{k'}$ is also a k-spanner for all $k \leq k'$. Let l be the minimum index for which $X_{k'}$ is *not* an l-spanner, that is $\mathcal{C} - Span_l(X_{k'}) \neq \emptyset$. By the choice of l, we have $l > k'$. Pick any $x \in \mathcal{C} - Span_l(X_{k'})$ and take $X_{k'+1} = X_{k'} \cup \{x\}$. Clearly, $|X_{k'+1}| \leq k' + 1$.

We now show that (9) holds for $k = k' + 1$. By the choice of l, for $j = k' + 1, k' + 2, ..., l - 1$, $X_{k'}$ is a j-spanner. Therefore $X_{k'+1}$ is also a j-spanner, and thus (9) holds. Consider any $j \geq l \geq k' + 1$. Let $A \in \mathcal{S}_j$ be the set for which $x \in A$. By induction, since $x \in \mathcal{C} - Span_j(X_{k'})$, we have $A \notin setscov_j(X_{k'})$. But now $x \in X_{k'+1}$, so $A \in setscov_j(X_{k'+1})$, and we get $|setscov_j(F_{k'+1})| \geq |setscov_j(F_{k'})| + 1 \geq k' + 1$. This completes the proof that our construction preserves invariant (9).

By (9), for each k we have $|setscov_k(X_k)| \geq k$, and thus X_k is a k-spanner. We can conclude then that $X_1, X_2, ..., X_n$ is an incremental spanner.

Incremental medians. We now show how to use incremental spanners to construct incremental medians. For $k = 1, 2, ..., n$, assign each customer $x \in \mathcal{C}$ to its closest facility $f \in F_k^*$ (that is, $d_{xf} = \delta_k$), breaking ties arbitrarily. Define C_k^f to be the set of customers assigned to f, and let $\mathcal{S}_k = \{C_k^f \mid f \in F_k^*\}$. Then each $\mathcal{S}_k$ contains k sets and forms a partition of $\mathcal{C}$. As we showed above, for these partitions $\mathcal{S}_1, \mathcal{S}_2, ..., \mathcal{S}_n$ there exists an incremental spanner $F_1, F_2, ..., F_n$.

We claim that $F_1, F_2, ..., F_n$ is a 2-competitive incremental medians sequence. Consider some fixed k. Since F_k is a k-spanner, for each customer $x \in \mathcal{C}$ there is $i \geq k$, $f \in F_i^*$ and $y \in F_k$ such that both $x, y \in C_i^f$. Thus $d(x, F_k) \leq d_{xy} \leq d_{xf} + d_{yf} = 2\delta_i \leq 2\delta_k$. This implies that $cost(F_k) \leq 2m\delta_k = 2opt_k$, and the claim follows.

Summarizing, we obtain the following result:

Theorem 4. *For any equable instance* $(\mathcal{C}, \mathcal{F})$ *of the medians problem there exists a 2-competitive incremental medians sequence.*

6   Final Comments

We improved both the lower and upper bounds for incremental medians, from 2 to 2.01 and from 8 to $2 + 4\sqrt{2} \approx 7.656$, respectively, thus proving that neither 2 nor 8 are the "right" bounds for this problem. (By optimizing the the parameters in Section 3 it is possible to improve the lower bound slightly, to about 2.01053.) In addition to its own independent interest, closing or significantly reducing the remaining gap would shed more light on the computational hardness of approximating incremental medians, as it would show to what degree the difficulty of the problem can be attributed to non-existence of incremental median sequences with small competitive ratios.

The expected values in the proof of Lemma 1 can be computed in polynomial-time, and thus our probabilistic construction can be derandomized using the method of conditional expectations. However, since our improvement is relatively minor, we did not pursue this direction of research, nor possible implications for upper bounds achievable in polynomial time.

We believe that some of the ideas in the paper can be used to prove even better bounds. In the upper bound proof in Section 4 we construct our sequence backwards, starting with all facilities, and gradually extracting smaller and smaller facility sets, two at a time. By extending the probabilistic construction to more than two steps at a time, we should be able to get a better bound. Even our two-step method still might have room for improvement, as the two choices for F and G considered in the proof of Theorem 2 are not "balanced", that is, the bounds on the cost of F and G in the two cases are not the same. Also, our construction of a 2-competitive incremental medians sequence for equable spaces is very different from previous constructions and we believe that its basic idea will be useful in improving the upper bound for general spaces.

Our lower bound argument uses only three steps, for $k = 1, 2, n$. It should be possible to improve our bound by using either $k > 2$ as the intermediate number of facilities or more (perhaps an unbounded number of) steps. Both ideas lead to difficulties that we were not able to overcome at this time. In a three-step strategy using $k = 1, k', n$ with $k' > 2$, an algorithm can place facilities $2, .., k'$ optimally (given the choice of the first facility), and thus increasing k' seems only to help the algorithm. A strategy that uses additional steps leads to a different problem. Average costs for the customers must decrease with k, and thus introducing additional steps creates shortcuts via optimal k'-medians for large k', reducing the algorithm's cost for small values of k.

The result from Section 5 may also be useful for lower bound proofs, as it shows that in "hard" instances, for a fixed k, the optimal customers' costs should be significantly different.

Acknowledgements. This research was supported by National Science Foundation Grants CCR-0208856, OISE-0340752, and CCF-0729071.

References

1. Arya, V., Garg, N., Khandekar, R., Meyerson, A., Munagala, K., Pandit, V.: Local search heuristic for k-median and facility location problems. In: STOC. Proc. 33rd Symp. Theory of Computing, pp. 21–29. ACM, New York (2001)
2. Arya, V., Garg, N., Khandekar, R., Meyerson, A., Munagala, K., Pandit, V.: Local search heuristics for k-median and facility location problems. SIAM Journal on Computing 33(3), 544–562 (2004)
3. Chrobak, M., Kenyon, C.: Competitiveness via doubling. SIGACT News, 115–126 (2006)
4. Chrobak, M., Kenyon, C., Noga, J., Young, N.: Online medians via online bidding. In: Correa, J.R., Hevia, A., Kiwi, M. (eds.) LATIN 2006. LNCS, vol. 3887, pp. 311–322. Springer, Heidelberg (2006)
5. Lin, G., Nagarajan, C., Rajamaran, R., Williamson, D.P.: A general approach for incremental approximation and hierarchical clustering. In: SODA. Proc. 17th Symp. on Discrete Algorithms, pp. 1147–1156 (2006)
6. Mettu, R.R, Plaxton, C.G.: The online median problem. In: Ramgopal, R. (ed.) FOCS. Proc. 41st Symp. Foundations of Computer Science, pp. 339–348. IEEE, Los Alamitos (2000)
7. Mettu, R.R., Plaxton, C.G.: The online median problem. SIAM J. Comput. 32, 816–832 (2003)

Minimum Weighted Sum Bin Packing

Leah Epstein[1] and Asaf Levin[2]

[1] Department of Mathematics, University of Haifa, 31905 Haifa, Israel
lea@math.haifa.ac.il
[2] Department of Statistics, The Hebrew University, Jerusalem, Israel
levinas@mscc.huji.ac.il

Abstract. We study MINIMUM WEIGHTED SUM BIN PACKING (MWSBP), which is a bin packing problem where the cost of an item is the index of the bin into which it is packed multiplied by its weight, and the goal is to minimize the total cost of the items. This is equivalent to a batch scheduling problem which we define, where the total weighted completion time is to be minimized. This problem is previously known to be NP-hard in the strong sense even for unit weight items. We design a polynomial time approximation scheme for it, and additionally, a dual polynomial time approximation scheme.

1 Introduction

Bin packing is a natural and well studied problem which has applications in problems of computer storage, bandwidth allocation, stock cutting, transportation and many other important fields.

Consider the following scenario. A processor receives a set of short tasks to run. Each task in this set has a given duration, which never exceeds one time unit, and has to be run non-preemptively. Moreover, each task has a non-negative weight. The processor is capable of processing one task at every time. At every integer time unit, the processor reports the output of running the tasks that were processed in the previous time unit, and starts a new batch of jobs. Clearly, each such batch of tasks must have a total processing time of at most one unit. Therefore, the tasks need to be partitioned into subsets, where for each subset, the total sum of processing times is at most one. Our goal is to minimize the weighted sum of completion times, where a completion time of a task is the time at which its output is reported. This definition of completion time to be the completion time of the whole batch is the common definition in batch scheduling problems. Note that if instead of the weighted sum of completion times we are interested in the maximum completion time of any task, this problem is equivalent to the standard bin packing problem, first studied in the early 1970's [12,2,1], and widely studied ever since. We define our problem as a variant of bin packing.

The MINIMUM WEIGHTED SUM BIN PACKING problem (MWSBP) is defined as follows. We are given a set I of n items denoted by $I = \{1, 2, \ldots, n\}$, where item i has a size $s_i \in (0, 1]$ and a weight $w_i \in [1, \infty)$. A feasible solution consists of partition of I into $I_1, I_2, \ldots, I_p$ such that for all $j = 1, 2, \ldots, p$, $\sum_{i \in I_j} s_i \leq 1$.

C. Kaklamanis and M. Skutella (Eds.): WAOA 2007, LNCS 4927, pp. 218–231, 2008.

The objective is to find a feasible solution so that $\sum_{j=1}^{p} j \cdot \sum_{i \in I_j} w_i$ is minimized.

The goal function is equivalent to minimizing $\sum_{j=1}^{p} \left(\sum_{i \in I_j \cup I_{j+1} \cup \cdots \cup I_p} w_i \right)$. The sets I_j are called bins and we identify a bin with the item set in it (thus I_j is a set of items).

The MIN-SUM SET COVER problem (MSSC) is a related problem studied by Feige, Lovász and Tetali [6] (see also Munagala, Babu, Motwani and Widom [10]). We are given a collection of subsets $S_1, S_2, \ldots, S_m$ of a ground set $S = \{1, 2, \ldots, n\}$. A feasible solution is an ordering π of a subset $\mathcal{S}'$ of $1, 2, \ldots, m$, such that $\bigcup_{\mathcal{X} \in \mathcal{S}'} \mathcal{X} = S$, and for each element j of the ground set we incur a cost i, where i is such that $j \in S_{\pi(i)}$ and $j \notin S_{\pi(k)}$ for all $k < i$. The goal is to find an ordering that minimizes the total cost. They showed that the greedy algorithm is a 4-approximation algorithm for MSSC. This greedy algorithm is equivalent to the well known greedy algorithm for the classical set cover problem [8,9]. They further showed that this approximation ratio is best possible unless $P = NP$. This result holds also for a weighted variant of MSSC where each item has a weight and for each element j we incur a cost $w_j \cdot i$ where i is defined in the same way as in MSSC (this can be easily deduced for integer weights by replacing each item with weight w_i by w_i distinct unit weight items, and then applying the greedy algorithm on the resulting instance. Since an optimal solution and the approximated solution for the new instance cover all copies of item i with the same set, the result holds). We note that the MWSBP is a special case of a weighted variant of the MSSC where the ground set is the set of items and the available subsets are all subsets of the items that fit into one bin. Although the number of subsets is exponential in the original input size to problem MWSBP, we can still apply the greedy algorithm where in each step we use an FPTAS for the Knapsack problem and obtain a $4 + \varepsilon$ approximation algorithm. Such an approach of using an approximated oracle in each greedy step for MSSC, was recently analyzed by Epstein, Halldórsson, Levin and Shachnai [4]. Hence, we conclude that prior to this study there is a known $4 + \varepsilon$ approximation algorithm for MWSBP where $\varepsilon > 0$ is an arbitrary small positive number.

The unweighted case of MWSBP was studied before in [5]. It is shown in that paper that the unweighted case is NP-hard in the strong sense, and a polynomial time approximation scheme for the unweighted case was designed. In the same paper, some natural fast bin-packing heuristics were analyzed, the best of which has an approximation ratio of at most $\frac{5}{3}$.

For an algorithm $\mathcal{A}$, we denote its cost by $\mathcal{A}$ as well. The cost of an optimal algorithm denoted by OPT. A ρ-*approximation algorithm* for a minimization problem is a polynomial time algorithm that returns a feasible solution with cost at most a factor ρ above the optimal cost. A *polynomial time approximation scheme* (PTAS) is a family of $(1 + \varepsilon)$-approximation algorithms over all $\varepsilon > 0$. A *dual polynomial time approximation scheme* (dual-PTAS) is a family of polynomial time algorithms such that for all $\varepsilon > 0$ the algorithm constructs a solution of cost at most OPT that uses bins of size $1 + \varepsilon$ instead of bins of size 1.

Our results. In Section 2 we show that Next Fit Decreasing has an approximation ratio of exactly 2, if the sequence of output bins is sorted optimally (i.e., in order of non-increasing sum of weights of items). Then, in Section 3 we design a dual-polynomial time approximation scheme, i.e., we design an approximation scheme where the cost of the returned solution is at most OPT, however the approximated solution is allowed to use bins with capacity $(1 + \varepsilon)$ whereas the optimal solution is allowed to use bins with unit capacity. Note that the value OPT does not need to be given in advance. In Section 4 we show a polynomial time approximation scheme (PTAS) for the problem. To design the PTAS, we use linear grouping similarly to [3], together with non-trivial pre-processing of the set of large items. The structure of our problem allows to design a PTAS (and not an APTAS as in [3]). Note that by the hardness result, a PTAS is basically the best possible approximation as an FPTAS does not exist. Note also that in presence of weights, the problem becomes more difficult to deal with than the unweighted problem of [5]. Therefore the structure of the PTAS is much more complex. Our dual PTAS however is relatively simple and has the interesting property that no rounding of the weights, or grouping by weight, is required.

2   Preliminaries

Many heuristics for the standard bin packing problem were suggested and analyzed. Among these heuristics, the most natural ones are Next Fit (NF), First Fit (FF) and Best Fit (BF). These algorithms assume an arbitrary ordering of the input. Next Fit uses one active bin into which it packs the input. Once the free space in this bin becomes too small to accommodate the next item, a new active bin is opened and the previous active bin is never used again. The two other algorithms keep all non-empty bins active, and try to pack every item in these bins before opening a new bin. Such algorithms belong to the class of Any Fit (AF) algorithms that consists of all algorithms that open a new bin only if there is no other option. Note that NF is not an AF algorithm. The algorithms Weighted Next Fit Decreasing (WNFD), Weighted First Fit Decreasing (WFFD), Weighted Best Fit Decreasing (WBFD), and the class Weighted Any Fit Decreasing (WAFD) are defined in the same way as NF, FF, BF and AF, only the input is not ordered arbitrarily but sorted in a non-increasing order of *the ratio* of the item size to the item weight.

To adapt these algorithms for our problem we note that the performance guarantee of an algorithm can only benefit from sorting the output bins in a non-increasing order of their total weight (the total weight of a bin is defined as the sum of weights of the items in it).

In [5] we showed that for each positive value of K, the approximation ratio of WAFD and of NF (with or without re-ordering the bins) and of WNFD without re-ordering the bins is at least K even when applied to unweighted instances (for unweighted instances, WAFD is equivalent to the algorithm Any Fit Decreasing

(AFD) and WNFD is equivalent to the algorithm Next Fit Decreasing (NFD)). We also showed that for unweighted case the approximation ratio of NFD when a re-ordering the bins is applied is at most 2 and at least $\frac{3}{2}$. We are able to show that this positive result holds also for the MWSBP and we improve the lower bound to get tight bounds on the performance of this algorithm for the weighted case.

Theorem 1. *The approximation ratio of* WNFD *when a re-ordering of the bins is applied is exactly* 2.

3   Dual-PTAS

Let $0 < \varepsilon \le \frac{1}{4}$ be a fixed constant, such that $\frac{1}{\varepsilon}$ is an even integer. We construct a solution of cost at most OPT that uses bins of size $1 + \varepsilon$ instead of bins of size 1. This scheme uses some of the ideas that the APTAS in the next section uses but it is significantly simpler. We describe it first to introduce some of the concepts used later. We use a parameter $\delta = \frac{\varepsilon}{2}$.

As a first step, we partition the set of items into *small items*, which are items of size at most δ, and *large items* which are all other items. We next perform a rounding of the *sizes* of large items (see [7]). For an item i of size $s_i > \delta$, we let s_i' be the smallest value of the form $\delta + b\delta^2$ for some integer $b \in \{1 \ldots, \frac{1}{\delta^2} - \frac{1}{\delta}\}$ such that $s_i \le s_i'$. For an item i of size $s_i \le \delta$, we let $s_i' = s_i$. We call the instance with sizes s_i' the rounded instance.

Lemma 1. *For every item* i, $\frac{s_i'}{s_i} \le 1 + \delta$.

Let OPT_R be the cost of an optimal solution which uses bins of size $1 + \delta$ instead of bins of size 1, and needs to pack the items of the set I with their original weights and the sizes s_i' for $1 \le i \le n$.

Lemma 2. $\text{OPT}_R \le \text{OPT}$.

As a result of the rounding, the number of different sizes of large items in now constant (less than $\frac{1}{\delta^2}$). We are going to restrict ourselves to solutions with the following property. Let $i, i' \in I$ be two large items such that $s_i' = s_{i'}'$ and $w_i \ge w_{i'}$. Let t and t' be the respective indices of bins where i and i' are packed. Then $t \le t'$. We call such a packing *sorted*. Let OPT_{RS} be the cost of a sorted optimal packing of the rounded instance into bins of size $1 + \delta$. Then we can show the following lemma.

Lemma 3. $\text{OPT}_{RS} \le \text{OPT}_R$.

We consider next packings into bins of size $1 + 2\delta$. We only consider packings of the following type. The items have rounded up sizes and are packed using a sorted packing. The sum of the rounded size of the large items in a bin is at most $1 + \delta$. Let $I_s = \{1, \ldots, \ell\}$ be the set of small items, sorted so that $\frac{s_1}{w_1} \ge \frac{s_2}{w_2} \ge \cdots \ge \frac{s_\ell}{w_\ell}$ (i.e., by Smith's ratio [11]). Then we require also that the set of small items packed in each prefix of bins $1, \ldots, j$ is a suffix of the small items (a set $k_j, k_j + 1, \ldots, \ell - 1, \ell$), and that the sum of all rounded sizes of items in the j-th bin, except the small item k_j (if it is packed into bin j), is at most $1 + \delta$. This can be seen visually as follows. We stack all large items (with rounded sizes) one on top of the other. Then we stack the small items, starting from the one of largest index. The "height" of the stack may exceed $1 + \delta$, but if the last item is removed, then the total (rounded) size does not exceed $1 + \delta$. Therefore, the total rounded size of the items in this stack is at most $1 + 2\delta$. Denote an optimal solution among solutions that fulfill these conditions by OPT_{RSC}.

Lemma 4. $\text{OPT}_{\text{RSC}} \leq \text{OPT}_{\text{RS}}$.

We showed that if we pack the rounded items into bins of size $1 + 2\delta$, we can restrict ourselves to packings where small items in each bin are a consecutive sequence and the set of small items in every prefix of bins is a suffix of the small items. All bins that contain small items, except possibly the last such bin, should contain a total size of items which is at least $1 + \delta$. Moreover, if we are given the locations of the large items of rounded size $\delta + b\delta^2$, these items can be distributed to the bins according to a list of the items sorted by weight, such that bins of smaller indices receive items of larger weight (with the same rounded size). Therefore, a packing for the rounded items can be specified by the number of items of rounded size $\delta + b\delta^2$ in each bin for $b = 1, \ldots, \frac{1}{\delta^2} - \frac{1}{\delta}$, and by the values k_j.

We next show how to construct a graph $G = (V, E)$, which allows us to find such an optimal packing of the input items. The packing is represented by a path in the graph. Edges in the graph represent bins, and are associated with costs. This results in a correspondence of paths to (possibly partial) packings. The cost of a path is equal to the costs of packings. Such a solution will be a valid packing of the rounded items into bins of size at most $1 + 2\delta$. However, since the original sizes are no larger than the rounded sizes, this solution implies a packing for the original items with the same cost and the same size of bins (i.e., $1 + 2\delta$).

A label of a vertex corresponds to a subset of unpacked items. For $v \in V$ we have a label which is a vector of length $\frac{1}{\delta^2} - \frac{1}{\delta} + 1$. The label is $label(v) = (n_1(v), \ldots, n_{\frac{1}{\delta^2} - \frac{1}{\delta}}(v), p(v))$. Among the first $\frac{1}{\delta^2} - \frac{1}{\delta}$ components, $n_b(v)$ is the number of remaining items of rounded size $\delta + b\delta^2$, and $p(v)$ is the largest index of an unpacked small item, thus $0 \leq p(v) \leq \ell \leq n$, where $p(v) = 0$ indicates that all small items are packed already. For $1 \leq b \leq \frac{1}{\delta^2} - \frac{1}{\delta}$, let $\nu_b \leq n$ be the number of items of rounded size $\delta + b\delta^2$ in the input. Then $0 \leq n_b \leq \nu_b$. Therefore, the number of different labels is at most $(n + 1)^{\frac{1}{\delta^2} - \frac{1}{\delta} + 1} < (n + 1)^{\frac{1}{\delta^2}}$. For each possible value of the label, we will have a vertex in the graph G. Therefore, G has a polynomial size (for a fixed value of ε).

An edge from v to u must correspond to the difference between the two labels, $label(v)$ and $label(u)$. This difference corresponds to a packing pattern of a single bin. A pattern for an outgoing edge of v, $pat(v) = (pat_1(v), \ldots, pat_{\frac{1}{\delta^2} - \frac{1}{\delta}}(v), q(v))$ is defined as follows. The pattern contains a number $pat_b(v)$ for $1 \leq b \leq \frac{1}{\delta^2} - \frac{1}{\delta}$, which indicates how many items of size $\delta+b\delta^2$ are in this bin. For all these values of b we require $n_b(v) - n_b(u) = pat_b(v)$. The last component $q(v)$ indicates the sequence of small items with consecutive indices to be packed in the bin, that is, the sequence of small items in this bin is defined to be the items of indices $q(v), \ldots, p(v)$. We require $1 \leq q(v) \leq p(v) + 1$, where $q(v) = p(v) + 1$ means that no small items are packed in the bin, and the special case $q(v) = 1$ and $p(v) = 0$ means that all small items have been packed previously. To comply with the requirements of the type of solutions we seek, if $q(v) > 1$, we require

that $\displaystyle\sum_{b=1}^{\frac{1}{\delta^2}-\frac{1}{\delta}} (\delta+b\delta^2)pat_b(v) + \sum_{i=q(v)+1}^{p(v)} s_i \leq 1+\delta$ and $1+\delta < \displaystyle\sum_{b=1}^{\frac{1}{\delta^2}-\frac{1}{\delta}} (\delta+b\delta^2)pat_b(v) +$

$\displaystyle\sum_{i=q(v)}^{p(v)} s_i \leq 1 + 2\delta$. If $q(v) = 1$ we still require the first and third inequalities.

A *packing path* in the graph is a path which starts at the vertex whose label is $(\nu_1, \ldots, \nu_{\frac{1}{\delta^2}-\frac{1}{\delta}}, \ell)$ and terminates in a vertex whose label is $(0, \ldots, 0, 0)$. The cost of any outgoing edge of a vertex v is the same. This is the total weight of all items that are associated with the label of v. Then, the cost of a path is identical to the cost of a packing, using the second definition of the cost of a solution to MWSBP. Since earlier bins contain items of larger weights (and the same rounded size), we may assume that the remaining n_b items of size $\delta + b\delta^2$ are the ones with smallest weight among the ν_b items of this rounded size, and we define the cost of an edge according to that. Assume that the weights of these items are $w^1(b) \leq w^2(b) \leq \ldots \leq w^{\nu(b)}(b)$. Then the cost of an outgoing edge of

v is $\displaystyle\sum_{b=1}^{\frac{1}{\delta^2}-\frac{1}{\delta}} \sum_{i=1}^{n_b(v)} w^i(b) + \sum_{i=1}^{p(v)} w_i$.

We are interested in a packing which is based on a path of minimum cost. As we saw above, this path can be translated into a packing with the same cost and bins of size $1 + 2\delta = 1 + \varepsilon$. Therefore, we proved the following theorem.

Theorem 2. *There is a dual polynomial time approximation scheme for problem* MWSBP.

4   PTAS

Let $0 < \varepsilon \leq \frac{1}{4}$ be a fixed constant, such that $\frac{1}{\varepsilon}$ is an even integer. We let $W = \sum_{j=1}^{n} w_j$. One of the ingredients of our PTAS is an approach which is similar to the approach of the APTAS for the classical bin packing problem due to Fernandez de la Vega and Lueker [3]. We say that an item i is *large* if its size is greater than ε^2, it is *heavy* if its size is at most ε^2 and its weight is at least $\varepsilon^3 W$, and otherwise, if its size is at most ε^2 and its weight is smaller than $\varepsilon^3 W$, it is *small*. Without loss of generality we make the following assumptions. Let m be the number of large items, and assume that these are the first m items (i.e., the large items are $L = \{1, 2, \ldots, m\}$). Moreover, we assume that $s_1 \geq s_2 \geq \cdots \geq s_m$. We further assume that there are ℓ heavy items and these are the next ℓ items (i.e., the heavy items are $H = \{m+1, m+2, \ldots, m+\ell\}$).

By definition, we have $\ell \leq \frac{1}{\varepsilon^4}$. Finally, we assume that the small items are sorted according to their Smith's ratio. That is, $\frac{s_{m+\ell+1}}{w_{m+\ell+1}} \geq \frac{s_{m+\ell+2}}{w_{m+\ell+2}} \geq \cdots \geq \frac{s_n}{w_n}$.

In some of the proofs we refer to the cost charged to an item, or the charging cost of this item. This cost for a given solution is defined to be the index of the bin where the item is packed, multiplied by its weight. Thus if a solution consists of bins $C_1, C_2, \ldots$, and item i is packed in bin C_j, then its charged cost is $j \cdot w_i$. Clearly, the total cost charged to all items is exactly the cost of the solution.

The first step of our scheme is to partition the set of large items into *classes* according to their weight. Without loss of generality we can assume that the minimum weight of an item is 1, and we define the first class as $L^1 = \{i : w_i < 1 + \varepsilon, 1 \leq i \leq m\}$. For all $j = 2, 3, \ldots$ we let $L^j = \{i : (1 + \varepsilon)^{j-1} \leq w_i < (1 + \varepsilon)^j, 1 \leq i \leq m\}$ to be the set of large items whose weights are in the interval $\left[(1 + \varepsilon)^{j-1}, (1 + \varepsilon)^j\right)$. We let r be the maximum index of a non-empty set L^j. The set L^j is called the j-th class.

A feasible solution to MWSBP is called a *nice solution* if for every bin S in the solution the following condition holds: Denote by $c(S)$ the largest index of a class of large items with an item in S, and denote by $d(S)$ the smallest index of a class of large items with an item in S. These values are defined only if S contains at least one large item, and the condition on such bins is that $c(S) - d(S) \leq 4\lceil\log_{1+\varepsilon} \frac{1}{\varepsilon}\rceil - 1$. I.e., the large items from S arise from at most $4\lceil\log_{1+\varepsilon} \frac{1}{\varepsilon}\rceil$ consecutive weight classes. We can show the following lemma.

Lemma 5. *Let C be a feasible solution to MWSBP . There exists a feasible nice solution C' that can be produced from C in polynomial running time, such that $C' \leq (1 + 4\varepsilon)C$.*

Recall that OPT the cost of an optimal solution to problem MWSBP, and denote by OPT_N the cost of the best nice solution. The following corollary of Lemma 5 shows that as far as we are interested in a polynomial time approximation scheme, we can bound the cost of the approximated solution by the cost of the best nice solution.

Corollary 1. $\text{OPT}_N \leq (1 + 4\varepsilon)\text{OPT}$.

The next step of our scheme is to apply linear grouping [3] of each class of large items into $\frac{1}{\varepsilon^4}$ sets (some of them may be empty if $|L^j| < \frac{1}{\varepsilon^4}$). I.e., we apply linear grouping to each class separately. More precisely, for each value of $j = 1, 2, \ldots, r$ such that $|L^j| \geq \frac{1}{\varepsilon^4}$, we partition L^j into $\frac{1}{\varepsilon^4}$ subsets $L_1^j, \ldots, L_{1/\varepsilon^4}^j$ such that the following two conditions hold: first $|L_1^j| \geq |L_2^j| \geq \cdots \geq |L_{1/\varepsilon^4}^j| \geq |L_1^j| - 1$ (i.e., the subsets are approximately of the same cardinality $\varepsilon^4|L^j|$), and second if $i \in L_k^j$ and $i' \in L_\ell^j$ such that $k < \ell$, then $i > i'$, and therefore $s_i \geq s_{i'}$ (i.e., we assign to L_1^j the largest $|L_1^j|$ items from L^j, afterwards we assign to L_2^j the largest $|L_2^j|$ items from the remaining ones in L^j, and continue in this way). Note that the first condition gives the exact cardinality of each L_i^j, which is either $\lfloor\varepsilon^4|L^j|\rfloor$ or $\lceil\varepsilon^4|L^j|\rceil$, and by the second condition the partition is well-defined up to the assignment of equal size items. If $|L^j| < \frac{1}{\varepsilon^4}$, then we partition L^j into $\frac{1}{\varepsilon^4}$ sets such that $L^j_1 = \emptyset$, the next $|L^j|$ sets contain one item each, and the rest of the sets are empty. We have $|L^j_2| \geq \cdots \geq |L^j_{1/\varepsilon^4}| \geq |L^j_2| - 1$ and the second property introduced above (sets with smaller indices have items which are no smaller than items in sets with larger indices) is kept starting from L^j_2.

We next consider the instance with rounded-up sizes defined in the following way. For all $k \geq 2$ and $i \in L^j_k$, we let $s'_i = \max_{i' \in L^j_k} s_{i'}$ be the *rounded-up size* of item i, and for a heavy item or a small item i we let $s'_i = s_i$. The *rounded-up instance* has item set $I \setminus (\cup_{j=1}^r L^j_1)$, where the size of $i \in I \setminus (\cup_{j=1}^r L^j_1)$ is s'_i. Note that if $|L^j| \leq \frac{1}{\varepsilon^4}$, then $s'_i = s_i$ for all $i \in L^j$.

For $i \geq 2$ and all $j = 1, 2, \ldots, r$, we denote $s^{i,j} = s'_k$ where $k \in L^j_i$. Denote by OPT'_N the cost of the cheapest nice solution to the rounded-up instance, and recall that OPT_N is the cost of the cheapest nice solution to the original instance. Then, the following lemma shows that it suffices that the solution we construct (which we explain later how to create it) has a cost close enough to $(1 + O(\varepsilon))\text{OPT}'_N$. We will thus be able to concentrate on a subset of feasible solutions.

Lemma 6. $\text{OPT}'_N \leq (1 + \varepsilon)\text{OPT}_N$.

Proof. We construct a feasible nice solution to the rounded-up instance whose cost is at most $(1+\varepsilon)\text{OPT}_N$. Given the cheapest nice solution OPT_N to the original instance, that packs the items into the sets $I_1, I_2, \ldots, I_p$, we create a new nice solution to the rounded-up instance as follows. The heavy and small items as well as large items belonging to classes L^j such that $|L^j| \leq \frac{1}{\varepsilon^4}$ are not moved and are kept in their respective subsets. However, the positions of the other large items are different. If I_q has $n_{j,k}$ items from L^j_k (for $j = 1, 2, \ldots, r$ such that $|L^j| \geq \frac{1}{\varepsilon^4}$ and for $k = 1, 2, \ldots$), then the subset I_q receives $n_{j,k}$ items from the set L^j_{k+1} of the rounded-up instance. We apply this procedure for all $q = 1, 2, \ldots, p$, all $j = 1, 2, \ldots, r$ and all $k \leq \frac{1}{\varepsilon^4} - 1$. If at some time we run out of items of some set L^j_{k+1}, and do not have an item to assign, it means that $|L^j_k| = |L^j_{k+1}| + 1$. There is only one such value of k for each value of j and this can only happen for one item. In this case, the space remains unused.

In this way we place all the large items of the rounded-up instance in the bins $I_1, I_2, \ldots, I_p$ (we can do it because of the first property of the partition and since there exists at least one item in L^j_{k+1} whose original size is $s^{k+1,j}$). The resulting solution is clearly feasible because for all j, k and a pair of items $i \in L^j_k$ and $i' \in L^j_{k+1}$, we have $s_i \geq s'_{i'}$ by the second property of the partition. The resulting solution is clearly nice because the original solution is nice and we only replace item by another item if they belong to the same class. By our partition into classes we note that the difference in the weight of two items from the same class is at most a factor of $(1 + \varepsilon)$. Therefore, the cost of the solution for the rounded-up instance is at most $(1 + \varepsilon)$ times the cost of the solution of the original instance, and the claim follows. $\square$

A feasible solution to the rounded-up instance is called *nice and easy* if it is nice, and additionally, for each bin in the solution that contains at least one large item, the total weight of the large items in the bin is at least ε^2 times the total weight of the items in the bin.

Lemma 7. *Given a nice solution to the rounded-up instance, C, there is a nice and easy solution C' to the rounded-up instance, whose cost is at most $(1+3\varepsilon)C$. Moreover, C' can be constructed in polynomial time from C.*

We denote by $\mathrm{OPT}'_{\mathrm{NE}}$ the cost of the cheapest nice and easy solution to the rounded-up instance. The following corollary follows from Lemma 7.

Corollary 2. $\mathrm{OPT}'_{\mathrm{NE}} \leq (1+3\varepsilon)\mathrm{OPT}'_{\mathrm{N}}$.

We call a solution *reasonable*, if the bins of this solution are sorted according to non-increasing total weight, that is, if bin S appears before bin S' in the solution, then the total weight of items in S, is not smaller than the total weight of items in S'.

Claim. Let P be a property of bins. Let OPT_P be an optimal solution among all solutions, where all bins have property P. Then OPT_P is reasonable.

Proof. Given two bins S and S' of OPT_P, such that S appears before S'. Let W_S and W'_S the total weight of items in these bins. If $W_S < W'_S$, swapping the locations of bins I and I' would result in a solution of smaller cost. This new solution still satisfies P since it is a property of bins.   □

Therefore, we can assume that all optimal solutions we are dealing with are reasonable. Moreover, we can sort any solution in polynomial time without increasing its cost, so that it becomes reasonable.

We next show that nice and easy solutions that are reasonable fulfill the following convenient properties.

Lemma 8. *Given a nice, easy and reasonable solution to the rounded-up instance, C, consider two bins of the solution, S and S', such that S appears before S' in the solution, and each one of the two bins contain at least one large item. Then we have $c(S') - c(S) \leq 4\lceil \log_{1+\varepsilon} \frac{1}{\varepsilon} \rceil$.*

Proof. Let W_S and W'_S be the total weights of items in bins S and S' respectively. Since C is reasonable, $W_S \geq W'_S$. Let a_S be a large item of maximum weight in S. Since the size of a large item is at least $\frac{1}{\varepsilon^2}$, S contains at most $\frac{1}{\varepsilon^2}$ large items. The total weight of these items is at least $\varepsilon^2 W_S$ (since C is easy) and thus an item of maximum weight has weight of $w_{a_S} \geq \frac{\varepsilon^2 W_s}{1/\varepsilon^2} = \varepsilon^4 W_s$. Similarly, let a'_S be a large item of maximum weight in S'. Clearly $w_{a'_S} \leq W'_S$. Thus we have $(1+\varepsilon)^{c(S')-1} \leq W_{a'_S} \leq W'_S \leq W_S \leq \frac{w_{a_S}}{\varepsilon^4} < \frac{1}{\varepsilon^4}(1+\varepsilon)^{c(S)}$. Taking the logarithm with base $1+\varepsilon$ we get $c(S') - 1 < 4\lceil \log_{1+\varepsilon} \frac{1}{\varepsilon} \rceil + c(S)$ and thus the claim follows because both sides of the last inequality are integer numbers.   □

The following corollary holds for a wide class of solutions, including OPT'_{NE}.

Corollary 3. *Given a nice, easy and reasonable solution to the rounded-up instance $\mathcal{C}$, consider a bin of the solution, S. Then every item i of a class j such that $j \geq d(S) + 8\lceil\log_{1+\varepsilon}\frac{1}{\varepsilon}\rceil$ is packed in a bin which precedes S in the solution $\mathcal{C}$.*

Proof. Consider a bin S' which appears after S in $\mathcal{C}$. By Lemma 8, $c(S') \leq c(S) + 4\lceil\log_{1+\varepsilon}\frac{1}{\varepsilon}\rceil$. By the definition of a nice solution, $c(S) \leq d(S) + 4\lceil\log_{1+\varepsilon}\frac{1}{\varepsilon}\rceil - 1$. We thus get $c(S') \leq d(S) + 8\lceil\log_{1+\varepsilon}\frac{1}{\varepsilon}\rceil - 1$. Since $j \geq d(S) + 8\lceil\log_{1+\varepsilon}\frac{1}{\varepsilon}\rceil$, item i cannot be packed in S'. Since this holds for every bin S' that appears after S, we conclude that i is packed into a bin that precedes S in the solution. □

We next show how to construct an infeasible solution SOL for the rounded-up instance whose cost is at most OPT'_{NE}, we later show how to adapt it into a feasible solution, increasing the cost of the solution by a small enough multiplicative factor. We create a graph $G = (V, E)$ in the following way. For each $v \in V$ we associate a label $label(v) = (n_2^1(v), n_3^1(v), \ldots, n_{1/\varepsilon^4}^1(v), \ldots, n_2^r(v), n_3^r(v), \ldots, n_{1/\varepsilon^4}^r(v),$ $a_{m+1}(v), \ldots, a_{m+\ell}(v), p(v))$, where $n_i^j(v)$ denotes the number of items from L_i^j that are not packed yet (for $i \geq 2$ and $j = 1, 2, \ldots, r$), $a_k(v)$ is a binary value indicating whether the k-th heavy item has been already packed or not ($a_k(v) = 1$ means that this item is still unpacked) where $k = m + 1, \ldots, m + \ell$, and $p(v)$ ($m + \ell \leq p(v) \leq n$) denotes that the small items $p(v) + 1, \ldots, n$ are already packed and the small items left are exactly $m + \ell + 1, m + \ell + 2, \ldots, p(v)$. For each possible value of the label, we will have a vertex in V. Note that since the label of a vertex may have a linear number of components, and each of them can have a polynomial number of possibilities, it seems that our graph will have exponential number of vertices. Therefore additional conditions need to be introduced. We next show that using the nice and easy properties of the resulting solution we are interested in we can decrease substantially the number of vertices in the graph. If $label(v)$ has zeros in all components besides (perhaps) the last $\ell + 1$ components, then it is put in the graph. Otherwise, we insert only vertices with label $label(v)$ that satisfies the following condition: Denote by $b(v)$ the maximum index of a class j such that $n_2^j(v) + n_3^j(v) + \cdots + n_{1/\varepsilon^4}^j(v) > 0$, and let $b(v) = 0$ if no such j exists. Then for all $j' \leq b(v) - 8\lceil\log_{1+\varepsilon}\frac{1}{\varepsilon}\rceil$ and for all $k = 1, 2, \ldots 1/\varepsilon^4$, $n_k^j(v) = |L_k^j|$. We denote by V the resulting vertex set.

Lemma 9. *$|V|$ is polynomial.*

We next describe the edge set E. A *pattern for an outgoing edge* of a vertex v, denoted by $pat(v)$, is defined as follows. Such a pattern defines a packing of a bin. A number $pat_i^j(v)$ is associated with the number of items from each L_i^j (for $j = 1, 2, \ldots, r$ and $i \geq 2$) that are packed in the bin. Further, there is a number $pat_j(v)$ for $j = m + 1, m + 2, \ldots, m + \ell$ that equals one if the j-th heavy item is packed in the bin, and otherwise it equals zero. Finally, an interval $[q(v), p(v)]$ which is associated with a set of small items that are placed in this bin. Note that we require $m + \ell + 1 \leq q(v) \leq p(v) + 1$, and if $q(v) = p(v) + 1$, then no small items are packed into this bin. We let $pat(v) = (pat_2^1(v), \ldots, pat_{\frac{1}{\varepsilon^4}}^1(v), \ldots, pat_{\frac{1}{\varepsilon^4}}^r(v), pat_{m+1}(v), \ldots, pat_{m+\ell}(v), q(v))$.

If $q(v) > m + \ell + 1$, i.e. the first small item is not packed into the bin, then we have the following constraint. We require the sum of the rounded-up sizes of all items that are placed in the bin to be strictly larger than 1, however if we remove the small item $q(v)$ then the total rounded-up sizes of the items in the bin is at most 1. That is, $\sum\limits_{j=1}^{r} \sum\limits_{i=2}^{\frac{1}{\varepsilon^4}} pat_i^j(v) s^{i,j} + \sum\limits_{j=m+1}^{m+\ell} pat_j(v) \cdot s_j' + \sum\limits_{j=q(v)}^{p(v)} s_j' > 1$

and $\sum\limits_{j=1}^{r} \sum\limits_{i=2}^{\frac{1}{\varepsilon^4}} pat_i^j(v) s^{i,j} + \sum\limits_{j=m+1}^{m+\ell} pat_j(v) \cdot s_j' + \sum\limits_{j=q(v)+1}^{p(v)} s_j' \le 1$. Therefore, if we are

given the number of items from each L_i^j that are placed in a bin, the set of heavy items in this bin, and $p(v)$, then $q(v)$ is defined uniquely.

For a pattern $pat(v)$ such that for $2 \le i \le \frac{1}{\varepsilon^4}$ and for all $j = 1, 2, \ldots, r$ we have $n_i^j(v) \ge pat_i^j(v)$ and for all $j = m+1, \ldots, m+\ell$ we have $pat_j(v) \le a_j(v)$, there is an edge from v to the vertex u which has the label

$$\left(n_2^1(v) - pat_2^1(v), n_3^1(v) - pat_3^1(v), \ldots, n_{\frac{1}{\varepsilon^4}}^1(v) - pat_{\frac{1}{\varepsilon^4}}^1(v), \ldots, n_2^r(v) - pat_2^r(v), \ldots, \right.$$

$$\left. n_{\frac{1}{\varepsilon^4}}^r(v) - pat_{\frac{1}{\varepsilon^4}}^r(v), a_{m+1}(v) - pat_{m+1}(v), \ldots, a_{m+\ell}(v) - pat_{m+\ell}(v), q(v) - 1 \right)$$ whose

cost is $\sum\limits_{j=m+\ell+1}^{p(v)} w_j + \sum\limits_{j=m+1}^{m+\ell} a_j(v) w_j + \sum\limits_{j=1}^{r} \sum\limits_{i=2}^{\frac{1}{\varepsilon^4}} n_i^j(v) \cdot (1+\varepsilon)^j$ (i.e., the total weight of items that are still unpacked in v when we round up the weight of a large item to the nearest power of $1 + \varepsilon$).

If such a vertex u does not exist then we do not insert the edge. Thus, to build the graph in polynomial time, we check for every vertex u whether an edge (u, v) should be inserted (i.e., if it corresponds to a pattern), and if so, the pattern of this edge and its cost are computed.

We next describe a source vertex s and a destination t. The label of s is

$$label(s) = \left(|L_2^1|, \ldots, \left| L_{\frac{1}{\varepsilon^4}}^1 \right|, |L_2^2|, \ldots, \left| L_{\frac{1}{\varepsilon^4}}^2 \right|, \ldots, |L_2^r|, \ldots, \left| L_{\frac{1}{\varepsilon^4}}^r \right|, 1, 1, \ldots, 1, n \right)$$

and the label of t is $(0, \ldots, 0, m+\ell)$. In the resulting graph we find a minimum cost path $\mathcal{P}$ from s to t.

If the k-th edge of the path $\mathcal{P}$ corresponds to a pattern $pat(v)$, then the k-th bin of the solution SOL is created as follows. For $i \ge 2$ and $j = 1, 2, \ldots, r$, we pack to this bin $pat_i^j(v)$ items from L_i^j that were not packed before. Finally, we add to this bin all heavy items such that $pat_j(v) = 1$. We add to this bin the set of small items $q(v), q(v) + 1, \ldots, p(v)$. Note that SOL might be infeasible (as long as there are enough small items, the total size of items packed in each bin will be larger than 1).

Note that the cost of SOL is at most the cost of the path $\mathcal{P}$. This holds because of the equivalent form of the goal function of MWSBP, and since the cost of an edge used a rounding up of the weights of the large items to their maximal possible values.

Lemma 10. *The cost of* SOL *is at most* $(1 + \varepsilon) \cdot \mathrm{OPT}'_{\mathrm{NE}}$.

We next show how to transform SOL into a feasible solution SOL' to the rounded-up instance while paying at most another $\varepsilon\text{OPT}'_N$. Note that SOL is an infeasible solution as it uses bins of size larger than 1.

Denote by $S_1, S_2, \ldots$ the bins used by SOL (according to their order in SOL). We partition the bins into *blocks* where each block has $\frac{1}{\varepsilon^2}$ bins. The i-th block consists of the bins $S_{\frac{i-1}{\varepsilon^2}+1}, \ldots, S_{\frac{i}{\varepsilon^2}}$. The last block may have a smaller number of bins.

We transform SOL into SOL' by adding to each block one bin that is placed in this block, and in this bin we pack one small item from each original bin of the block that contains at least one small item. The position of this extra bin within the block is as follows: In the first block we place the extra bin in the $\frac{1}{\varepsilon}+1$-th position, and in the other blocks we add this additional bin as the first bin of the block. From each bin in the block, we move a small item of minimum index to the new additional bin. Then, the new bin is packed with at most $\frac{1}{\varepsilon^2}$ small items, and therefore the total size assigned to such a new bin is at most 1. Since we remove one small item, then by definition of $q(v)$, we get that the same holds for all bins. We conclude that the solution is feasible. We next bound its cost.

Lemma 11. *The cost of* SOL' *is at most* $(1 + 2\varepsilon)$SOL.

Proof. The total cost of SOL is exactly the total charging costs of all items according to SOL. We similarly partition the cost of the items according to SOL'. We first consider the increase of the cost charged to items that is caused by insertion of new bins. The worst case for each block occurs for bins that are located just after a new bin. In the first block, the bin in position $\frac{1}{\varepsilon}+1$ is shifted to position $\frac{1}{\varepsilon}+2$. In other blocks, the bin is position $\frac{j}{\varepsilon^2}+1$ is shifted to position $\frac{j}{\varepsilon^2}+j+2$ $(j \geq 1)$. The ratio between the positions of the bin in the first case is at most $\frac{1+2\varepsilon}{1+\varepsilon} \leq 1+\varepsilon$. In the second case, the worst case occurs for the second block, for which $j = 1$ and we get the ratio $\frac{1+3\varepsilon^2}{1+\varepsilon^2} \leq 1+2\varepsilon^2 < 1+\varepsilon$ for $\varepsilon \leq \frac{1}{4}$.

If an item i is *not* packed in the first block, then its charged cost in SOL' is at most $(1 + 2\varepsilon^2)$ times its charged cost in SOL. We saw that the increase in the location of bins is bounded by this factor, and the only items that are shifted are small items that are moved to locations with smaller indices. Therefore, it remains to bound the increase of the total charged cost of the items in the first block.

First, note that if j is either large or heavy item then its charged cost in SOL' is at most $(1 + \varepsilon)$ times its charged cost in SOL (the two charged costs are the same if j is packed in the first $\frac{1}{\varepsilon}$ bins and otherwise the two charged costs differ by at most a factor of $1 + \varepsilon$). Therefore, there are at most $\frac{1}{\varepsilon}$ small items that are packed in bins $1, \ldots, \frac{1}{\varepsilon}$ that are postponed until the $\frac{1}{\varepsilon} + 1$-th bin, each of them has weight of at most $\varepsilon^3 W$ (because these are small items, and thus not heavy), and therefore the total charging cost of these items increased by an additive factor of at most εW. The other items from the first $\frac{1}{\varepsilon}$ bins have the same charging cost in the two solutions. Since W is clearly a lower bound on the cost of any solution, and on SOL in particular, we conclude that the total cost of SOL$'$ is at most $(1+\varepsilon)$ times the total cost of SOL plus εSOL, i.e., at most $(1+2\varepsilon)$SOL as we claimed.   □

Even though we restricted our search of solutions, the resulting solution for the rounded-up input is not necessary nice and easy, nor can we claim that it is reasonable. We can apply transformations on SOL$'$ as we performed on OPT, changing it into a nice solution and afterwards to a nice and easy solution.

The proof of the following lemma allows us to assume that the resulting solution is nice, easy and reasonable.

Lemma 12. *There exists a nice and easy reasonable solution* SOL$''$ *that can be computed in polynomial time such that the cost of* SOL$''$ *is at most* $(1+10\varepsilon)$SOL$'$.

Proof. Using Lemmas 5 and 7, we see that a nice and easy solution of cost at most $(1+3\varepsilon)(1+4\varepsilon)SOL'$ can be constructed from SOL$'$ in polynomial time. We further sort the bins to obtain a reasonable solution, this does not change the first property, and may only reduce the cost of the solution. Using $\varepsilon \le \frac{1}{4}$ we get a cost of at most $(1+3\varepsilon)(1+4\varepsilon)SOL' \le (1+10\varepsilon)SOL'$.   □

Though we phrased SOL$''$ as a solution to the rounded-up instance, it is clear that it would be possible to give a feasible solution to the instance with item set $I \setminus \cup_{j=1}^{r} L_1^j$ and the size of an item i is s_i (because by decreasing the size of each item, any feasible solution remains feasible with the same cost). It remains to show what we do with the items of $L_1^1 \cup \cdots \cup L_1^r$. We apply this replacement later, so the solutions we consider at this point are still to the rounded-up problem. We insert the items of $L_1^1 \cup \cdots \cup L_1^r$ into the solution of the rounded-up problem, SOL$''$.

For each $j = 1, 2, \ldots, r$ we create another $|L_1^j|$ bins that are placed immediately after the last item of $L_2^j \cup L_3^j \cup \cdots \cup L_{\frac{1}{\varepsilon^4}}^j$ is packed (pushing some bins forward), each of this additional bins contain exactly one item from L_1^j. Note that insertion of bins for some values of j could push bins for a smaller value of j forward so that they would no longer be in the position just after the last items of the same class in the rounded-up instance. We denote by SOL$'''$ the resulting solution. It is clear that SOL$'''$ is a feasible solution to the rounded-up instance together with the $L_1^1 \cup \cdots \cup L_1^r$ items, that can be modified without additional cost into a feasible solution of the original instance.

Lemma 13. *The cost of* SOL$'''$ *is at most* $(1+2\varepsilon^2) \cdot (1+48\varepsilon^2(1+\varepsilon)\lceil \log_{1+\varepsilon} \frac{1}{\varepsilon} \rceil) < 1 + 150\varepsilon^2 \log_{1+\varepsilon} \frac{1}{\varepsilon}$ *times the cost of* SOL$''$.

We sequentially apply Corollary 1, Lemma 6, Corollary 2, and Lemmas 10, 11, 12 we have SOL$'' \le (1+128\varepsilon)$OPT for $\varepsilon \le \frac{1}{4}$. Since by Lemma 13 SOL$''' \le$. We let $\varepsilon = \frac{\delta^2}{1000000}$ for some $\delta \le 1$. We need to upper bound $\varepsilon^2 \log_{1+\varepsilon} \frac{1}{\varepsilon} = \varepsilon^2 \frac{\ln \frac{1}{\varepsilon}}{\ln(1+\varepsilon)}$. We have $\frac{\varepsilon}{\ln(1+\varepsilon)} < 2$ for $\varepsilon < 1$. We also have $\ln \frac{1}{\varepsilon} \le \frac{1}{\sqrt{\varepsilon}}$ for $\varepsilon \le \frac{1}{4}$, so the expression is upper bounded by $2\sqrt{\varepsilon} = \frac{\delta}{500}$. Thus we have SOL$''' \le (1+\frac{128\delta^2}{1000000})(1+\frac{150\delta}{500})$OPT $< (1+\delta)$OPT.

We get an approximation ratio of at most $1 + \delta$ for every $\delta \leq 1$ such that $\frac{1}{\delta}$ is an integer. The running time of our algorithm is polynomial in the number of items for any constant value of δ. Thus we have proved the following.

Theorem 3. *There is a polynomial time approximation scheme for problem* MWSBP.

References

1. Coffman, E.G., Garey, M.R., Johnson, D.S.: Approximation algorithms for bin packing: A survey. In: Hochbaum, D. (ed.) Approximation algorithms, PWS Publishing Company (1997)
2. Csirik, J., Woeginger, G.J.: On-line packing and covering problems. In: Fiat, A., Woeginger, G.J. (eds.) Online Algorithms: The State of the Art, pp. 147–177 (1998)
3. de la Vega, W.F., Lueker, G.S.: Bin packing can be solved within $1 + \varepsilon$ in linear time. Combinatorica 1, 349–355 (1981)
4. Epstein, L., Halldórsson, M.M., Levin, A., Shachnai, H.: Weighted sum coloring in batch scheduling of conflicting jobs. In: Díaz, J., Jansen, K., Rolim, J.D.P., Zwick, U. (eds.) APPROX 2006. LNCS, vol. 4110, pp. 116–127. Springer, Heidelberg (2006)
5. Epstein, L., Levin, A.: Min-sum bin packing (manuscript)
6. Feige, U., Lovász, L., Tetali, P.: Approximating min sum set cover. Algorithmica 40(4), 219–234 (2004)
7. Hochbaum, D.S., Shmoys, D.B.: Using dual approximation algorithms for scheduling problems: theoretical and practical results. Journal of the ACM 34(1), 144–162 (1987)
8. Johnson, D.S.: Approximation algorithms for combinatorial problems. Journal of Computer and System Sciences 9, 256–278 (1974)
9. Lovàsz, L.: On the ratio of optimal integral and fractional covers. Discrete Mathematics 13, 383–390 (1975)
10. Munagala, K., Babu, S., Motwani, R., Widom, J.: The pipelined set cover problem. In: Eiter, T., Libkin, L. (eds.) ICDT 2005. LNCS, vol. 3363, pp. 83–98. Springer, Heidelberg (2004)
11. Smith, W.E.: Various optimizers for single-stage production. Naval Research and Logistics Quarterly 3, 59–66 (1956)
12. Ullman, J.D.: The performance of a memory allocation algorithm. Technical Report 100, Princeton University, Princeton, NJ (1971)

Approximation Schemes for Packing
Splittable Items with Cardinality Constraints

Leah Epstein[1] and Rob van Stee[2,*]

[1] Department of Mathematics, University of Haifa, 31905 Haifa, Israel
lea@math.haifa.ac.il
[2] Department of Computer Science, University of Karlsruhe, D-76128 Karlsruhe,
Germany
vanstee@ira.uka.de

Abstract. We continue the study of bin packing with splittable items and cardinality constraints. In this problem, a set of items must be packed into as few bins as possible. Items may be split, but each bin may contain at most k (parts of) items, where k is some fixed constant. Complicating the problem further is the fact that items may be larger than 1, which is the size of a bin. We close this problem by providing a polynomial-time approximation scheme for it. We first present a scheme for the case $k = 2$ and then for the general case of constant k.

Additionally, we present *dual* approximation schemes for $k = 2$ and constant k. Thus we show that for any $\varepsilon > 0$, it is possible to pack the items into the optimal number of bins in polynomial time, if the algorithm may use bins of size $1 + \varepsilon$.

1 Introduction

In bin packing problems, a set of *items* is given and the goal is to pack them into the smallest possible number of containers, called *bins*. The items are typically given as numbers between 0 and 1, which is the bin size. In this paper we consider items that may be larger than 1. Items are allowed to be *split* and distributed among an arbitrary number of bins.

Clearly, if we allow items to be split and have no other constraints, a simple Next Fit-type algorithm can generate an optimal solution. However, we require that at most k (parts of) different items are packed together in a single bin. This is called a *cardinality constraint*, and it makes the problem NP-hard in the strong sense for any fixed $k \geq 2$ [3,6].

This problem was introduced by Chung et al. [3], who discussed the problem of allocating memory to parallel processors. The goal is that each processor has sufficient memory and not too much memory is being wasted. If processors have memory requirements that vary wildly over time, any memory allocation where a single memory can only be accessed by one processor will be inefficient. A solution to this problem is to allow memory sharing between processors. However,

---

* Research supported by the Alexander von Humboldt Foundation.

C. Kaklamanis and M. Skutella (Eds.): WAOA 2007, LNCS 4927, pp. 232–245, 2008.

if there is a single shared memory for all the processors, there will be a lot of contention which is also undesirable. It is currently infeasible to build a large, fast shared memory and in practice, such memories are time-multiplexed. For n processors, this increases the effective memory access time by a factor of n.

Chung et al. [3] suggested a new architecture where each memory may be accessed by at most *two* processors, avoiding the disadvantages of the two extreme earlier models. This leads to the bin packing problem described above, where in their paper $k = 2$: the bins are the memories and the items to be packed represent the memory requirements of the processors.

In this paper, we study approximation algorithms in terms of the *absolute approximation ratio* or the *absolute performance guarantee*. Let $\mathcal{B}(\mathcal{I})$ (or $\mathcal{B}$, if the input $\mathcal{I}$ is clear from the context), be the cost of algorithm $\mathcal{B}$ on the input $\mathcal{I}$. An algorithm $\mathcal{A}$ is an $\mathcal{R}$-approximation (with respect to the absolute approximation ratio) if for every input $\mathcal{I}$, $\mathcal{A}(\mathcal{I}) \leq \mathcal{R} \cdot \text{OPT}(\sigma)$, where OPT is an optimal algorithm for the problem. The absolute approximation ratio of an algorithm is the infimum value of $\mathcal{R}$ such that the algorithm is an $\mathcal{R}$-approximation.

The *asymptotic* approximation ratio for an algorithm $\mathcal{A}$ is defined to be $\mathcal{R}_{\mathcal{A}}^{\infty} = \limsup_{n \to \infty} \sup_{\mathcal{I}} \{ \frac{\mathcal{A}(\mathcal{I})}{\text{OPT}(\mathcal{I})} | \text{OPT}(\mathcal{I}) = n \}$. This ratio is relevant if we are particularly interested in the performance of algorithms on large inputs, that cannot be packed in few bins. Fernandez de la Vega and Lueker [4] designed an APTAS for standard bin packing. Their work was followed by the work of Karmarkar and Karp [10] who developed an AFPTAS.

Regarding the absolute approximation ratio, for the classical bin packing problem a simple reduction from the PARTITION problem (see problem SP12 in [7]) shows that no polynomial-time algorithm has an absolute performance guarantee better than $\frac{3}{2}$ unless P=NP. This reduction is no longer valid for our problem, where items may be split.

Chung et al. [3] showed that the problem with splittable items is NP-hard in the strong sense for $k = 2$. They use a reduction from the 3-PARTITION problem (see problem [SP15] in [7]). In a recent paper [6], we showed that this problem is NP-hard in the strong sense for any fixed value of k.

Chung et al. [3] also gave a 3/2-approximation for the case $k = 2$. Graham and Mao [8] analyzed the asymptotic approximation ratio of several algorithms, giving upper bounds of 1.498 for $k = 2$, 3/2 for $k = 3$ and $2 - 2/k$ for $k \geq 4$. In [6], we gave a simple algorithm with an absolute approximation ratio of $2 - 1/k$ for $k \geq 2$, and an algorithm with absolute approximation ratio of 7/5 for $k = 2$.

Bin packing with cardinality constraints (and regular, non-splittable items) was introduced and studied in an offline environment as early as in 1975 by Krause, Shen and Schwetman [12,13]. They showed that the performance guarantee of the well known First Fit algorithm is at most $2.7 - \frac{12}{5k}$. Additional results were offline approximation algorithms of performance guarantee 2. Kellerer and Pferschy [11] designed an improved offline approximation algorithm with performance guarantee 1.5 and finally a PTAS was designed in [2] (for a more general problem).

On the other hand, Babel et al. [1] designed a simple *online* algorithm with asymptotic approximation ratio 2 for any value of k. They also designed improved algorithms for $k = 2, 3$. Finally, Epstein [5] gave an optimal online bounded space algorithm (i.e., an algorithm which can have a constant number of active bins at every time) for this problem. Its asymptotic worst-case ratio is an increasing function of k and tends to $1 + h_\infty \approx 2.69103$, where h_∞ is the best possible performance guarantee of an online bounded space algorithm for regular bin packing (without cardinality constraints). Additionally, she improved the online upper bounds for $3 \le k \le 6$.

A related problem was studied recently by Shachnai, Tamir and Yehezkely [14]. They considered an offline bin packing problem where items may be split arbitrarily. They consider two models: one where splitting items comes at a cost, as each part of a split item increases by a constant additive factor, and one where there is an upper bound on the total number of splits. They showed that both these problems do not admit a PTAS unless P = NP. They designed approximation schemes for both problems. Their problem is different from our problem since in their case all items have size at most 1. In their case it is possible to exploit the existence of simple structures of optimal solutions, which are more complicated in our case.

Our results. Our first main result is a polynomial-time approximation scheme. Recall that for standard bin packing, this is impossible unless P = NP. We first present our scheme for the special case of $k = 2$ and then show how to extend it to the general case. The main difficulty here is that we have less structure in the packing, making it harder to search all potential packings.

We also present a dual PTAS for this problem, first for $k = 2$ and then for general k. That is, given bins of size $1 + \varepsilon$ for an arbitrary $\varepsilon > 0$, we give an algorithm to pack these items into at most N bins, where N is the number of bins (of size 1) in an optimal solution. The difficulty of designing such a dual PTAS lies in the packing of large items. Since they can be arbitrarily large, the number of items does not imply any upper bounds on the optimal cost, and no known rounding techniques apply in this case.

Note that a dual PTAS for standard bin packing is a component in the PTAS for scheduling on identical machines, which was given by Hochbaum and Shmoys [9].

Due to space constraints, we have omitted almost all proofs.

2   PTAS for $k = 2$

2.1   The Structure of the Optimal Packing

Before we begin our analysis, we make some observations regarding the packing of OPT. A packing can be represented by a graph where the items are nodes and edges correspond (one-to-one) to bins. If there is a bin which contains (parts of) two items, there is an edge between these items. A bin with only one item corresponds to a loop on that item. The paper [3] showed that for any given packing, it is possible to modify the packing such that there are no cycles in the associated graph. Thus the graph consists of a forest together with some loops.

We now consider items that are larger than $1/\varepsilon$. We modify the input as follows. Any item of size $x > 1/\varepsilon$ is replaced by $\lfloor \varepsilon x \rfloor$ items of size $1/\varepsilon$ and one additional item of size $x - \frac{1}{\varepsilon}\lfloor \varepsilon x \rfloor$ (if this last amount is nonzero). Denote the original input by I and the modified input by I'. We have the following lemma.

Lemma 1. $\text{OPT}(I') \leq (1 + \varepsilon)\text{OPT}(I)$.

This lemma implies that for an input with items larger than $1/\varepsilon$, we can begin by splitting these items into pieces of size $1/\varepsilon$. Then if we find a solution which approximates $\text{OPT}(I')$, this solution approximates $\text{OPT}(I)$ nearly as well.

The optimal packing for the modified input I' consists of a forest and some loops. The trees can be arbitrarily large, where the size of a tree is the number of its nodes. However, given an optimal solution with large trees (possibly with loops), we can split these trees into trees (with loops) of constant size. Denote by $\text{OPT}'(I')$ an optimal solution for the case where there is an additional constraint that all trees that are created in the packing must have size at most $1/\varepsilon^2$. We then have the following lemma.

Lemma 2. $\text{OPT}'(I') \leq (1 + 2\varepsilon)\text{OPT}(I')$.

2.2   Description of the PTAS

We look for solutions with trees of size at most $1/\varepsilon^2$. We start by using techniques introduced by Lueker and Fernandez de la Vega [4]. Let n be the number of items in I'. Assume first that $n \geq 1/\varepsilon^2$. The easier case $n < 1/\varepsilon^2$ is treated below. We sort the items in order of nonincreasing size and put the items into groups of $\lceil n\varepsilon^2 \rceil$ successive items (possibly less items for the last group). Say that this gives $p + 1$ groups.

We now modify I' as follows. We remove the first group (the one with the largest items). For each other group, we round the item sizes inside this group up to the size of the largest item in the group. This creates an input I'' which does not require more bins to be packed than I' (since we can map every item in I'' to an item of I' that is no smaller than it), and does not require less bins than I' without the first group (since we rounded up the item sizes).

We are going to consider all possible packings for I'', that is, all possible forests with trees of size at most $1/\varepsilon^2$.

From a packing of I'' to a packing of I'. Given a packing (represented by a forest), we are going to change it as follows: wherever an item of group i is needed, we are going to take an arbitrary unpacked item from this group. Since it is smaller than the rounded version, it definitely fits in the space that is allocated to it. This leaves the $\lceil n\varepsilon^2 \rceil$ largest items of I' (the items in group 1) unpacked and we pack them into separate bins (into chains where possible). The size of each such item (in I') is at most $1/\varepsilon$ so this requires at most $(n\varepsilon^2 + 1)/\varepsilon \leq 2n\varepsilon$ extra bins (using that $n \geq 1/\varepsilon^2$).

The packing that we will finally use is the one that uses the least amount of bins *and* that gives a feasible packing. The important thing to note is that we are going to try all possible forests that pack all the items, thus we also try the one that corresponds to the packing of $\mathrm{OPT}'(I')$. Our modified packing then requires only $2n\varepsilon$ more bins than $\mathrm{OPT}'(I')$.

From a tree to a packing of I''. An important ingredient of our algorithm is still missing. We need to actually allocate items to bins based on the tree representation.

To begin with, we may assume that the trees in $\mathrm{OPT}'(I')$ are *minimal* in the sense that any partition of the items in a tree into two sets requires more bins to pack the items than the original tree.

Second, note that in any tree, items can be packed starting from the leaves without wasting any space, so we do not have any empty space in bins apart from possibly one bin that contains part of the root. This also means that given a tree, the number of bins required to pack this tree follows immediately from the total size of the items packed.

Third, to pack the items into bins we do not need the loops explicitly: we can ignore them, and it will be clear from the size of an item whether or not we need to pack some bins that contain only a part of this item. Thus from now on we work with real trees.

We define a type of a tree to be a pair (j, E) where j is the number of vertices $(1 \leq j \leq 1/\varepsilon^2 + 1)$. We assume that these vertices are always numbered $1, \ldots, j$ and E is a subset of $j - 1$ edges. A *pattern* consists of two parts. The first one is a type of a tree (defined above), and the second is a vector of length j, where component i (for $1 \leq i \leq j$) is the group to which node i belongs (a number between 1 and $p + 1$). The number of patterns is constant, since there is a constant amount of trees with at most $1/\varepsilon^2 + 1$ nodes, and $p + 1 \leq 1/\varepsilon^2$ possible groups for each node. For a given pattern, items are packed starting from the leaves. During the process, we sometimes *assign* items to bins without immediately packing them into those bins. A pattern is *valid* if this process leads to all items being packed without violating any cardinality constraints and the representation of the final packing is the original tree.

Packing a leaf is done as follows: first fill up the bins it is assigned to, if any. Then open as many new bins as you need to pack this item. If the final bin is not filled completely, assign the item at the other side of the edge leading to this leaf to that bin. (If this violates the cardinality constraint of k, or if the final bin is not used at all, the pattern is not valid.) Remove the leaf and the edge that leads to it (possibly creating a new leaf).

Note that in this process, items may be assigned to multiple bins before finally being packed into them. When a leaf completely fits into bins it was already assigned to, and it is not the last node packed in the tree, we have in effect found a smaller tree and we know that this pattern is not valid since the tree was not minimal. This proves the following lemma.

Lemma 3. *Given a tree representation of the optimal packing with minimal trees, it is possible to assign the items to bins such that for each tree, there is at most one bin which is not completely full.*

Enumerating the forests. There is a constant number of patterns, so certainly a constant number of valid patterns. Each valid pattern can be picked at most n times, thus we have at most $(n+1)^{O(1)}$ assignments to check. Once we know how many instances of each pattern you have, we can check whether this gives a valid packing (the right number of items of each group), and finally select the one that uses the least number of bins. This can be done in polynomial time.

Our PTAS (for the case $n \geq 1/\varepsilon^2$) is summarized in Figure 1.

---

1. Any item of size $x > 1/\varepsilon$ is replaced by $\lfloor \varepsilon x \rfloor$ items of size $1/\varepsilon$ and one additional item of size $x - \frac{1}{\varepsilon}\lfloor \varepsilon x \rfloor$ (if this last amount is nonzero).
2. Put the items into a constant number of groups and round them as in Lueker and Fernandez de la Vega [4]. All items in a group have the same size. The largest group is packed separately in new bins.
3. Determine the set of valid patterns (trees plus specification of groups of nodes).
4. Determine the forest (combination of patterns) which uses the least number of bins and which packs all the items.
5. Replace the rounded items by the original items.

Fig. 1. The PTAS for $k = 2$ and $n \geq 1/\varepsilon^2$

A trivial lower bound on the amount of bins needed to pack the entire input is $n/2$. Putting it all together, we find that our algorithm uses the following number of bins:

$$\text{OPT}'(I') + 2n\varepsilon \leq (1 + 4\varepsilon)\text{OPT}'(I') \leq (1 + 10\varepsilon)\text{OPT}(I') \leq (1 + 16\varepsilon)\text{OPT}(I),$$

where we have applied $\text{OPT}'(I') \geq n/2$, Lemma 2, and Lemma 1 in this order, as well as $\varepsilon \leq 1/2$.

Finally we consider the case where $n < 1/\varepsilon^2$. For a constant number of items, there exists only a constant number of forests including the allocation of items to nodes. A given forest can be filled up as described above (starting from the leaves), using the exact sizes of items. Hence in this case we do not split large items or large trees and find the optimal solution in polynomial time.

Altogether, this proves the following theorem.

Theorem 1. *There exists a polynomial-time approximation scheme for cardinality constrained bin packing of splittable items where each bin is allowed to have at most two items or parts of items.*

3   PTAS for Constant k

Our PTAS for constant k will be essentially the same as the one in Figure 1 for $k = 2$. However, we need to implement Step 3 for this more general case. For this we use a modified graph representation. If a bin contains parts $x_1, \ldots, x_k$, we order them in some way and create edges only between successive items in this ordering. Thus the edges in this case are e.g. $(x_1, x_2), \ldots, (x_{k-1}, x_k)$. Each item is still represented by a single node, thus one node might be involved in several of such chains, where each chain represents one bin.

So in this case, we no longer have a one-to-one correspondence of edges and bins. Instead, there are now at least as many edges as there are bins. Note that the order of the parts inside a bin is irrelevant. Thus we can reorder the parts in a chain arbitrarily. This gives a different graph for the same packing into bins. It is now more difficult to construct a packing into bins from a given graph. We have the following two important lemmas.

Lemma 4. *If the graph of a packing contains a cycle, it is possible to modify the packing such that this cycle is removed without increasing the number of bins.*

Lemma 5. *There exists an optimal packing such that each item of size at most i/k is split into at most i parts, for all $i > 1$.*

From a tree to a packing of rounded items. Lemmas 1 and 2 also hold for general k. However it is now not so clear how to construct the bins from a given forest.

To do this, we further modify our graph representation, and let each item be represented by x nodes if and only if it is split into x parts in the packing. The parts of one item are connected by a simple chain, as are the parts that are in one bin. We can then start packing bins from the leaves of the tree and repeatedly remove leaves similar to before. There are now two cases, since the edge that connects the leaf to the tree leads either to a copy of the leaf (same item) or to a different item.

If a leaf leads to another item, then the remaining unpacked part of the leaf item must be small enough that it can be packed entirely inside the bin it is assigned to, so do that. (If not, there would be an edge leading to a copy of the leaf.) If it does not fit, we know that the tree is not valid. Also, assign the item at the other end of this edge to this bin. If there are already k items in the bin, the tree is not valid.

If a leaf leads to another part of the same item, then the bin this leaf is assigned to (or a new bin, if it is not assigned to anything) can be filled up by this leaf. This holds because no future unpacked items can be assigned to this bin (otherwise there would be an edge to that item from this leaf).

Using this packing process, it can be seen that Lemma 3 also holds for this case. In order to apply this process, we do not only need to know the group to which each node belongs but also *which* of the items of that size is packed there. Again, let the type of a tree be a pair (j, E) where j is the number of *nodes* in the tree (as mentioned above, there is one node in the tree for every part of an item) and E is a set of $j - 1$ edges.

We now need a vector (a, b) for each node (to get a pattern for the tree). Thus, a is the group ($a \in \{1, \ldots, p+1\}$, where $p+1$ is the number of groups like in Section 2.2) and b is the number of the item of this group ($b \in \{1, \ldots, 1/\varepsilon^2\}$, as there are at most $1/\varepsilon^2$ items in a tree, there are certainly at most $1/\varepsilon^2$ items of any one of the groups). In a valid tree, the nodes of type (a, b) for any fixed a

and b must be in a chain, since they represent parts of one item. The maximum length of such a chain is bounded by the following Lemma.

Lemma 6. *The length of a chain representing one item in a valid pattern is bounded by $1/\varepsilon^2 + 1/\varepsilon$.*

Proof. There can be at most $1/\varepsilon^2$ nodes in the chain that have an edge to another item, because otherwise there would be two nodes having edges to the same item, giving a cycle.

There can be at most $1/\varepsilon$ nodes in the chain that do not have an edge to another item, since each such node has a bin to itself and such a bin (apart from at most one) will be fully packed in an optimal solution by Lemma 3. The size of an item is at most $1/\varepsilon$. □

We can now determine in polynomial time how often each pattern is used in an optimal packing, as in the previous PTAS. We now have that a trivial lower bound on the cost of packing n items is n/k. Thus our PTAS only works for constant k, and requires at most $(1 + (2k + 4)\varepsilon)\mathrm{OPT}(I)$ bins.

Again, for a constant number of items, only a constant amount of forests needs to be checked, and we can find an optimal solution. Thus we have the following theorem.

Theorem 2. *For any constant $k \geq 2$, there exists a polynomial-time approximation scheme for cardinality constrained bin packing of splittable items where each bin is allowed to have at most k items or parts of items.*

4   A Dual PTAS for $k = 2$

We have already seen that the optimal packing can be represented by a forest together with some loops. Moreover, in each tree, the only items that have degree more than two have size more than 1. Items of size in $(\frac{1}{2}, 1]$ have at most two neighbors. We call such items *medium*. Items of size in $(0, \frac{1}{2}]$ have at most one neighbor. We call such items *small*.

Our algorithm tries to find a good way to cut items, i.e., split them into parts. The cuts are performed in two stages. As a first step we cut a single piece off medium and large items. Our algorithm performs an enumeration on such possible cuts. Clearly, these are not the only cuts that an optimal algorithm may perform on these items for its packing. However, by Lemmas 7 and 8, proved in [6], no further cuts are required for items of size at most 1.

Lemma 7. *There exists an optimal packing in which all items of size at most $1/2$ are leaves.*

Lemma 8. *There exists an optimal packing in which any item of size in $((i-1)/2, i/2]$ has at most i neighbors for all $i \geq 2$.*

When we perform cuts on items, our algorithm considers the two resulting parts to be two independent items and thus allows to cut them further (for parts that have size more than 1) while creating a packing. The enumeration considers a set of cut options which cover sufficiently many packings to find a very good one. The options include the "empty cut", i.e. the case that this item is not cut at all.

We do this initial cutting in order to simplify the tree structure. We would like to work with trees that contain at most one large item, and each tree is a star rooted at a large item or a part of a large item. We now show that by cutting off a piece of size at most 1 from each item that is medium or large, and treating this piece as an independent item, we get a packing which has this property without increasing the number of bins required to pack the input. Note that these techniques are useful only for the dual PTAS and not for the PTAS since the modification of the input is done by cutting some items. We later use the fact that we can slightly increase the sizes of bins in order to efficiently enumerate the possible cutting points.

Lemma 9. *It is possible to modify the input in such a way that the optimal packing for the new input requires the same number of bins as the old input, and there exists an optimal packing for the new input such that all medium items have degree 1.*

Lemma 10. *It is possible to modify the input in such a way that the optimal packing for the new input requires the same number of bins as the old input, and there exists an optimal packing for the new input such that each tree contains at most one large item.*

We conclude that by modifying the input appropriately, there exists an optimal packing which consists of stars with large items in the middle (where such a large item that is a root of a star might be smaller by at most 1 than the corresponding large item in the original input), single edges, and loops. We will look for a packing that has this structure. Denote the number of input items by n.

4.1   Description of the Algorithm

Our dual PTAS works as follows. We use a parameter δ which is based on ε, and which is the inverse of some odd integer. Specifically, we let $K = \min\{i \mid i \geq 2/\varepsilon, 2 \nmid i\}$ and $\delta = 1/K$. We begin by rounding item sizes (of all items that are not large) up to the nearest multiple of δ (possibly to 0). There are $K + 1$ possible sizes of such items. For a given tree, we can fill the bins starting with these items. This means that each cut of an item will now occur at an integer multiple of δ. This also holds for a tree that contains no small items (items of size at most $1/2$) but does contain medium items. By the above, if a tree contains no items of size at most 1, it consists of only a loop (a single item).

Denote the number of items of size $i\delta$ by M_i for $i = (K + 1)/2, \ldots, K$. For each size, we guess how many items of this size are cut at each integer multiple of δ that is at most $1/2$. Note that we do not need to consider cuts above $1/2$, since cutting an item of size $i\delta$ at the point $j\delta$ or at the point $(i - j)\delta$ gives the same parts. Thus the possible cutting points are $i\delta$ for $i = 0, \ldots, (K - 1)/2$.

1. Let $K = \min\{i | i \geq 2/\varepsilon, 2 \nmid i\}$ and $\delta = 1/K$.
2. Round each item size which is no larger than 1 up to the nearest multiple of δ. Let the number of items of size $i\delta$ be M_i for $i = (K+1)/2, \ldots, K$.
3. For each medium item size, *guess* how many items of this size are cut at $j\delta$ for $j = 0, \ldots, (K-1)/2$.
4. *Guess* how many items of size $j\delta$ are cut off from large items for $j = 0, \ldots, K$.
5. Create a graph with L layers, plus source and sink. The construction of the graph is shown in Figure 3. This graph represents all possible packings for the current set of guesses. Find a path with minimal cost from the source to the sink. This is the cost of packing the input with these guesses.
6. Use the packing of this guess to create a packing for the original instance.

Fig. 2. The dual PTAS for $k = 2$

Our dual PTAS is summarized in Figure 2. Each guessing step can be emulated via an exhaustive enumeration of all the possibilities for this piece of information. So our algorithm runs all the possibilities, and among them chooses the best solution achieved. Denote the number of large items by L. For convenience of notation, we will also denote this number by $M_{(K-1)/2}$. We guess how many pieces of each size of at most 1 that is an integer multiple of δ are cut off. Note that a large item may stop being large when some part of it is cut off. However, in our algorithm, we still group it among the large items (and in particular, allow it to be cut further). The cuts can be represented by a vector of size $(K + 1)^2/4 + (K + 1)$, which tells us how many items of each size $(K+1)\delta/2, \ldots, K\delta$ are cut off at each point, and how many pieces of each size are cut off from the large items.

Construction of the graph. For every possible set of cuts, we do the following. We construct a layered graph which represents possible packings. The graph starts at a single source node, then there are L layers which correspond to the L large items, and finally there is a sink. We maintain a *summary vector* which describes how many **unpacked** (parts of) items there are of every size $i\delta$ ($i = 0, \ldots, K$). This vector is denoted by $s(u)$ for a node u. Additionally, we maintain a *cutoff vector* which contains unpacked parts of size less than 1 **of large items**. This vector is denoted by $c(u)$ for a node u. We concatenate both vectors into a single *packing vector* of length $2(K+1)$ which contains all relevant information needed to find the optimal packing for these parts. Note that the parts which were cut off large items are listed twice, once in the main list of unpacked items, so that they can be packed, and once in the list of parts of large items, to make sure that the pieces that were cut off are matched to the large items.

For two nonnegative integer vectors a and b of length ℓ, we say that $a \geq b$ if $a_i \geq b_i$ for $i = 1, \ldots, \ell$. We say that $a \to b$ if there exists a unique j such that $a_j = b_j + 1$ and $a_i = b_i$ for $i \in \{1, \ldots, \ell\} \setminus \{j\}$. We describe the construction of the layered graph in Figure 3.

The cost of an edge (u, v) that is mentioned in Step 4 of Figure 3 can be computed as follows. This step creates a star rooted at a given large item (the

1. Layer 0 and layer $L + 1$ contain a single node. The node in layer 0 is labeled with the packing vector, while the node in layer L is labeled with the all-zero vector.
2. Sort the large items in some way. Each large item is associated with a layer between 1 and L. Each of these layers contains one node for *every* (nonnegative, integer) vector that is smaller than the original packing vector.
3. For a node u, denote the cutoff vector by $c(u)$ and the summary vector by $s(u)$. For any node u in layer i ($i = 0, \ldots, L - 1$), there is an arc to every node v in layer $i + 1$ such that $c(u) \to c(v)$ and $s(u) \geq s(v)$.
4. The cost of arc (u, v), where v is in layer i ($i = 1, \ldots, L$), is the cost of packing the ith large item excluding a piece of size specified by the nonzero entry in $c(u) - c(v)$ (this size may be 0), together with the items specified by $s(u) - s(v)$.
5. For every node u in layer L, there is an arc to the single node in layer $L + 1$. The cost of this arc is the cost of packing all items in $s(u)$.

Fig. 3. Construction of the layered graph for one set of guesses (cuts)

i-th item in the list of large items is associated with layer i). The size of the large item that needs to be packed, is given by its original size minus the size of the part of item which corresponds to the nonzero entry of $c(u) - c(v)$. This item is to be packed with items specified by $s(u) - s(v)$. The only item that we cut further at this point is the large item associated with the current layer. Moreover, that is the only item that may be combined with other items. Thus, if we denote the sizes of items specified by $s(u) - s(v)$ by $a_1, \ldots, a_p$ and the size of the part of the large item that needs to be packed by X, then the number of bins is $\max\{p, \lceil X + \sum_{i=1}^{p} a_i \rceil\}$.

The cost of an edge (u, v) that is mentioned in Step 5 of Figure 3 can be computed as follows. The items to pack here are specified by $s(u)$. These items are not split further, they are packed in bins containing one or two of these items. We apply the First-Fit-Decreasing algorithm with the restriction that no bin can contain more than two items. By Lemma 11, this gives an optimal packing.

Lemma 11. *FFD is an optimal algorithm for cardinality constrained bin packing for $k = 2$.*

Proof. We modify the input as follows. For an item $x > 0$ let $x' = (x + 1)/3$. Then $1/3 < x' \leq 2/3$. Three modified items clearly do not fit together, and for two items $x' + y' \leq 1 \iff x + y \leq 1$.

Thus the number of bins required to pack the modified input is the same as for the original input. We now have an input where all items are larger than $1/3$. It is known [15] that for such an input, FFD gives an optimal solution.   □

Packing the original input. Once we have found the set of cuts that allows the best packing, it is easy to find the packing for the original input items. Say large item 1 (in our ordering) is packed into bins together with parts of size $k_1\delta, k_2\delta, \ldots, k_{a_1}\delta$. Using the original vector that represents the set of cuts, we find the first i such that there exists an item of size $i\delta < 1$ which is cut at $k_1\delta$, or at $(i-k_1)\delta$, and the part of size $k_1\delta$ that is created by this cut is so far unpacked. We then mark this part as packed and continue. (For each item size less than 1, we keep track of how many first and second parts are packed of each size.)

The correct part of this item of size less than 1 is put in bin 1. Bin 1 is filled with some part of large item 1 (namely, $1 + 2\delta - k_1\delta$). Then we find an unpacked part for bin 2 in the same manner, etc. At the end we have some part of the large item left, exactly how large this is is determined by what piece was cut off from the first large item. If this part has a positive size, it is packed in consecutive bins, and we move to the next large item. Finally, we find parts that are paired up in the same manner. Each bin contains only two parts, and we rounded up to the nearest multiple of δ, so we can use bins of size $1 + 2\delta$ to pack the unrounded parts.

Lemma 12. *The running time of this algorithm is* $n^{O(1/\varepsilon^2)}$.

Lemma 13. *This algorithm uses at most* OPT(L) *bins of size* $1 + 2\delta$ *to pack the input* L.

Proof. The optimal solution of the original instance (in bins of size 1) can be adapted to pack the rounded items (to the nearest multiple of δ) in the same number of bins of size $1 + 2\delta$, using only cuts at multiples of δ. Denote this packing by P. The PTAS tries all possible packings of this form for the rounded items and thus tries the packing P at some point. Therefore, it manages to pack the original items in bins of size $1 + 2\delta$, needing at most the optimal number of bins for these items. $\square$

Theorem 3. *For any* $\varepsilon > 0$, *there exists a polynomial-time algorithm for cardinality constrained bin packing of splittable items where each bin is allowed to have at most two items or parts of items. This algorithm gives packs the items in the optimal number of bins, but uses bins of size* $1 + \varepsilon$.

5 A Dual PTAS for Constant k

We give an algorithm for packing the input items into the optimal number of bins, but where the bins have size $1 + \varepsilon$. In fact we will pack the items in bin of size $1 + k\delta$, where δ depends on ε and k. Therefore, we only have a dual PTAS for the case where k is constant. We choose ε, so that δ is the inverse of some odd integer. Let $M = 1/\delta + k$. All items of size more than $1 + k\delta = M\delta$ are called large.

We will again use the fact that there is an optimal packing which is a forest (Lemma 4). We modify the input in two steps.

Sizes of items and parts. A first step would be a revision of sizes of items and parts of items. We take an optimal packing, and replace any part of size x with a part of size $\lceil \frac{x}{\delta} \rceil \delta$. As a result, the total size of parts in a bin can increase by an additive factor of at most $k\delta$. Therefore from this time on, we use bins of size $1 + k\delta$. One problem is that the sizes of items may have increased in an unbounded way. We would like to change the size of an item of size y to exactly $\lceil \frac{y}{\delta} \rceil \delta$. Therefore, we repeatedly pick a packed piece of an item whose size increased too much, and decrease its size by δ. This is done until all items are back to the desired size.

All parts in the packing now have sizes that are multiples of δ. Note that it may happen that the number of bins used decreases, if there are bins where all the pieces in it have their size reduced to 0.

Large items. As in the previous Section (Lemma 10), we would like to pack the large items one by one and not combine them together into bins. Note that we showed in Section 4 that Lemma 3 still holds in this case. It is straightforward to adapt the proof of Lemma 10 for the case where the bins have size $1 + k\delta$ and large items have size more than $1 + k\delta$ (instead of 1). Thus we find that for each large item, we can cut off one part of size at most $1 + k\delta$, and moreover this part is not cut further later.

Non-large items. We now consider the non-large items (size at most $1 + k\delta$). We need to allow these items (except non-large parts cut off from large items) to be cut at every integer multiple of δ. This is sufficient since in the optimal packing all parts have sizes that are integer multiples of δ. Moreover, by Lemma 5, it is sufficient to let non-large items be cut at most $k' = k(1 + k\delta) < 2k$ times. Therefore our scheme need only check such packings.

Description of the dual PTAS. We begin by rounding up all items into integer multiples of δ. To convert our packing into a packing of the original instance, for each item of original size y we need to decrease the size of at most one of its parts by $\lceil \frac{y}{\delta} \rceil \delta - y$ (this amount may be zero). From now we only discuss the rounded items. Note that these items are the same as used in the adapted optimal packing described above.

After rounding, the non-large items in the input can be represented by a vector indicating how many items exist of each non-large size, out of M possible non-large sizes. For each size, the number of parts cut off from those items of a particular smaller size can also be represented by a vector. We need to try all possibilities for these cutoff vectors. For each possibility, we will enumerate all possible packings of the items of size at most $1 + k\delta$ into bins of size $1 + k\delta$ such that no bin is empty. Here we use the fact that there is only a constant number of different packings of one bin (patterns), and a packing can be specified by giving how often each pattern is used.

For each such packing, we will construct a layered graph similar to the one in the previous section, with one layer for each large item. Each node now represents a subset of the bins of the current packing. The cost of an edge between two nodes is determined by the difference packing vector and the size of the large item of the current layer.

6   Conclusions

In this paper, we provided approximation schemes for bin packing of splittable items with cardinality constraints for all values of k. We also provided dual approximation schemes. It should be noted that our upper bounds are absolute, i.e. there is no additive term.

References

1. Babel, L., Chen, B., Kellerer, H., Kotov, V.: Algorithms for on-line bin-packing problems with cardinality constraints. Discrete Applied Mathematics 143(1-3), 238–251 (2004)
2. Caprara, A., Kellerer, H., Pferschy, U.: Approximation schemes for ordered vector packing problems. Naval Research Logistics 92, 58–69 (2003)
3. Chung, F., Graham, R., Mao, J., Varghese, G.: Parallelism versus memory allocation in pipelined router forwarding engines. Theory of Computing Systems 39(6), 829–849 (2006)
4. de la Vega, W.F., Lueker, G.S.: Bin packing can be solved within 1+epsilon in linear time. Combinatorica 1(4), 349–355 (1981)
5. Epstein, L.: Online bin packing with cardinality constraints. SIAM Journal on Discrete Mathematics 20(4), 1015–1030 (2006)
6. Epstein, L., van Stee, R.: Improved results for a memory allocation problem. In: WADS 2007. Workshop on Algorithms and Data Structures, pp. 362–373 (2007)
7. Garey, M.R., Johnson, D.S.: Computers and Intractability: A Guide to the theory of NP-Completeness. W. H. Freeman and Company, New York (1979)
8. Graham, R.L., Mao, J.: Parallel resource allocation of splittable items with cardinality constraints (manuscript)
9. Hochbaum, D.S., Shmoys, D.B.: Using dual approximation algorithms for scheduling problems: Theoretical and practical results. Journal of the ACM 34(1), 144–162 (1987)
10. Karmarkar, N., Karp, R.M.: An efficient approximation scheme for the one-dimensional bin-packing problem. In: Proceedings of the 23rd Annual Symposium on Foundations of Computer Science, pp. 312–320 (1982)
11. Kellerer, H., Pferschy, U.: Cardinality constrained bin-packing problems. Annals of Operations Research 92, 335–348 (1999)
12. Krause, K.L., Shen, V.Y., Schwetman, H.D.: Analysis of several task-scheduling algorithms for a model of multiprogramming computer systems. Journal of the ACM 22(4), 522–550 (1975)
13. Krause, K.L., Shen, V.Y., Schwetman, H.D.: Errata: Analysis of several task-scheduling algorithms for a model of multiprogramming computer systems. Journal of the ACM 24(3), 527–527 (1977)
14. Shachnai, H., Tamir, T., Yehezkely, O.: Approximation schemes for packing with item fragmentation. In: Erlebach, T., Persinao, G. (eds.) WAOA 2005. LNCS, vol. 3879, pp. 334–347. Springer, Heidelberg (2006)
15. Simchi-Levi, D.: New worst-case results for the bin-packing problem. Naval Research Logistics 41(4), 579–585 (1994)

A Randomized Algorithm for Two Servers in Cross Polytope Spaces

Wolfgang Bein[1,*], Kazuo Iwama[2], Jun Kawahara[2], Lawrence L. Larmore[1], and James A. Oravec[1]

[1] Center for the Advanced Study of Algorithms, School of Computer Science,
University of Nevada Las Vegas, Nevada 89154, USA
{bein,larmore,oravec}@cs.unlv.edu
[2] School of Informatics, Kyoto University,
Kyoto 606-8501, Japan
{iwama,jkawahara}@kuis.kyoto-u.ac.jp

Abstract. It has been a long-standing open problem to determine the exact randomized competitiveness of the 2-server problem, that is, the minimum competitiveness of any randomized online algorithm for the 2-server problem. For deterministic algorithms the best competitive ratio that can be obtained is 2 and no randomized algorithm is known that improves this ratio for general spaces. For the line, Bartal *et al.* [2] give a $\frac{155}{78}$ competitive algorithm, but their algorithm is specific to the geometry of the line.

We consider here the 2-server problem over Cross Polytope Spaces $M_{2,4}$. We obtain an algorithm with competitive ratio of $\frac{19}{12}$, and show that this ratio is best possible. This algorithm gives the second non-trivial example of metric spaces with better than 2 competitive ratio.

The algorithm uses a design technique called the knowledge state technique – a method not specific to $M_{2,4}$.

1 Background

In the k-server problem, there are k mobile identical servers in a metric space $\mathcal{M}$. At any time, a point $r \in M$ can be "requested," and must be "served" by moving one of the k servers to the point r. The cost of that service is defined to be the distance the server is moved; for a sequence of requests the goal is to serve the requests at small cost. An *online algorithm* for the server problem decides, at each request, which server to move, but does not know the sequence of future requests. We analyze an online algorithm for the server problem in terms of its *competitive ratio*, which essentially gives the ratio of its cost over the cost of an optimal (offline) algorithm which has knowledge of the entire request sequence before making any decisions. More precisely, we say that an online algorithm $\mathcal{A}$ for the server problem is C-*competitive*, if there is a constant K, such that, given

---

* Research of the first author (Bein) done while visiting Kyoto University as Kyoto University Visiting Professor. Research of the first author (Bein) and the fourth author (Larmore) supported by NSF grant CCR-0312093.

C. Kaklamanis and M. Skutella (Eds.): WAOA 2007, LNCS 4927, pp. 246–259, 2008.

any request sequence ϱ, $cost_A(\varrho) \leq C \cdot cost_{OPT}(\varrho) + K$. For a randomized online algorithm, we state competitiveness in terms of expected cost. The competitive ratio of A is the smallest C for which A is C-competitive.

The server problem was first proposed by Manasse, McGeoch and Sleator [14] and the problem has been studied widely since then. They also introduced the now well-known k-server conjecture, which states that, for each k, there exists an online algorithm for k servers which is k-competitive in any metric space. The conjecture was immediately proved true for $k = 2$, but for larger k remains open except in special cases, including lines [8], trees [9], and spaces with at most $k + 2$ points [11]. Even some simple-looking special cases have not been settled, for example the 3-server problem in the circle and in the Euclidean plane [8,9,12]. In general, the best currently known upper bound is $2k - 1$, given by Koutsoupias and Papadimitriou [12]. Thus there is a rich literature for *deterministic* online algorithms for this problem.

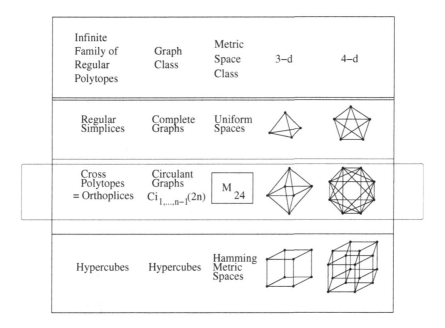

Fig. 1. The Class $M_{2,4}$

Randomization is a powerful for many online problems [7]. Yet, very little is known for randomized algorithms for the k-server problem. It seems to be quite hard to determine the exact randomized competitiveness of the k-server problem, that is, the minimum competitiveness of any randomized online algorithm for the server problem. Even in the case $k = 2$ it is not known whether its competitiveness is lower than 2, the known value of the deterministic competitiveness. This is surprising and it is quite intuitive that a "better than 2-competitive" algorithm should exist. In fact, $1 + e^{-\frac{1}{2}} \approx 1.6065$, is the greatest lower bound with a published proof (see [10]) on the competitiveness of any randomized on-line algorithm.[1] There has been some progress for special cases. The randomized competitiveness is known to be $\frac{3}{2}$ for all uniform spaces and is also known for three-point spaces [13]. For the special case of the line, Bartal *et al.* [2] have given a randomized algorithm with a competitive ratio of $\frac{155}{78} \approx 1.987$.

Our Contribution. In this paper we give a randomized online algorithm for the 2-server problem in Cross Polytope Spaces with optimal competitive ratio of $\frac{19}{12}$. Cross Polytope Spaces, denoted by $\mathcal{M}_{24}$, have been studied extensively as early as the 19th century, see Schläfli [15], as well as Figure 1. They consist of all metric spaces such that

- all distances are 1 or 2,
- $d(x, y) + d(y, z) + d(z, x) \leq 4$.

By an abuse of terminology we will sometimes simply say "the metric space $\mathcal{M}_{2,4}$" to denote this class of metric spaces.

In terms of the server problem, $\mathcal{M}_{24}$ generalizes uniform spaces and thus paging. It is also useful to gain insight into the 2-server problem over more general spaces. Our technique can, in principle, be used to design algorithms for spaces $M_{\ell,k}$, $\ell \leq \frac{k}{2}$, where distances are $1, \ldots, \ell$ and the perimeter of every triangle is at most k.

Our algorithm is not derived in an ad hoc way, instead it is constructed by using a design technique called the knowledge state technique. It is worth mentioning that it would be hard to come up with the actual behavioral algorithm, which we call the "wireframe algorithm", if one were not to use this technique. Yet the algorithm can be easily implemented and uses little memory, though the derivation and the proof of competitiveness is only via the technique.

In the next section we briefly describe the knowledge state technique, and then give a knowledge state description of the algorithm together with a proof of competitiveness. This description is in what is called the mixed model of computation – a generalization of a distributional description of a randomized online algorithm. As mentioned, the technique makes it easier to contrive the $\frac{19}{12}$-competitive algorithm. In this form however, the algorithm would be hard to implement as it is not described in the usual behavioral way. Thus in Section 3 we translate this description into the behavioral (and easily implementable) wireframe algorithm. We are also able to show that our algorithm has a competitive ratio, which is best possible; we show the lower bound in Section 4.

2   Knowledge States

We remind the reader that many randomized algorithms are given in distributional form, including a number of well known paging algorithms, *e.g.* [1,3]. For the 2-server problem, such an algorithm is essentially a state transition diagram, where each state is a probabilistic distribution of configurations (each

---

[1] A lower bound very slightly larger than $1 + e^{-\frac{1}{2}}$ is given in [10], but without proof.

configuration is a set of two points in the space – the locations of the servers); a transition from one state to the next state is a deterministic transition to a new distribution. Figure 2 illustrates such a step. Here the algorithm has both servers initially at configuration (x, z). Serving request r the algorithm transitions to a distribution with mass $\frac{1}{2}$ at (r, x) and mass $\frac{1}{2}$ at (r, z). Unfortunately the number of configurations in each state (and hence the number of states) can increase arbitrarily. One way to help avoid this is to allow non-deterministic transitions. We note that we have a great degree of freedom in designing our state transition diagram. As it turns out, our algorithm needs only eight states to achieve the optimal competitive ratio.

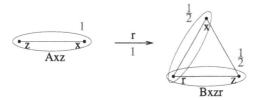

Fig. 2. A Step of a Distributional Algorithm

We describe our algorithm and the lower bound result in terms of knowledge states. We describe knowledge states briefly in this section and refer the reader to [5,6] for a more detailed description of this concept. It incorporates non-deterministic transitions as well as estimates on the offline cost.

As mentioned above, we use a variation of the *distribution model* to describe our randomized algorithm. That is, at each step the state of the algorithm will be described by a probability distribution on the set of all possible configurations at that step. The distribution model is equivalent to the behavioral model for randomized online algorithms against an oblivious adversary; see, for example, [7]. In the standard distribution model, the algorithm deterministically chooses a distribution at each step, but in this paper we allow the algorithm to use randomization to choose the distribution. We call such a step a *Las Vegas Step*; the reader might preview Figure 4. This variation, called the *mixed model* of randomized algorithms, is a generalization of both the behavioral model and the distributional model.

Let $\mathcal{X}$ denote the set of all configurations. (Naturally, for the 2-server problem, a configuration is simply a 2-tuple (a, b) of points in the metric space, which describes the location of two servers.) We say that a function $\omega : \mathcal{X} \to \mathbf{R}$ is *Lipschitz* if $\omega(w) \le \omega(u) + d(u, w)$ for all $u, w \in \mathcal{X}$. An *estimator* is a non-negative Lipschitz function $\mathcal{X} \to \mathbf{R}$. If $S \subseteq \mathcal{X}$, we say that S *supports an estimator* ω if, for any $w \in \mathcal{X}$ there exists some $u \in S$ such that $\omega(w) = \omega(u) + d(u, w)$. If ω is supported by a finite set, then there is a unique minimal set S which supports ω, which we call the *estimator support* of ω. We call the

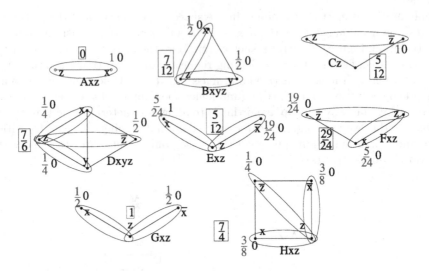

Fig. 3. The Knowledge States

cardinality of the support the *order* of the estimator. We say that an estimator ω has *zero minimum* if $\min_{u \in \mathcal{X}} \omega(u) = 0$.[2]

A *knowledge state algorithm* [4,5,6] is a mixed online algorithm that computes an *estimator* at each step. The estimators used throughout this paper will have very low order, i.e. the estimator can be described by giving values on very few configurations. Furthermore, distributions of a knowledge state algorithm are only concentrated on the estimator support, *i.e.* they are zero on all configurations other than the configurations in the estimator support.

More formally, if $\mathcal{A}$ is a knowledge-state algorithm, then:

1. At any given step, $\mathcal{A}$ keeps track of a pair (ω, π), where π is a finite distribution on $\mathcal{X}$, and $\omega : \mathcal{X} \to \mathbb{R}$ is the current estimator. The distribution is positive only on configurations which are in the support of the estimator ω. We call that pair the *current knowledge state*.
2. If $S = (\omega, \pi)$ is the knowledge state and the next request is r, then $\mathcal{A}$ uses randomization to pick a new knowledge state $S' = (\omega', \pi')$.

We now describe Item 1 for our specific situation. Thus, let M be a metric space in the class $\mathcal{M}_{24}$. We will call a finite set of points $S \subset M$ a *constellation*. To define the knowledge states for our algorithm, we only need a total of eight constellations, where each constellation has no more than four points.

---

[2] We remind the reader of the concept of a *work function*. Work functions are estimator functions. For example, work functions were used by Lund and Reingold [13] to describe an "opt-graph", which describes all possible moves of an optimal adversary. In short, work functions provide information about the optimal cost of serving the past request sequence. For a request sequence ϱ, by $\omega^{\varrho}(u)$, we denote the minimum cost of serving ϱ and ending in configuration u.

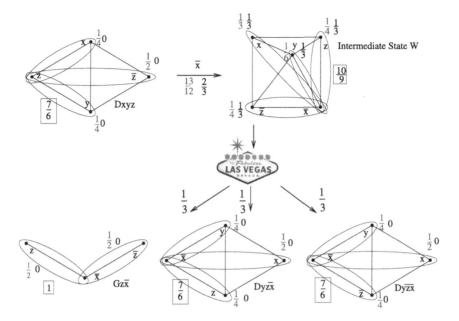

Fig. 4. One Move of the Knowledge State Algorithm

Each constellation is used to define a knowledge state of order no more than 3. In fact, for M_{24} it will suffice to consider a very small and simple class of knowledge states: these knowledge states are shown in Figure 3. In Figure 3, a line between two points indicates a distance of 1 between the two points, and the absence of a line means that the points are 2 apart. Note also that for any point $x \in M$, we denote an antipodal point (*i.e.* a point a distance of 2 away) by $\bar{x}$. The ovals encircling two points are the support configurations of the estimators and distributions; the red numbers (the numbers to the left in the pairs of numbers) give the values of the distribution and the black numbers (the numbers to the right in the pairs of numbers) give the value on the support of the estimator. We refer to these knowledge states as $A_{xz}, B_{xyz}, C_z, D_{xyz}, E_{xz}, F_{xz}, G_{xz}$ and H_{xz}. When we only refer to the configurations we will use the same notation except we will use lower case letters; thus the constellations are referred to as $a_{xz}, b_{xyz}, c_z, d_{xyz}, e_{xz}, f_{xz}, g_{xz}$ and h_{xz}. We finally note that the numbers in the boxes denote a potential, which is used later.

We now turn to Item 2 and describe how, using randomization, a new knowledge state is chosen. Given $S = (\omega, \pi)$ and r there are subsequent knowledge states $S_i = (\omega_i, \pi_i)$ and subsequent nonactive weights λ_i for $i = 1, \ldots m$, $\sum_{i=1}^{m} \lambda_i = 1$. Then for each i, $\mathcal{A}$ chooses S' to be S_i with probability λ_i. Again, for $M_{2,4}$, Figure 3 shows all eight possible subsequents.

We will now discuss how we can see if a knowledge state algorithm is competitive. Given the subsequents, a real number $adjust_{\mathcal{A}}(S, r)$ is computed such that $(\omega \wedge r)(u) \geq adjust_{\mathcal{A}}(S, r) + \sum_{i=1}^{m} \lambda_i \omega_i(u)$ for each $x \in \mathcal{X}$, where we define function $\omega \wedge r$ as $(\omega \wedge r)(w) = \min \{\omega(u) + d(u, w) \mid u \ni r\}$. We will use a standard potential argument to prove competitiveness, and thus we will need to associate a potential Φ with each knowledge state. We now define the *update condition* for a given step. To this end, fix competitive ratio $C > 1$. Let S be the current knowledge state, let $\{S_i\}$ be the subsequents for the current step, and λ_i be the probability that S_i will be chosen in this current step. Let $cost_A$ to be the expected cost of the algorithm A. Then the update condition is that

$$\Phi(S) \;\geq\; cost_A - C \cdot adjust + \sum_i \lambda_i \, \Phi(S_i) \;. \tag{1}$$

We will make use of the following lemma from [5,6]:

Lemma 1. *If the update condition holds at every step of a knowledge state algorithm then the algorithm is C-competitive.*

Figure 4 shows the step where the knowledge state S is $Dxyz$ and $\bar{x}$ is requested. For S we have an estimator with support $\omega(\{z, x\}) = \omega(\{z, y\}) = \omega(\{z, \bar{z}\}) = 0$

Table 1. The Knowledge State Algorithm for $M_{2,4}$

KS State	Request	Resulting KS	Φ_0	Φ_1	offset	$cost_A$	slack
Axz	$\bar{x}$	Cx	0	$\frac{5}{12}$	1	1	$\frac{1}{6}$
Axz	r	$Bxzr$	0	$\frac{7}{12}$	1	1	0
$Bxyz$	x	Axz	$\frac{7}{12}$	0	0	$\frac{1}{2}$	$\frac{1}{18}$
$Bxyz$	$\bar{z}$	$Bxy\bar{z}$	$\frac{7}{12}$	$\frac{7}{12}$	1	1	$\frac{7}{12}$
$Bxyz$	$\bar{x}$	$Dyz\bar{x}$	$\frac{7}{12}$	$\frac{7}{6}$	1	1	0
$Bxyz$	r	$\frac{1}{3}Bxyr + \frac{1}{3}Bxzr + \frac{1}{3}Byzr$	$\frac{7}{12}$	$\frac{7}{12}$	$\frac{2}{3}$	1	$\frac{1}{18}$
Cz	r	Gzr	$\frac{5}{12}$	1	1	1	0
$Dxyz$	x	$E\bar{z}x$	$\frac{7}{6}$	$\frac{5}{12}$	0	$\frac{3}{4}$	0
$Dxyz$	$\bar{z}$	Cz	$\frac{7}{6}$	$\frac{5}{12}$	0	$\frac{1}{2}$	$\frac{1}{4}$
$Dxyz$	$\bar{x}$	$\frac{1}{3}Dyz\bar{x} + \frac{1}{3}Dy\bar{z}\bar{x} + \frac{1}{3}Gz\bar{x}$	$\frac{7}{6}$	$\frac{10}{9}$	$\frac{2}{3}$	$\frac{13}{12}$	$\frac{1}{36}$
$Dxyz$	r	$\frac{1}{2}Bxzr + \frac{1}{2}By\bar{z}r$	$\frac{7}{6}$	$\frac{7}{12}$	$\frac{1}{2}$	1	$\frac{3}{8}$
Exz	x	Fzx	$\frac{5}{12}$	$\frac{29}{24}$	1	$\frac{19}{24}$	0
Exz	$\bar{x}$	$A\bar{x}z$	$\frac{5}{12}$	0	0	$\frac{5}{12}$	0
Exz	$\bar{z}$	$A\bar{x}z$	$\frac{5}{12}$	0	0	$\frac{5}{12}$	0
Exz	r	$B\bar{x}zr$	$\frac{5}{12}$	$\frac{7}{12}$	1	1	$\frac{5}{12}$
Fxz	x	Ezx	$\frac{29}{24}$	$\frac{5}{12}$	0	$\frac{19}{24}$	0
Fxz	$\bar{z}$	Cz	$\frac{29}{24}$	$\frac{5}{12}$	0	$\frac{5}{24}$	$\frac{7}{12}$
Fxz	$\bar{x}$	$Hz\bar{x}$	$\frac{29}{24}$	$\frac{7}{4}$	1	$\frac{25}{24}$	0
Fxz	r	$\frac{1}{2}Axr + \frac{1}{2}Gzr$	$\frac{29}{24}$	$\frac{1}{2}$	$\frac{1}{2}$	$\frac{31}{24}$	$\frac{17}{24}$
Gxz	x	Axz	1	0	0	1	0
Gxz	$\bar{z}$	Cz	1	$\frac{5}{12}$	1	1	$\frac{7}{6}$
Gxz	r	Axr	1	0	$\frac{3}{2}$	1	$\frac{13}{12}$
Hxz	r	$\frac{1}{2}Bxzr + \frac{1}{2}B\bar{x}\bar{z}r$	$\frac{7}{4}$	$\frac{7}{12}$	$\frac{1}{2}$	1	$\frac{23}{24}$
Hxz	x	$E\bar{z}x$	$\frac{7}{4}$	$\frac{5}{12}$	0	1	$\frac{1}{3}$
Hxz	$\bar{z}$	Cz	$\frac{7}{4}$	$\frac{5}{12}$	0	$\frac{3}{4}$	$\frac{7}{12}$

and distribution $\frac{1}{4}$ on $\{z, x\}$, $\frac{1}{4}$ on $\{z, y\}$ and $\frac{1}{2}$ on $\{z, \bar{z}\}$. In this situation the knowledge state algorithm chooses knowledge states $G_{z\bar{x}}$, $D_{yz\bar{x}}$, and $D_{y\bar{z}\bar{x}}$ with equal probability of $\frac{1}{3}$. (See the single numbers on the edges under the Las Vegas sign in Figure 4.) Next, we will argue that the update condition, *i.e.* inequality 1, does indeed hold for this step. To argue this we first focus on the "intermediate state" W depicted to the right of $Dxyz$. First note by using elementary arithmetic that the weighted average of the three subsequent states $G_{z\bar{x}}$, $D_{yz\bar{x}}$, and $D_{y\bar{z}\bar{x}}$ gives exactly the intermediate state W, both its distribution as well as its estimator function.

Turning now to $\omega \wedge \bar{x}$ it is easily calculated that the resulting estimator support set consists of $\{\{\bar{x}, x\}, \{\bar{x}, \bar{z}\}, \{\bar{x}, z\}, \{\bar{x}, y\}\}$ with value 1 on all the elements in the support set. Note now that if $\omega \wedge \bar{x}$ is lowered by $\frac{2}{3}$, this function is equal to the estimator of W. In other words, if $adjust_A(S, \bar{x}) = \frac{2}{3}$, then $(\omega \wedge \bar{x}) - adjust_A(S, \bar{x})$ is the estimator of W. (The value $adjust_A(S, \bar{x})$ appears as the second value under the arrow from D_{xyz} to W in Figure 4.)

We now analyze the cost of the algorithm for the step. It is the cost of the move from the distribution of D_{xyz} to the distribution of W. We remind the reader that this can be done solving a transportation problem. An instance of the *transportation problem* is a weighted directed bipartite graph with distributions on both parts. More formally, an instance is an ordered quintuple $(A, B, cost, \alpha, \beta)$ where U and V are finite non-empty sets, α is a distribution on U, β is a distribution on V, and $cost$ is a real-valued function on $U \times V$. A *solution* to this instance is a distribution γ on $U \times V$ such that

1. $\gamma(\{u\} \times V) = \alpha(u)$ for all $u \in U$.
2. $\gamma(U \times \{v\}) = \beta(v)$ for all $v \in V$.

Then $cost(\gamma) = \sum_{u \in U} \sum_{v \in V} \gamma(u, v) cost(u, v)$, and γ is a *minimal* solution if $cost(\gamma)$ is minimized over all solutions, in which case we call $cost(\gamma)$ the *minimum transportation cost*. The left part of Figure 5 shows the instance of the transportation problem which results from the situation in Figure 4. A solution of the problem is given by the following: Move $\frac{1}{4}$ from $\{z, x\}$ to $\{\bar{x}, x\}$, move $\frac{1}{6}$ from $\{z, y\}$ to $\{\bar{x}, y\}$, move $\frac{1}{4}$ from $\{z, \bar{z}\}$ to $\{\bar{x}, \bar{z}\}$, move $\frac{1}{4}$ from $\{z, \bar{z}\}$ to $\{\bar{x}, z\}$, each at cost 1; and $\frac{1}{12}$ from $\{z, y\}$ to $\{\bar{x}, x\}$ at cost 2. Thus the total cost of the algorithm in this step is $\frac{13}{12}$. (This number also appears as the first value under the arrow from D_{xyz} to W in Figure 4.)

Finally we fix competitiveness $C = \frac{19}{12}$. We are now ready to check the update condition 1 for this step. We have $\Phi(S) = \frac{7}{6}$; $cost_A = \frac{13}{12}$, $C \cdot adjust = \frac{19}{12} \cdot \frac{2}{3}$ and $\sum_i \lambda_i \Phi(S_i) = \frac{10}{9}$. Thus $\Phi(S) - cost_A + C \cdot adjust - \sum_i \lambda_i \Phi(S_i) \geq 0$.

A complete listing of moves of the algorithm is given in columns one to three of Table 1.

We have:

Theorem 1. *The knowledge state algorithm of Table 1 is C-competitive with* $C = \frac{19}{12}$.

Proof. Update condition 1 is verified for every step in Table 1.

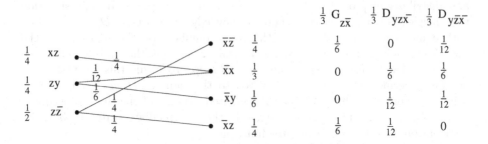

Fig. 5. The Distributional Transportation Problem

3   The Wireframe Algorithm

The knowledge state algorithm described in the previous section was analyzed in the mixed model of computation. We recall that this implies that there is a competitive behavioral online algorithm. The following lemma is well-known. (It is, for example, implicit in Chapter 6 of [7].)

Lemma 2. *The mixed model and the behavioral model of randomized online algorithms are equivalent, in the following sense. If A_1 is an algorithm of one of the models, there exist an algorithm A_2 of the other model, such that, given any request sequence ϱ, the cost of A_2 for ϱ is no greater than the cost of A_1.*

We will now translate our algorithm into behavioral form, a form in which it is easy to implement the algorithm into an actual working computer program. The resulting behavioral algorithm is called the "wireframe algorithm." At each step, in addition to the position of the two servers, the algorithm also keeps track of certain points in an octahedron. This can be best illustrated by a wireframe of an octahedron; see Figure 6. In the situation depicted in Figure 6 (see top octahedron) the server positions are at points y and z and the algorithm keeps track of constellation d_{xyz}. (See the dashed lines.) Note that constellation d_{xyz} can be best thought of as the wireframe shown in Figure 6 (top part). Next, the Figure (lower part) shows the behavior of the algorithm if $\bar{x}$ is requested. With probability $\frac{1}{6}$ the algorithm moves the server at point y to point $\bar{x}$ *and* and the server at point z to x and goes into state (*i.e.* wireframe) $d_{yz\bar{x}}$. With remaining probability $\frac{1}{6}$ the algorithm does exactly the same server movements (y to $\bar{x}$ and z to x) and goes into state (*i.e.* wireframe) $d_{yz\bar{x}}$. Furthermore, with equal probabilities $\frac{1}{3}$ the wireframe algorithm moves the server at point z to point $\bar{x}$ *and* goes into state (*i.e.* wireframe) $d_{yz\bar{x}}$ or $d_{y\bar{z}\bar{x}}$.

Note the following:

- The algorithm has no concept of knowledge states, it merely remembers where the servers are and keeps track of only very limited extra information. Upon a request, depending on this extra information, the algorithm then decides how to move the servers and how to update its information.

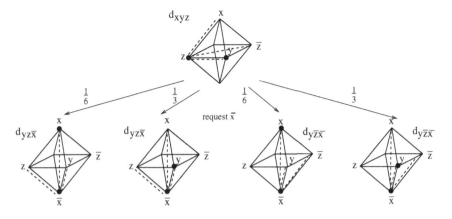

Fig. 6. One Move of the Wireframe Algorithm

- A server algorithm is called *lazy* if, in a step, it only moves one server to serve a request and it does not move any other server. Note that the algorithm is non-lazy.
- In the situation described in Figure 6 the request to $\bar{x}$ is served by moving the server at point z to point $\bar{x}$ with probability $\frac{2}{3} = \frac{1}{3} + \frac{1}{3}$, but the constellation memorized by the algorithm after the move is either $d_{yz\bar{x}}$ or $d_{y\bar{z}\bar{x}}$. With probability $\frac{1}{3} = \frac{1}{6} + \frac{1}{6}$, the algorithm moves the server at point y to point $\bar{x}$ *and* and the server at point z to x. (Again, with the caveat that two different constellations are possible after the move.)

For the behavioral move just described we will use the notation:

$$\boxed{d_{xyz}, yz \,|\, \bar{x}\, |\, \tfrac{1}{6}\{d_{yz\bar{x}}, x\bar{x}\} + \tfrac{1}{3}\{d_{yz\bar{x}}, y\bar{x}\} + \tfrac{1}{6}\{d_{y\bar{z}\bar{x}}, x\bar{x}\} + \tfrac{1}{3}\{d_{y\bar{z}\bar{x}}, y\bar{x}\}.}$$

Using this notation, the wireframe algorithm is completely described in Table 2.

The derivation of the behavioral algorithm from the mixed algorithm is routine (it is described in general in [5,6]); we will briefly discuss how the step of Figure 6 results from translating the transition of Figure 4. The translation uses the solution to transportation problem of Figure 5. Note that there is probability mass of $\frac{1}{4}$ at $\{(z,y)\}$. Mass $\frac{1}{12}$ is moved to $\{(\bar{x},x)\}$ and mass $\frac{1}{6}$ is moved to $\{(\bar{x},y)\}$. Thus, *given* that the constellation is $\{(z,y)\}$ the conditional probabilities for $\{(\bar{x},x)\}$ and $\{(\bar{x},y)\}$ are $\frac{1}{3}$ and $\frac{2}{3}$ respectively. Following Figure 5 to the right, we see that the mass $\frac{1}{3}$ at $\{(\bar{x},x)\}$ is equally divided between constellations $d_{yz\bar{x}}$ and $d_{y\bar{z}\bar{x}}$. The same is true for the mass at $\{(\bar{x},y)\}$. We conclude that the algorithm chooses with probability $\frac{1}{6}$ servers at $\{(x\bar{x})\}$ with $d_{yz\bar{x}}$, with probability $\frac{1}{3}$ servers at $\{(y\bar{x})\}$ with $d_{yz\bar{x}}$, with probability $\frac{1}{6}$ servers at $\{(x\bar{x})\}$ with $d_{y\bar{z}\bar{x}}$, and with probability $\frac{1}{3}$ servers with $\{(y\bar{x})\}$ with $d_{y\bar{z}\bar{x}}$.

In summary we have:

Theorem 2. *The wireframe algorithm of Table 2 is C-competitive with $C = \frac{19}{12}$.*

Table 2. The Moves of the Wireframe Algorithm

a_{xz}, xz	$\bar{x}$	$1\{c_x, x\bar{x}\}$
a_{xz}, xz	r	$\frac{1}{2}\{b_{xzr}, xr\} + \frac{1}{2}\{b_{xzr}, zr\}$
b_{xyz}, xz	x	$1\{a_{xz}, xz\}$
b_{xyz}, yz	x	$1\{a_{xz}, xz\}$
b_{xyz}, xz	$\bar{x}$	$1\{d_{yz\bar{x}}, x\bar{x}\}$
b_{xyz}, yz	$\bar{x}$	$\frac{1}{2}\{d_{yz\bar{x}}, y\bar{x}\} + \frac{1}{2}\{d_{yz\bar{x}}, z\bar{x}\}$
b_{xyz}, xz	$\bar{z}$	$1\{b_{xy\bar{z}}, x\bar{x}\}$
b_{xyz}, yz	$\bar{z}$	$1\{b_{xy\bar{z}}, y\bar{x}\}$
b_{xyz}, xz	r	$\frac{1}{3}\{b_{xyr}, xr\} + \frac{1}{3}\{b_{xzr}, xr\}$ $+ \frac{1}{6}\{b_{xzr}, zr\} + \frac{1}{6}\{b_{yzr}, zr\}$
b_{xyz}, yz	r	$\frac{1}{3}\{b_{xyr}, yr\} + \frac{1}{6}\{b_{xzr}, zr\}$ $+ \frac{1}{3}\{b_{yzr}, yr\} + \frac{1}{6}\{b_{yzr}, zr\}$
$c_{xz}, z\bar{z}$	r	$\frac{1}{2}\{g_{zr}, zr\} + \frac{1}{2}\{g_{zr}, \bar{z}r\}$
d_{xyz}, xz	x	$1\{e_{\bar{z}x}, yz\}$
d_{xyz}, yz	x	$1\{e_{\bar{z}x}, yz\}$
$d_{xyz}, \bar{z}z$	x	$\frac{7}{12}\{e_{\bar{z}x}, yz\} + \frac{5}{12}\{e_{\bar{z}x}, y\bar{z}\}$
d_{xyz}, xz	$\bar{z}$	$1\{c_z, z\bar{z}\}$
d_{xyz}, yz	$\bar{z}$	$1\{c_z, z\bar{z}\}$
$d_{xyz}, \bar{z}z$	$\bar{z}$	$1\{c_z, z\bar{z}\}$
d_{xyz}, xz	r	$1\{b_{xzr}, xr\}$
d_{xyz}, yz	r	$\frac{1}{2}\{b_{xzr}, zr\} + \frac{1}{2}\{b_{y\bar{z}r}, \bar{z}r\}$
$d_{xyz}, \bar{z}z$	r	$1\{b_{y\bar{z}r}, yr\}$
d_{xyz}, xz	$\bar{x}$	$\frac{1}{2}\{d_{yz\bar{x}}, x\bar{x}\} + \frac{1}{2}\{d_{yz\bar{x}}, x\bar{x}\}$
d_{xyz}, yz	$\bar{x}$	$\frac{1}{6}\{d_{yz\bar{x}}, x\bar{x}\} + \frac{1}{3}\{d_{yz\bar{x}}, y\bar{x}\}$ $+ \frac{1}{6}\{d_{yz\bar{x}}, x\bar{x}\} + \frac{1}{3}\{d_{yz\bar{x}}, y\bar{x}\}$
$d_{xyz}, \bar{z}z$	$\bar{x}$	$\frac{1}{3}\{g_{z\bar{x}}, z\bar{x}\} + \frac{1}{3}\{g_{z\bar{x}}, \bar{z}\bar{x}\}$ $+ \frac{1}{6}\{d_{yz\bar{x}}, z\bar{x}\} + \frac{1}{3}\{d_{yz\bar{x}}, \bar{z}\bar{x}\}$
e_{xz}, xz	x	$1\{f_{zx}, zx\}$
$e_{xz}, \bar{x}z$	x	$1\{f_{zx}, \bar{x}x\}$
e_{xz}, xz	$\bar{z}$	$1\{a_{\bar{x}z}, \bar{x}z\}$
$e_{xz}, \bar{x}z$	$\bar{z}$	$1\{a_{\bar{x}z}, \bar{x}z\}$
e_{xz}, xz	r	$1\{b_{\bar{x}zr}, zr\}$
$e_{xz}, \bar{x}z$	r	$\frac{12}{19}\{b_{\bar{x}zr}, \bar{x}r\} + \frac{7}{19}\{b_{\bar{x}zr}, zr\}$
e_{xz}, xz	$\bar{x}$	$1\{a_{\bar{x}z}, \bar{x}z\}$
$e_{xz}, \bar{x}z$	$\bar{x}$	$1\{a_{\bar{x}z}, \bar{x}z\}$

f_{xz}, xz	x	$1\{e_{zx}, zx\}$
$f_{xz}, \bar{z}z$	x	$\frac{14}{19}\{e_{zx}, zx\} + \frac{5}{19}\{e_{zx}, \bar{z}x\}$
f_{xz}, xz	$\bar{z}$	$1\{c_z, z\bar{z}\}$
$f_{xz}, \bar{z}z$	$\bar{z}$	$1\{c_z, z\bar{z}\}$
f_{xz}, xz	$\bar{x}$	$1\{h_{z\bar{x}}, x\bar{x}\}$
$f_{xz}, \bar{z}z$	$\bar{x}$	$\frac{1}{19}\{h_{z\bar{x}}, x\bar{x}\} + \frac{9}{19}\{h_{z\bar{x}}, z\bar{x}\}$ $+ \frac{9}{19}\{h_{z\bar{x}}, \bar{z}\bar{x}\}$
f_{xz}, xz	r	$1\{a_{xr}, xr\}$
$f_{xz}, \bar{z}z$	r	$\frac{7}{19}\{a_{xr}, xr\} + \frac{6}{19}\{g_{zr}, zr\}$ $+ \frac{6}{19}\{g_{zr}, \bar{z}r\}$
g_{xz}, xz	x	$1\{a_{xz}, zx\}$
$g_{xz}, \bar{x}z$	x	$1\{a_{xz}, zx\}$
g_{xz}, xz	$\bar{z}$	$1\{c_z, z\bar{z}\}$
$g_{xz}, \bar{x}z$	$\bar{z}$	$1\{c_z, z\bar{z}\}$
g_{xz}, xz	r	$1\{a_{xr}, xr\}$
$g_{xz}, \bar{x}z$	r	$1\{a_{xr}, xr\}$
h_{xz}, xz	r	$\frac{2}{3}\{b_{xzr}, xr\} + \frac{1}{3}\{b_{xzr}, zr\}$
$h_{xz}, \bar{x}z$	r	$\frac{1}{3}\{b_{xzr}, zr\} + \frac{2}{3}\{b_{\bar{x}\bar{z}r}, \bar{x}r\}$
$h_{xz}, \bar{z}z$	r	$1\{b_{\bar{x}\bar{z}r}, \bar{z}r\}$
h_{xz}, xz	x	$1\{e_{\bar{z}x}, xz\}$
$h_{xz}, \bar{x}z$	x	$1\{e_{\bar{z}x}, xz\}$
$h_{xz}, \bar{z}z$	x	$\frac{1}{6}\{e_{\bar{z}x}, xz\} + \frac{5}{6}\{e_{\bar{z}x}, x\bar{z}\}$
h_{xz}, xz	$\bar{z}$	$1\{c_z, z\bar{z}\}$
$h_{xz}, \bar{x}z$	$\bar{z}$	$1\{c_z, z\bar{z}\}$
$h_{xz}, \bar{z}z$	$\bar{z}$	$1\{c_z, z\bar{z}\}$

4   The Lower Bound

Indeed, the competitiveness of the wireframe algorithm is best possible:

Theorem 3. *Let $\mathcal{A}$ be any randomized online algorithm for the 2-server problem for $M_{2,4}$. Let C be the competitiveness of $\mathcal{A}$. Then $C \geq \frac{19}{12}$.*

Proof. We only give a sketch; the formal proof will be given in the full paper. We refer to Figure 7. Consider the cross-polyhedron $\{x, \bar{x}, y, \bar{y}, z, \bar{z}\}$. Without loss of generality, the initial server position is $\{(x, y)\}$. We call this situation START. Now consider:

1. Adversary requests z; pays 1.
2. $\mathcal{A}$ serves request; pays 1.
3. Without loss of generality, $\mathcal{A}$ is at x with probability $\leq \frac{1}{2}$.
4. Adversary requests $\bar{y}$; pays 1.
5. $\mathcal{A}$ serves request; pays 1.

It can easily be argued that the probability that $\mathcal{A}$ has a server at y is not smaller than $\frac{1}{2}$ and that the probability that there is a server at x is not larger than $\frac{1}{4}$.

6. Adversary requests x, pays 0.
7. $\mathcal{A}$ serves the request; pays $\geq \frac{3}{4}$.

We call this situation the MIDDLE, see Figure 7.
    Let

$$p = \text{probability there is a server at } \bar{y}$$
$$q = \text{probability there is a server at } z$$
$$r = \text{probability there is a server at } y$$

Then $p + q + r = 1$.

Case i: $q + 2r \geq \frac{5}{12}$ (Return to START)

a) Adversary repeatedly requests $\bar{y}, x$, *i.e.* hammers at $\{(\bar{y}, x)\}$; pays 0.
b) $\mathcal{A}$ must move server from y and z to $\bar{y}$; pays $q + 2r$.

Situation has returned to START: Adversary has paid 2, $\mathcal{A}$ has paid no less than $2 + \frac{3}{4} + \frac{5}{12} = \frac{38}{12}$. Thus the ratio is no less than $\frac{19}{12}$.

Case ii: $q + 2r < \frac{5}{12}$ (Cycle back to MIDDLE)

a) Adversary requests y, pays 1.
b) $\mathcal{A}$ serves request and pays $1 - r$. Thus, the probability that there is a server at x after this move is $\leq 1 - p$.
c) Adversary requests x, pays 0.
d) $\mathcal{A}$ serves request, pays $\geq p$.

We have now returned to MIDDLE, with the roles of y and $\bar{y}$ interchanged. The "middle loop" consists of the two requests between the two times of MIDDLE. Analysis of the middle loop: Adversary pays 1. $\mathcal{A}$ pays no less than $(1 - r) + p = (p + q) + p = 2p + q$. But $2p + q = 2(p + q + r) - q - 2r \geq 2 - \frac{5}{12} = \frac{19}{12}$.

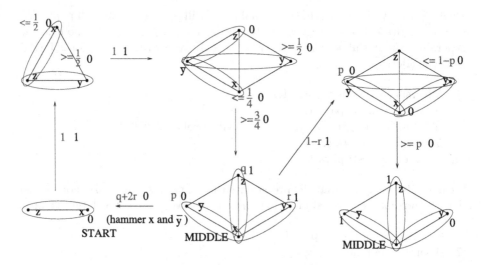

Fig. 7. The Lower Bound Graph

5   Concluding Remarks

The knowledge states from Figure 3 were found by trial and error using computer experimentation. We mention that there exists a slightly simpler knowledge state algorithm for $M_{2,4}$, which is of order 2, with competitiveness $\frac{7}{4}$. We note that for that case we also have calculated, through computer experimentation, the minimum value of C in the sense that no lower competitiveness for any order 2 knowledge state algorithm for $M_{2,4}$ exists. This value is $C = \frac{173+\sqrt{137}}{112}$.

These results, as well as our results for the server problem in uniform spaces (equivalent to the caching problem), indicate a natural trade-off between competitiveness and memory of online randomized algorithms.

References

1. Achlioptas, D., Chrobak, M., Noga, J.: Competitive analysis of randomized paging algorithms. Theoret. Comput. Sci. 234, 203–218 (2000)
2. Bartal, Y., Chrobak, M., Larmore, L.L.: A randomized algorithm for two servers on the line. In: Bilardi, G., Pietracaprina, A., Italiano, G.F., Pucci, G. (eds.) ESA 1998. LNCS, vol. 1461, pp. 247–258. Springer, Heidelberg (1998)
3. Bein, W., Larmore, L.L., Noga, J.: Equitable revisited. In: Proceedings of the 15th Annual European Symposium on Algorithms. LNCS, vol. 4698, pp. 419–426. Springer, Heidelberg (2007)
4. Bein, W., Larmore, L.L., Reischuk, R.: Knowledge states for the caching problem in shared memory multiprocessor systems. In: Proceedings of the 7th International Symposium on Parallel Architectures, Algorithms and Networks, pp. 307–312. IEEE, Los Alamitos (2004)
5. Bein, W., Larmore, L.L., Reischuk, R.: Knowledge state algorithms: Randomization with limited information (2007), Arxiv: archive.org/cs/0701142

6. Bein, W., Larmore, L.L., Reischuk, R.: Knowledge states: A tool for randomized online algorithms. In: Proceedings of the 41st Annual Hawaii International Conference on System Sciences (CD-ROM), p. 10. IEEE Computer Society Press, Los Alamitos (2008)
7. Borodin, A., El-Yaniv, R.: Online Computation and Competitive Analysis. Cambridge University Press, Cambridge (1998)
8. Chrobak, M., Karloff, H., Payne, T.H., Vishwanathan, S.: New results on server problems. SIAM J. Discrete Math. 4, 172–181 (1991)
9. Chrobak, M., Larmore, L.L.: An optimal online algorithm for k servers on trees. SIAM J. Comput. 20, 144–148 (1991)
10. Chrobak, M., Larmore, L.L., Lund, C., Reingold, N.: A better lower bound on the competitive ratio of the randomized 2-server problem. Inform. Process. Lett. 63, 79–83 (1997)
11. Koutsoupias, E., Papadimitriou, C.: Beyond competitive analysis. In: Proc. 35th Symp. Foundations of Computer Science (FOCS), pp. 394–400. IEEE, Los Alamitos (1994)
12. Koutsoupias, E., Papadimitriou, C.: On the k-server conjecture. J. ACM 42, 971–983 (1995)
13. Lund, C., Reingold, N.: Linear programs for randomized on-line algorithms. In: Proc. 5th Symp. on Discrete Algorithms (SODA), pp. 382–391. ACM/SIAM (1994)
14. Manasse, M., McGeoch, L.A., Sleator, D.: Competitive algorithms for server problems. J. Algorithms 11, 208–230 (1990)
15. Schläfli, L.: Theorie der vielfachen Kontinuität. Birkhäuser, Basel (1857)

Deterministic Algorithms for Rank Aggregation and Other Ranking and Clustering Problems

Anke van Zuylen* and David P. Williamson**

School of Operations Research and Information Engineering,
Cornell University, Ithaca, NY 14853
Fax: (607) 255-9129
avz2@cornell.edu
dpw@cs.cornell.edu

Abstract. We consider ranking and clustering problems related to the aggregation of inconsistent information. Ailon, Charikar, and Newman [1] proposed randomized constant factor approximation algorithms for these problems. Together with Hegde and Jain, we recently proposed deterministic versions of some of these randomized algorithms [2]. With one exception, these algorithms required the solution of a linear programming relaxation. In this paper, we introduce a purely combinatorial deterministic pivoting algorithm for weighted ranking problems with weights that satisfy the triangle inequality; our analysis is quite simple. We then shown how to use this algorithm to get the first deterministic combinatorial approximation algorithm for the partial rank aggregation problem with performance guarantee better than 2. In addition, we extend our approach to the linear programming based algorithms in Ailon et al. [1] and Ailon [3]. Finally, we show that constrained rank aggregation is not harder than unconstrained rank aggregation.

Keywords: derandomization, rank aggregation, feedback arc set in tournaments.

1   Introduction

We consider the problem of ranking or clustering a set of elements, based on input information for each pair of elements. The objective is to find a solution that minimizes the deviation from the input information. For example, we may want to cluster webpages based on similarity scores, where for each pair of pages we have a score between 0 and 1, and we want to find a clustering that minimizes the sum of the similarity scores of pages in different clusters plus the sum of (one minus the similarity score) for pages in the same cluster. Another example arises in meta-search engines for Web search, where we want to get robust rankings that are not sensitive to the various shortcomings and biases of individual search engines by combining the rankings of the individual search engines [4].

---

* Supported by NSF grant CCF-0514628.
** Supported by NSF grant CCF-0514628.

C. Kaklamanis and M. Skutella (Eds.): WAOA 2007, LNCS 4927, pp. 260–273, 2008.

More formally, in the *weighted minimum feedback arc set problem in tourna-ments*, we are given a set of elements V, nonnegative weights $w_{(i,j)}$ and $w_{(j,i)}$ for each pair of distinct elements i and j, and we want to find a permutation π that minimizes the weight of pairs of elements out of order with respect to the per-mutation, i.e. $\sum_{\pi(i)<\pi(j)} w_{(j,i)}$. We say the weights satisfy *probability constraints* if for any pair i, j, $w_{(i,j)} + w_{(j,i)} = 1$, or the *triangle inequality* if for any triplet i, j, k, $w_{(i,j)} + w_{(j,k)} \geq w_{(i,k)}$. We will sometimes refer to this problem as the *ranking* problem. In the *rank aggregation problem*, the input is a collection of orderings of V, and $w_{(i,j)}$ is the fraction of orderings in which i is ordered before j; note that these weights obey both the probability constraints and triangle inequality. In the *constrained* version of the ranking problem, we are also given a partial order P as input and the output permutation π must be consistent with P, i.e. if $(i,j) \in P$ then $\pi(i) < \pi(j)$.

In the *weighted clustering problem*, we are given a set of elements V, and values $w^+_{\{i,j\}}$ and $w^-_{\{i,j\}}$ for every distinct pair of elements i, j. We want to find a clustering minimizing $\sum_{i,j \text{ in different clusters}} w^+_{\{i,j\}} + \sum_{i,j \text{ in same cluster}} w^-_{\{i,j\}}$. We say the weights satisfy probability constraints if for every $i, j \in V$, $w^+_{\{i,j\}} + w^-_{\{i,j\}} = 1$. We will say the weights satisfy the triangle inequality if for every triple i, j, k, $w^-_{\{i,j\}} + w^-_{\{j,k\}} \leq w^-_{\{i,k\}}$ and $w^+_{\{i,j\}} + w^-_{\{j,k\}} \leq w^+_{\{i,k\}}$. The problem where exactly one of $w^+_{\{i,j\}}$ and $w^-_{\{i,j\}}$ is 1 (and the other 0) is called *correlation clustering*. The clustering problem corresponding to rank aggregation, in which we want to aggregate a collection of clusterings of the same set of elements, is called *consensus clustering*. We can also have a constrained version of the weighted clustering problem by giving as input sets of pairs of items P^+ and P^-, where pairs in P^+ must be in the same output cluster, while pairs in P^- must be in different output clusters.

Both rank aggregation and consensus clustering are NP-hard [4,5], so the more general problems of ranking or clustering with weights that satisfy the triangle inequality or probability constraints, or both, are also NP-hard.

Ailon, Charikar and Newman [1] give the first constant-factor approximation algorithms for the unconstrained ranking and clustering problems with weights that satisfy either triangle inequality constraints, probability constraints, or both. Their algorithms are randomized and based on Quicksort: the algorithms recursively generate a solution by choosing a random vertex as "pivot" and ordering all other vertices with respect to the pivot vertex according to some criterion. For example, in the first type of algorithm they give for the ranking problem, a vertex j is ordered before the pivot k if $w_{(j,k)} \geq w_{(k,j)}$ or ordered after k otherwise. Next, the algorithm recurses on the two instances induced by the vertices before and after the pivot.

In the case of rank aggregation and consensus clustering, a folklore result is that returning the best of the input rankings or clusterings is a 2-approximation algorithm. Ailon, Charikar and Newman also show that one can obtain better approximation factors for rank aggregation and consensus clustering by returning the best of their algorithm's solution and the best input ranking/clustering.

For instance, for rank aggregation, they obtain a randomized $\frac{11}{7}$-approximation algorithm using their first type of algorithm, and a randomized $\frac{4}{3}$-approximation algorithm using their second, LP-based, algorithm.

There has been a good deal of follow-up work since the Ailon et al. paper. Ailon and Charikar [6] extend the pivot-based approximation algorithms for clustering to hierarchical clustering. Coppersmith, Fleischer, and Rudra [7] give a simple greedy 5-approximation algorithm for the ranking problem when weights obey the probability constraints. Van Zuylen, Hegde, Jain and Williamson [2] give deterministic variants of the pivoting algorithms in Ailon et al. and Ailon and Charikar and extend them to the constrained versions of these problems. All but one of their algorithms require solving an LP relaxation of the problem, and their techniques do not extend to the improved results in the Ailon et al. paper for rank aggregation and consensus clustering. They do give a combinatorial approximation algorithm for the ranking problem when weights obey the probability constraints, with a performance guarantee of 4 in the unconstrained case, and 6 for constrained problems.

Kenyon-Mathieu and Schudy [8] show that there exists a polynomial-time approximation scheme for unconstrained weighted feedback arc set in tournaments with weights satisfying $b \leq w_{(i,j)} + w_{(j,i)} \leq 1$ for all $i, j \in V$ for some $b > 0$. Note that this includes problems satisfying the probability constraints and hence includes the rank aggregation problem as a special case. Their approximation scheme assumes the availability of a solution with cost that is not more than a constant factor α from optimal. To get a $(1 + \epsilon)$-approximate solution, the running time of their algorithm is doubly exponential in $\frac{1}{\epsilon}, \frac{1}{b}$ and α.

Ailon [3] considers the *partial rank aggregation* problem, which was introduced by Fagin, Kumar, Mahdian, Sivakumar and Vee [9,10]. Unlike full rank aggregation, the input rankings do not have to be permutations of the same set of elements. Instead, input rankings are allowed to be top-m rankings, i.e. permutations of only a subset of the elements (in which case we make the natural assumption that the unranked elements all share the position after the last ranked element), or the rankings may be p-ratings, i.e. mappings from V to $\{1, \ldots, p\}$, as is the case for example in movie rankings. More precisely, a partial ranking of a set of elements V is a function $\pi : V \to \{1, \ldots, |V|\}$. If π is bijective, it is a *full ranking*. We will say the distance between two partial rankings π_1 and π_2 is the number of pairs i, j such that $\pi_1(i) < \pi_1(j)$, and $\pi_2(i) > \pi_2(j)$. The goal of partial rank aggregation is, given ℓ partial rankings of V, to output a permutation of the elements of V that minimizes the sum of the distances from the ℓ input rankings. Note that the output is required to be a permutation, and cannot be a partial ranking. Ailon [3] generalizes and improves some of the results from Ailon et al. to partial rank aggregation. He shows that perturbing the solution to the linear programming relaxation and using these perturbed values as probabilities gives a randomized $\frac{3}{2}$-approximation algorithm for partial rank aggregation. Since his analysis only uses the fact that the weights satisfy the triangle inequality, this also yields $\frac{3}{2}$-approximation algorithm for ranking with triangle inequality constraints on the weights.

1.1   Our Results

Our goals in obtaining the results for this paper were twofold. First, we wanted to do an implementation study of the various pivoting algorithms. But none of the deterministic pivoting algorithms thus far are especially practical. The PTAS of Kenyon-Mathieu and Schudy is of theoretical interest only. Although the deterministic algorithms of Van Zuylen et al. are polynomial-time, with the exception of their combinatorial algorithm for ranking with probability constraints, they require solving a linear program with $O(n^2)$ variables and $O(n^3)$ constraints for $n = |V|$, which with standard LP packages is likely to be slow for even moderate values of n. Thus we give purely combinatorial, deterministic pivoting algorithms. For weights obeying the triangle inequality, we give a 2-approximation algorithm, whose analysis is particularly simple. In the case of rank aggregation and consensus clustering, we give an $\frac{8}{5}$-approximation algorithm. The $\frac{8}{5}$-approximation algorithm extends to the partial rank aggregation problem as well. This gives the first combinatorial algorithm for partial rank aggregation with an approximation guarantee less than 2. The running time of our combinatorial algorithms is $O(n^3)$ compared to $O(n^2)$ for their randomized counterparts in Ailon et al.

Second, we wished to give deterministic algorithms matching the best randomized algorithms of Ailon et al. and Ailon in the case of rank aggregation and partial aggregation, and correlation and consensus clustering. It is a fundamental question whether everything computable in randomized polynomial time is computable in deterministic polynomial time (something that the recent PRIMES in P result by Agrawal, Kayal, and Saxena [11] provided some additional evidence for). The techniques from Ailon et al. and Ailon are not amenable to standard techniques of derandomization, but we show (in the current paper and [2]) that we can amortize in place of the expectation and make the randomized algorithm deterministic. In particular, we show how to derandomize the $\frac{4}{3}$-approximation algorithm of Ailon et al. for rank aggregation and consensus clustering, the $\frac{5}{2}$-approximation algorithm of Ailon et al. for ranking and clustering with probability constraints, and the $\frac{3}{2}$-approximation algorithm of Ailon for partial rank aggregation. These algorithms invoke an interesting two-step derandomization, in which we first choose a pivot so as to minimize a ratio of expectations; then we apply the method of conditional expectations to decide how to order (cluster) the elements with respect to the pivot.

Finally, we show that if the weights satisfy the triangle inequality, then any approximation result that holds for unconstrained ranking or clustering problems also holds for constrained problems, by showing a sequence of local moves that remove any violations of the constraints and do not increase the cost of the solution.

In the remainder of the paper, we will only discuss our results for ranking, and not for clustering. It is straightforward to translate these results to the clustering setting (see Ailon et al. [1] and Van Zuylen et al. [2]).

Table 1. The table summarizes the best known approximation guarantees. Italicized entries are expected approximation guarantees from randomized algorithms. 'A' refers to Ailon [3], 'ACN' refers to Ailon, Charikar and Newman [1], 'KS' refers to Kenyon-Mathieu and Schudy [8], 'ZHJW' refers to Van Zuylen, Hegde, Jain, and Williamson [2] and 'ZW' refers to this paper.

Ranking				
	prob. constr.	triangle ineq.	full rank agg.	partial rank agg
Combin.	4(ZHJW), *3(ACN)*	2(ZW), *2(ACN)*,	$\frac{8}{5}$(ZW), $\frac{11}{7}$*(ACN)*	$\frac{8}{5}$(ZW)
LP based	$\frac{5}{2}$(ZW), $\frac{5}{2}$*(ACN)*	$\frac{3}{2}$(ZW), $\frac{3}{2}$*(A)*	$\frac{4}{3}$(ZW), $\frac{4}{3}$*(ACN)*	$\frac{3}{2}$(ZW), $\frac{3}{2}$*(A)*
PTAS	$1+\varepsilon$(KS)		$1+\varepsilon$(KS)	

Clustering			
	prob. constr.	triangle ineq.	consensus clustering
Combinatorial	6(ZHJW), *3(ACN)*	2(ZW), *2(ACN)*	$\frac{8}{5}$(ZW), $\frac{11}{7}$*(ACN)*
LP based	$\frac{5}{2}$(ZW), $\frac{5}{2}$*(ACN)*	2(ZHJW), *2(ACN)*	$\frac{4}{3}$(ZW), $\frac{4}{3}$*(ACN)*

2   Combinatorial Pivoting Algorithms

Given an instance of the weighted feedback arc set problem in tournaments, suppose we form a tournament $G = (V, A)$ by including arc (i, j) only if $w_{(i,j)} \geq w_{(j,i)}$ (breaking ties arbitrarily). This is called the majority tournament in Ailon et al. [1]. Clearly, if the tournament is acyclic, then it corresponds to an optimal permutation: the cost for pair i, j in any solution is at least $\min\{w_{(i,j)}, w_{(j,i)}\}$, and this lower bound is met for every pair.

Ailon, Charikar and Newman [1] propose a simple algorithm to obtain a permutation that costs at most 3 times the optimum if the weights satisfy the triangle inequality, or at most 2 times the optimum if the weights satisfy both triangle inequality and probability constraints. Their algorithm, FAS-Pivot, is given below. We use the following notation: We denote by $G(V')$ the subgraph of G induced by $V' \subset V$. If π_1 and π_2 are permutations of disjoint sets V_1, V_2, we let π_1, π_2 denote the concatenation of the two permutations.

FAS-Pivot$(G = (V, A))$

Pick a pivot $k \in V$.
$V_L = \{i \in V : (i, k) \in A\}, V_R = \{i \in V : (k, i) \in A\}$.
Return FAS-Pivot$(G(V_L))$, k, FAS-Pivot$(G(V_R))$.

In Ailon, Charikar and Newman's algorithm, a pivot is chosen randomly. In our deterministic versions of this algorithm, we will propose different ways of choosing the pivot, depending on the information we have about the input.

For a pair i, j with (i, j) in the majority tournament A, we will let $w_{ij} = w_{(i,j)}$ and $\bar{w}_{ij} = w_{(j,i)}$. Note that this implies that $w_{ij} = \max(w_{(i,j)}, w_{(j,i)})$ and $\bar{w}_{ij} = \min(w_{(i,j)}, w_{(j,i)})$. Then if a pair i, j is ordered according to A, the cost incurred by the algorithm is $\bar{w}_{ij}$, and for each pair not ordered according to A, the cost is w_{ij}. We will call the first type of arcs "forward arcs" and the second "backward arcs".

Ailon, Charikar and Newman bound the expected cost for the backward arcs if the pivot is chosen randomly. Subsequently, Van Zuylen, Hegde, Jain and Williamson showed that one can obtain a deterministic version of this algorithm by first solving a linear programming relaxation, and then carefully choosing the pivot vertex based on the solution to the linear program. We show that if the weights satisfy the triangle inequality, then we can give a deterministic algorithm that does not require us to solve a linear program and that achieves the same guarantees as in Van Zuylen et al. for this case. The idea of our algorithm is to use $\bar{w}_{ij}$ as a "budget" for vertex pair i, j, and to show that we can always choose a pivot so that the cost of the backward arcs created by pivoting on this vertex is at most twice the budget for these arcs.

Theorem 1. *There exists a deterministic combinatorial 2-approximation algorithm for weighted feedback arc set in tournaments with triangle inequality.*

Proof. We use the algorithm described above, but specify how to choose a good pivot. For a given pivot k, we let $T_k(V) \subset A$ be the set of arcs that become backward by pivoting on k when the set of vertices in the recursive call is V. Our choice of pivot is then:

$$\text{Pick } k \in V \text{ minimizing}^1 \frac{\sum_{(i,j) \in T_k(V)} w_{ij}}{\sum_{(i,j) \in T_k(V)} \bar{w}_{ij}} \qquad (1)$$

As was observed in [2], (i,j) is a backward arc if (k,i) and (j,k) in A, in other words, exactly when (i,j) is in a directed triangle in A with the pivot k. Therefore $T_k(V)$ contains exactly the arcs that are in a directed triangle with k in $G(V)$. The cost incurred for the arcs in $T_k(V)$ if k is the pivot is equal to $\sum_{(i,j) \in T_k(V)} w_{ij}$, and we have a lower bound on the cost in any feasible solution for these vertex pairs of $\sum_{(i,j) \in T_k(V)} \bar{w}_{ij}$.

Let T be the set of directed triangles in $G(V)$, and for a triangle $t = (i,j), (j,k), (k,i)$, let $w(t) = w_{ij} + w_{jk} + w_{ki}$ and let $\bar{w}(t) = \bar{w}_{ij} + \bar{w}_{jk} + \bar{w}_{ki}$. If we sum $\sum_{(i,j) \in T_k(V)} w_{ij}$ over all $k \in V$, i.e. $\sum_{k \in V} \sum_{(i,j) \in T_k(V)} w_{ij}$, then we count w_{ij} exactly once for every pivot k such that $(i,j), (j,k), (k,i)$ is a directed triangle, hence $\sum_{k \in V} \sum_{(i,j) \in T_k(V)} w_{ij} = \sum_{t \in T} w(t)$. Similarly, $\sum_{(i,j) \in T_k(V)} \bar{w}_{ij} = \sum_{t \in T} \bar{w}(t)$.

Now, note that for $t = (i,j), (j,k), (k,i)$, by the triangle inequality on the weights, $w_{ij} = w_{(i,j)} \leq w_{(i,k)} + w_{(k,j)} = \bar{w}_{ki} + \bar{w}_{jk}$, or more generally $w_a \leq \sum_{a' \in t: a' \neq a} \bar{w}_{a'}$ for any $a \in t$. Hence $w(t) \leq 2\bar{w}(t)$. Thus we have that

$$\sum_{k \in V} \sum_{(i,j) \in T_k(V)} w_{ij} = \sum_{t \in T} w(t) \leq 2 \sum_{t \in T} \bar{w}(t) = \sum_{k \in V} \sum_{(i,j) \in T_k(V)} \bar{w}_{ij}.$$

---

[1] Throughout this work, we define a ratio to be 0 if both numerator and denominator are 0. If only the denominator is 0, we define it to be ∞.

Hence, there exists some k such that $\sum_{(i,j)\in T_k(V)} w_{ij} \leq 2\sum_{(i,j)\in T_k(V)} \bar{w}_{ij}$, and the cost incurred for the backward arcs when pivoting on k is not more than 2 times the lower bound on the cost for these vertex pairs.   □

As in Ailon, Charikar and Newman [1], we can do better in the case of rank aggregation. In fact, we will extend the ideas from Ailon, Charikar and Newman [1], and Ailon [3] to give a combinatorial $\frac{8}{5}$-approximation algorithm for partial rank aggregation.

In the partial rank aggregation problem, we are given ℓ partial rankings $\pi_1, \ldots, \pi_\ell$ where $\pi_k : V \to \{1, \ldots, |V|\}$ for $k = 1, \ldots, \ell$. In the (full) rank aggregation problem, $\pi_1, \ldots, \pi_\ell$ are bijective. We let $w_{(i,j)} = \frac{1}{\ell}\sum_{k=1}^{\ell} \mathbf{1}_{(\pi_k(i)<\pi_k(j))}$ and note that the weights for the partial rank aggregation problem satisfy the triangle inequality.

A well-known 2-approximation for full rank aggregation outputs one of the input permutations at random: the expected cost for pair i, j is $2w_{(i,j)}w_{(j,i)}$ which is not more than $2\bar{w}_{ij}$. It follows that returning the best input permutation is also a 2-approximation algorithm.

Ailon [3] proposes the algorithm RepeatChoice for partial rank aggregation. Let $\pi_1, \ldots, \pi_\ell$ be the input rankings; π will be our final output ranking. We start by setting $\pi(i) = 1$ for all $i \in V$. Then we repeatedly choose an input ranking π_k uniformly at random without replacement; we check each $i, j \in V$ and if $\pi(i) = \pi(j)$ but $\pi_k(i) < \pi_k(j)$, we modify π so that now $\pi'(i) < \pi'(j)$. We can do this by setting $\pi'(h) = \pi(h)$ if $\pi(h) \leq \pi(i)$ and $\pi'(h) = \pi(h) + 1$ if $h = j$ or $\pi(h) > \pi(i)$.

Note that π may not yet be a full ranking: We will say that $i \equiv j$ if $\pi_k(i) = \pi_k(j)$ for every input ranking π_k. At the end of the RepeatChoice procedure, we arbitrarily break the ties between i, j, $i \equiv j$, so that π is a full ranking. For ease of exposition, we will henceforth assume that there are no pairs i, j such that $\pi_k(i) = \pi_k(j)$ for all $k = 1, \ldots, \ell$, although our results also hold if such pairs do exist.

The probability that i is ranked before j is $\frac{w_{(i,j)}}{w_{(i,j)}+w_{(j,i)}}$ which incurs a cost of $w_{(j,i)}$. Since i is either ranked before j, or j before i, the expected cost for pair i, j is $\frac{2w_{(j,i)}w_{(i,j)}}{w_{(i,j)}+w_{(j,i)}}$. If we define the majority tournament $G = (V, A)$ as above, and let $w_{ij} = \max\{w_{(i,j)}, w_{(j,i)}\}$ and $\bar{w}_{ij} = \max\{w_{(i,j)}, w_{(j,i)}\}$ as before, then the total expected cost for the permutation returned by RepeatChoice is $\sum_{(i,j)\in A} \frac{2w_{ij}\bar{w}_{ij}}{w_{ij}+\bar{w}_{ij}} \leq 2\sum_{(i,j)\in A} \bar{w}_{ij}$. Ailon shows that this algorithm can be derandomized.

Ailon, Charikar and Newman show that the best of their algorithm's solution and the best input permutation is a $\frac{11}{7}$-approximation algorithm for (full) rank aggregation. We show that a similar guarantee can be given for our deterministic algorithm, and moreover that this result also holds for partial rank aggregation, i.e. the best of our algorithm's solution, and the solution given by RepeatChoice gives a combinatorial $\frac{8}{5}$-approximation algorithm for partial rank aggregation.

Theorem 2. *There exists a deterministic combinatorial $\frac{8}{5}$-approximation algorithm for partial rank aggregation.*

Proof. We again use the algorithm described above, but specify a different way of choosing a good pivot. Let $T_k(V) \subset A$ be the set of arcs that become backward by pivoting on k when the set of vertices in the recursive call is V. Let $\alpha_{ij} = w_{(i,j)} + w_{(j,i)} = w_{ij} + \bar{w}_{ij}$, and note that $\bar{w}_{ij} = \alpha_{ij} - w_{ij}$. In our partial rank aggregation algorithm, we use the following rule to choose a pivot vertex.

$$\text{Pick } k \in V \text{ minimizing } \frac{\sum_{(i,j) \in T_k(V)} \left(\frac{8}{5} w_{ij} - \frac{6}{5} \frac{w_{ij}^2}{\alpha_{ij}} \right)}{\sum_{(i,j) \in T_k(V)} (\alpha_{ij} - w_{ij})} \tag{2}$$

We charge our pivoting algorithm $\frac{2}{5}$ times the cost of the solution it generates, plus $\frac{3}{5}$ times the expected cost of the permutation returned by RepeatChoice. We will show that the total cost charged is not more than $\frac{8}{5}$ times the lower bound given by $\sum_{(i,j) \in A} \bar{w}_{ij} = \sum_{(i,j) \in A} (\alpha_{ij} - w_{ij})$. Taking the better solution from the pivoting solution and the (derandomized) RepeatChoice solution then gives a $\frac{8}{5}$-approximation algorithm.

The expected cost incurred by pair i, j in RepeatChoice is $2 \frac{w_{(i,j)} w_{(j,i)}}{w_{(i,j)} + w_{(j,i)}} = 2 \frac{w_{ij}(\alpha_{ij} - w_{ij})}{\alpha_{ij}}$. The cost if i, j is ranked according to the majority tournament is $\bar{w}_{ij} = \alpha_{ij} - w_{ij}$, and if it is not ordered according to the majority tournament, the cost is w_{ij}. Hence the charge for a forward arc is

$$\frac{2}{5}(\alpha_{ij} - w_{ij}) + \frac{6}{5} \frac{w_{ij}(\alpha_{ij} - w_{ij})}{\alpha_{ij}} \leq \frac{2}{5}(\alpha_{ij} - w_{ij}) + \frac{6}{5}(\alpha_{ij} - w_{ij}) = \frac{8}{5}(\alpha_{ij} - w_{ij})$$

The charge for a backward arc is

$$\frac{2}{5} w_{ij} + \frac{6}{5} \frac{w_{ij}(\alpha_{ij} - w_{ij})}{\alpha_{ij}} = \frac{8}{5} w_{ij} - \frac{6}{5} \frac{w_{ij}^2}{\alpha_{ij}}.$$

We will show that there always exists a pivot k such that the ratio in (2) is at most $\frac{8}{5}$. This implies that the combined charge for the arcs that become backward in one iteration can be bounded by $\frac{8}{5}$ times the lower bound on their combined cost in any feasible solution. Since the charge for a forward arc between a vertex pair is also at most $\frac{8}{5}$ times the lower bound for the vertex pair, the total charge at the end of the algorithm is at most $\frac{8}{5} \sum_{(i,j) \in A} (\alpha_{ij} - w_{ij}) = \sum_{(i,j) \in A} \bar{w}_{ij}$, which is at most $\frac{8}{5}$ times the optimal cost.

To show that a pivot with ratio at most $\frac{8}{5}$ exists, we use the same techniques as before. Let T again be the set of directed triangles in A, and for a triangle $t = (i,j), (j,k), (k,i)$, let $w(t) = w_{ij} + w_{jk} + w_{ki}, \alpha(t) = \alpha_{ij} + \alpha_{jk} + \alpha_{ki}$ and $z(t) = \frac{w_{ij}^2}{\alpha_{ij}} + \frac{w_{jk}^2}{\alpha_{jk}} + \frac{w_{ki}^2}{\alpha_{ki}}$. Note that

$$\sum_{k \in V} \sum_{(i,j) \in T_k(V)} \left(\frac{8}{5} w_{ij} - \frac{6}{5} \frac{w_{ij}^2}{\alpha_{ij}} \right) = \frac{8}{5} \sum_{t \in T} w(t) - \frac{6}{5} \sum_{t \in T} z(t),$$

and

$$\sum_{k \in V} \sum_{(i,j) \in T_k(V)} (\alpha_{ij} - w_{ij}) = \sum_{t \in T} \alpha(t) - \sum_{t \in T} w(t).$$

Note that by the triangle inequality constraints, $w(t) = w_{(i,j)} + w_{(j,k)} + w_{(k,i)} \leq w_{(i,j)} + w_{(j,k)} + (w_{(k,j)} + w_{(j,i)}) = \alpha_{ij} + \alpha_{jk}$. Similarly, we get that $w(t) \leq \alpha_{ij} + \alpha_{ki}$ and $w(t) \leq \alpha_{jk} + \alpha_{ki}$. Adding these constraints, we get that $w(t) \leq \frac{2}{3}\alpha(t)$.

By these observations and Claim 3 below, we can conclude that $\frac{8}{5}\sum_{t \in T} w(t) - \frac{6}{5}\sum_{t \in T} z(t) \leq \frac{8}{5}\left(\sum_{t \in T}\alpha(t) - \sum_{t \in T} w(t)\right)$, and hence there exists some $k \in V$ such that the ratio in (2) is at most $\frac{8}{5}$.   □

Claim 3. For $w = (w_1, w_2, w_3)$, and $\alpha = (\alpha_1, \alpha_2, \alpha_3)$ such that $0 \leq w_i \leq \alpha_i \leq 1$ for $i = 1, 2, 3$, and $\sum_{i=1}^{3} w_i \leq \frac{2}{3}\sum_{i=1}^{3}\alpha_i$:

$$16\sum_{i=1}^{3} w_i - 6\sum_{i=1}^{3}\frac{w_i^2}{\alpha_i} - 8\sum_{i=1}^{3}\alpha_i \leq 0$$

Proof. The proof uses standard techniques, and for space reasons is deferred to the full version.

Lemma 4. *The algorithms in Theorem 1 and 2 can be implemented in $O(n^3)$ time.*

Proof. We maintain a list of the directed triangles in G for which all three vertices are currently contained in a single recursive call, and for each vertex we maintain the total cost for the vertex pairs that get a backward arc if pivoting on that vertex and the total budget for these pairs (i.e. the numerator and denominator of (1) resp. (2)). If we disregard the time needed to obtain and update this information, then a single recursive call takes $O(n)$ time, and there are at most $O(n)$ iterations, giving a total of $O(n^2)$. Initializing the list of triangles and the numerator and denominator of (1) or (2) for each vertex takes $O(n^3)$ time. Over all recursive calls combined, the time needed to update the list of directed triangles, and the numerator and denominator of (1) or (2) is $O(n^3)$: After each pivot, we need to remove all triangles that either contain the pivot vertex, or contain (i,j) where i and j are separated into different recursive calls, and for each triangle removed from the list, we need to update the numerator and denominator of (1) or (2) for the three vertices in the triangle. Assuming the list of triangles is linked to the vertices and arcs contained in it and vice versa, finding a triangle that contains a certain vertex or arc, removing it, and updating the numerator and denominator for the vertices contained in it, can be done in constant time. Finally, note that each triangle is removed from the list exactly once.   □

3   Two-Step Derandomization of LP-Based Pivoting Algorithms

We now show how to extend the ideas from [2] to derandomize the randomized rounding algorithm in Ailon et al. [1], and the perturbed version in Ailon [3]. In particular, this allows us to obtain a deterministic $\frac{5}{2}$-approximation algorithm for ranking with probability constraints, and a $\frac{3}{2}$-approximation algorithm for partial rank aggregation. Combined with the ideas from Theorem 2, this also allows us to obtain a deterministic $\frac{4}{3}$-approximation algorithm for full rank aggregation.

The linear program we will use is the following:

$$\min \quad \sum_{i<j} \left(x_{(i,j)} w_{(j,i)} + x_{(j,i)} w_{(i,j)} \right)$$

$$\begin{aligned}
&\text{s.t.} & x_{(i,j)} + x_{(j,k)} + x_{(k,i)} &\geq 1 && \text{for all distinct } i, j, k \\
&(LP) & x_{(i,j)} + x_{(j,i)} &= 1 && \text{for all } i \neq j \\
&& x_{(i,j)} &\geq 0 && \text{for all } i \neq j
\end{aligned}$$

Given an optimal solution x to this LP, we will write $c_{ij} = c_{ji} = x_{(i,j)} w_{(j,i)} + x_{(j,i)} w_{(i,j)}$.

Given an optimal solution x to the linear programming relaxation, in Ailon et al. [1] a vertex i is ordered to the left of the pivot k with probability $x_{(i,k)}$, and to the right of the pivot with probability $x_{(k,i)} = 1 - x_{(i,k)}$. In Ailon [3], the probabilities are perturbed by a function h that satisfies $h(1-x) = 1-h(x)$. Since we can always take h to be the identity, we assume without loss of generality that the probabilities are always given by $h(x)$.

FASLP-Pivot(V, x)

Pick a (random) pivot $k \in V$.
Set $V_L = \emptyset, V_R = \emptyset$.
For all $i \in V, i \neq k$,
        with probability $h(x_{(i,k)})$: add i to V_L,
        else (with probability $h(x_{(k,i)})$): add i to V_R.
Return FASLP-Pivot(V_L, x), k, FASLP-Pivot(V_R, x).

We will say an arc (i, j) is a forward arc if the vertices i, j were in the same recursive call in which one of them was the pivot, and we will say an arc (i, j) is backward if the vertices i, j were in the same recursive call, in which some vertex $k \neq i, j$ was the pivot, and i was added to V_L and j was added to V_R. Note the difference from our previous definition of forward and backward arcs.

Let $T_k(V)$ be the set of arcs that become backward in a recursive call on V when k is the pivot. Note that $T_k(V)$ is a random set, since V_L, V_R are random sets. In particular, $(i, j) \in T_k(V)$ with probability $h(x_{(i,k)})h(x_{(k,j)})$, and the expected cost for the arcs that become backward arcs is $\mathbb{E}\left[\sum_{(i,j) \in T_k(V)} w_{(j,i)} \right] = \sum_{i \in V \setminus \{k\}} \sum_{j \in V \setminus \{k\}} h(x_{(i,k)})h(x_{(k,j)}) w_{(j,i)}$.

We derandomize the algorithm in two steps. First we choose a pivot k such that ratio of the expected cost for the arcs in $T_k(V)$ and $\mathbb{E}\left[\sum_{(i,j) \in T_k(V)} c_{ij} \right]$ is as small as possible. Then we use the method of conditional expectation [12] to assign the vertices in $V \setminus \{k\}$ to V_L or V_R.

We define the following notation: Let V_L, V_R, V' be a partition of $V \setminus \{k\}$, and let $\mathbb{E}\big[B_k(V)|V_L, V_R\big]$ be the expected total cost incurred in an iteration of FASLP-Pivot for the backward and forward arcs when pivoting on k conditioned on the vertices in V_L and V_R being ordered to the left and right of k respectively (and the vertices in V' are ordered left or right with probability $h(x_{(i,k)})$ and $h(x_{(k,i)})$). Let $\mathbb{E}\big[C_k(V)|V_L, V_R\big]$ be the expected total LP contribution for the vertex pairs that are in forward or backward arcs in an iteration of FASLP-Pivot when pivoting on k, again conditioned on V_L, V_R. Note that the conditional expected cost for backward arcs is

$$
\mathbb{E}\Big[\sum_{(i,j)\in T_k(V)} w_{(j,i)}|V_L, V_R\Big] = \sum_{i\in V_L}\sum_{j\in V_R} w_{(j,i)} + \sum_{i\in V'}\sum_{j\in V'} h(x_{(i,k)})h(x_{(k,j)})w_{(j,i)}
$$
$$
+ \sum_{i\in V_L}\sum_{j\in V'} h(x_{(k,j)})w_{(j,i)} + \sum_{i\in V'}\sum_{j\in V_R} h(x_{(i,k)})w_{(j,i)},
$$

and the conditional expected LP budget for backward arcs can be computed similarly. Hence we can easily compute these conditional expectations and we get

$$
\mathbb{E}\Big[B_k(V)|V_L, V_R\Big] = \mathbb{E}\Big[\sum_{(i,j)\in T_k(V)} w_{(j,i)}|V_L, V_R\Big] + \sum_{i\in V_L} w_{(k,i)} + \sum_{i\in V_R} w_{(i,k)}
$$
$$
+ \sum_{i\in V'} \Big(h(x_{(i,k)})w_{(k,i)} + h(x_{(k,i)})w_{(i,k)}\Big),
$$

and $\mathbb{E}\Big[C_k(V)|V_L, V_R\Big] = \mathbb{E}\Big[\sum_{(i,j)\in T_k(V)} c_{ij} \mid V_L, V_R\Big] + \sum_{i\in V\setminus\{k\}} c_{ik}$.

DerandFASLP-Pivot(V, x)

Pick $k \in V$ minimizing $\dfrac{\mathbb{E}\Big[\sum_{(i,j)\in T_k(V)} w_{(j,i)}\Big]}{\mathbb{E}\Big[\sum_{(i,j)\in T_k(V)} c_{ij}\Big]}$.

Set $V_L = \emptyset, V_R = \emptyset$.
For $i \in V\setminus\{k\}$

    If $\quad \dfrac{\mathbb{E}\Big[B_k(V)| V_L \cup \{i\}, V_R\Big]}{\mathbb{E}\Big[C_k(V) \mid V_L \cup \{i\}, V_R\Big]} \leq \dfrac{\mathbb{E}\Big[B_k(V) \mid V_L, V_R \cup \{i\}\Big]}{\mathbb{E}\Big[C_k(V) \mid V_L, V_R \cup \{i\}\Big]}$

        add i to V_L,

    else

        add i to V_R.

Return DerandFASLP-Pivot(V_L, x), k, DerandFASLP-Pivot(V_R, x).

Lemma 5. *If $h(x_{(i,j)})w_{(j,i)} + h(x_{(j,i)})w_{(i,j)} \leq \alpha c_{ij}$ and there always exists a pivot k with ratio at most α, then DerandFASLP-Pivot is an α-approximation algorithm.*

Proof. In our analysis of DerandFASLP-Pivot, it will be convenient to consider an "intermediate" derandomization of FASLP-Pivot. Let DFASLP-Pivot be the algorithm that chooses a pivot as in DerandFASLP-Pivot, but then randomly assigns vertices to V_L and V_R as in FASLP-Pivot.

We think of $c_{ij} = x_{(i,j)}w_{(j,i)} + x_{(j,i)}w_{(i,j)}$ as the "LP budget" for vertex pair i,j. We say a pair i,j gets decided in an iteration of DFASLP-Pivot if it either gets a forward arc (i.e. one of i,j is the pivot) or a backward arc (i.e. one of them gets assigned to V_L and one to V_R). Under the assumptions in the lemma, the expected cost for the pairs that get decided in an iteration of DFASLP-Pivot is at most α times the expected LP budget for these pairs: the total expected cost for the pairs that get decided in an iteration of DFASLP-Pivot with pivot k is
$$\mathbb{E}\Big[B_k(V) \mid V_L = \emptyset, V_R = \emptyset\Big] =$$

$$\sum_{i \in V\setminus\{k\}} \big(h(x_{(i,k)})w_{(k,i)} + h(x_{(k,i)})w_{(i,k)}\big) + \mathbb{E}\Big[\sum_{(i,j)\in T_k(V)} w_{(j,i)}\Big],$$

and by the assumptions of the lemma this expected cost is at most

$$\alpha\Big(\sum_{i \in V\setminus\{k\}} c_{ik} + \mathbb{E}\Big[\sum_{(i,j)\in T_k(V)} c_{ij}\Big]\Big) = \alpha\mathbb{E}\Big[C_k(V) \mid V_L = \emptyset, V_R = \emptyset\Big].$$

Hence DFASLP-Pivot is an expected α-approximation algorithm. By standard conditional expectation arguments, we know that if we consider some vertex $i \in V\setminus(V_L\cup V_R\cup\{k\})$ and $\mathbb{E}\Big[B_k(V) \mid V_L, V_R\Big] \leq \alpha\mathbb{E}\Big[C_k(V) \mid V_L, V_R\Big]$, then we can add i to either V_L or V_R and maintain the invariant that $\mathbb{E}\Big[B_k(V) \mid V_L, V_R\Big] \leq \alpha\mathbb{E}\Big[C_k(V) \mid V_L, V_R\Big]$. Therefore, DerandFASLP-Pivot returns a a partition V_L, V_R of $V\setminus\{k\}$ such that the cost of ordering the vertices in V_L and V_R to the left and right of k respectively, is at most α times the total LP budget of the pairs that get decided in that iteration. □

Note that one can show that there always exists a pivot with ratio at most α, by showing that $\sum_{k\in V} \mathbb{E}\Big[\sum_{(i,j)\in T_k(V)} w_{(j,i)}\Big] \leq \alpha\sum_{k\in V} \mathbb{E}\Big[\sum_{(i,j)\in T_k(V)} c_{ij}\Big]$ for any feasible LP solution. Using similar observations as in the proof of Theorem 1, it is possible to reduce this inequality to a certain inequality on triples of vertices. Using Lemma 13 in [1] this gives us Corollary 6 and using an inequality from [3] this implies corollary 7.

Corollary 6. *DerandFASLP-Pivot with $h(x) = x$ is a $\frac{5}{2}$-approximation algorithm for weighted feedback arc set in tournaments with probability constraints.*

Corollary 7. *DerandFASLP-Pivot is a $\frac{3}{2}$-approximation algorithm for partial rank aggregation and for ranking with weights that obey triangle inequality if we use $h(x) = \begin{cases} \frac{3}{4}x, & 0 \leq x \leq \frac{1}{3} \\ \frac{3}{2}x - \frac{1}{4}, & \frac{1}{3} < x \leq \frac{2}{3} \\ \frac{3}{4}x + \frac{1}{4}, & \frac{2}{3} < x \leq 1 \end{cases}$ as proposed in [3].*

We can obtain a $\frac{4}{3}$-approximation algorithm for rank aggregation by using the techniques from Theorem 2 to show that the best of DerandFASLP-Pivot and picking a random input permutation is within $\frac{4}{3}$ of optimal. Similar to the technique in the proof of Theorem 2 we replace the weight $w_{(i,j)}$ by $\frac{2}{3}w_{(i,j)} + \frac{1}{3}(2w_{(i,j)}w_{(j,i)})$, and show that DerandFASLP-Pivot returns a solution for which the cost with respect to these new weights is at most $\frac{4}{3}$ times optimal. Since the cost with respect to these new weights is a convex combination of the cost of the solution with respect to the original weights, and the cost of a randomly chosen input permutation, we get that the best of the algorithm's solution and the best input permutation is a $\frac{4}{3}$-approximation algorithm. For space reasons the details of the proof are deferred to the full version.

Theorem 8. *There exists a deterministic $\frac{4}{3}$-approximation algorithm for rank aggregation.*

4   Constrained Problems

We now consider ranking problems where we are also given a partial order P, and the output permutation π must be consistent with P; in other words, if $(i,j) \in P$ then $\pi(i) < \pi(j)$. We make the natural assumption that the weights are consistent with P, i.e. if $(i,j) \in P$ then $w_{(j,i)} = 0$. It is possible to use similar techniques as in [2] to ensure that the algorithms in Section 2 return a feasible solution. However, a stronger result is given by the following lemma, which says that if the weights satisfy the triangle inequality, then any permutation that is not consistent with P is not a local minimum. We thank Frans Schalekamp for suggesting that this may be the case. This means that all results in this paper, except for the result in Corollary 6, also hold for constrained problems.

Lemma 9. *Given weights that satisfy the triangle inequality, a permutation π, and a partial order P such that $w_{(j,i)} = 0$ for $(i,j) \in P$, then we can find a permutation π' that is consistent with P and costs not more than π.*

Proof. Let $(i,j) \in P$ and suppose $\pi(j) < \pi(i)$. We call such (i,j) violated. Let $K(i,j)$ be the set of vertices k such that $\pi(j) < \pi(k) < \pi(i)$, and let (i^*,j^*) be a violated pair such that for any vertex $k \in K(i^*,j^*)$ it is the case that $(j^*,k) \notin P$ and $(k,i^*) \notin P$. (Note that by transitivity of P, if a violated pair exists, then there exists a violated pair that satisfies this condition.)

Consider the permutation π' we obtain by moving j^* to the position just after i^* with probability $p = \frac{1}{2}$ or otherwise moving i^* to the position just before j^*. Note that (i^*,j^*) is not violated in π' and no new violations are created.

The expected difference in the cost of permutations π' and π is given by

$$w_{(j^*,i^*)} - w_{(i^*,j^*)} + \frac{1}{2} \sum_{k \in K(i^*,j^*)} (w_{(j^*,k)} - w_{(k,j^*)} + w_{(k,i^*)} - w_{(i^*,k)})$$

$$\leq w_{(j^*,i^*)} - w_{(i^*,j^*)} + \frac{1}{2} \sum_{k \in K(i^*,j^*)} (2w_{(j^*,i^*)}) = -w_{(i^*,j^*)} \leq 0,$$

where the first inequality follows from the triangle inequality, and the last equality follows since $w_{(j^*,i^*)} = 0$. Hence either moving j^* to the position just after i^* or moving i^* to the position just before j^* does not increase the cost of the permutation, and has fewer violations. $\qquad\square$

References

1. Ailon, N., Charikar, M., Newman, A.: Aggregating inconsistent information: Ranking and clustering. In: STOC 2005, pp. 684–693 (2005)
2. van Zuylen, A., Hegde, R., Jain, K., Williamson, D.P.: Deterministic pivoting algorithms for constrained ranking and clustering problems. In: SODA 2007, pp. 405–414 (2007)
3. Ailon, N.: Aggregation of partial rankings, p-ratings and top-m lists. In: SODA 2007, pp. 415–424 (2007)
4. Dwork, C., Kumar, S.R., Naor, M., Sivakumar, D.: Rank aggregation methods for the web. In: WWW 2001, pp. 613–622 (2001)
5. Wakabayashi, Y.: The complexity of computing medians of relations. Resenhas 3(3), 323–349 (1998)
6. Ailon, N., Charikar, M.: Fitting tree metrics: Hierarchical clustering and phylogeny. In: FOCS 2005, pp. 73–82 (2005)
7. Coppersmith, D., Fleischer, L., Rudra, A.: Ordering by weighted number of wins gives a good ranking for weighted tournaments. In: SODA 2006, pp. 776–782 (2006)
8. Kenyon-Mathieu, C., Schudy, W.: How to rank with few errors: A PTAS for weighted feedback arc set on tournaments. In: STOC 2007, pp. 95–103 (2007)
9. Fagin, R., Kumar, R., Mahdian, M., Sivakumar, D., Vee, E.: Comparing partial rankings. SIAM J. Discret. Math. 20(3), 628–648 (2006)
10. Fagin, R., Kumar, R., Sivakumar, D.: Comparing top k lists. SIAM J. Discret. Math. 17(1), 134–160 (2003)
11. Agrawal, M., Kayal, N., Saxena, N.: PRIMES is in P. Ann. of Math (2) 160(2), 781–793 (2004)
12. Alon, N., Spencer, J.: The Probabilistic Method. Wiley Interscience, Chichester (1992)

Online Rectangle Filling*

Haitao Wang, Amitabh Chaudhary, and Danny Z. Chen

Department of Computer Science and Engineering
University of Notre Dame, Notre Dame, IN 46556, USA

Abstract. We study an online geometric problem arising in channel-aware scheduling of wireless networks, which we call *online rectangle filling*. We present an online algorithm for this problem with a competitive ratio of 1.848. We also prove a lower bound of 1.6358 for the competitive ratio of the problem. In addition, we give an $O(n^2)$-time optimal offline algorithm, where n is the size of the input. All three results are significant improvements on the previous results. Our techniques are based on new observations of the combinatorial structures of the problem.

1 Introduction

We consider the following online problem motivated by channel-aware scheduling in wireless networks: Given an online sequence of nonnegative real numbers $h(1), h(2), \ldots,$ representing the maximum transmission capacities of a wireless channel at each time step, compute a sequence of transmission rates $u(1), u(2), \ldots,$ that satisfies two constraints: (1) the transmission rate $u(i)$ at each time step i is no more than the channel capacity $h(i)$, and (2) if at any step we decide to change the transmission rate, a penalty of one time step in which the transmission rate is zero is incurred (this is required for the transmitter and receiver to coordinate and reset a new transmission rate). The objective is to maximize the throughput, i.e., $\sum_{i=1,2,\ldots} u(i)$. The decision is made online with a lookahead of one time step, i.e., at each step i, we decide on $u(i)$ having seen the capacities $h(1), \ldots, h(i+1)$, but before we see the capacity $h(i+2)$ and beyond. The penalty on changing the transmission rate implies that for any i, either $u(i) = u(i+1)$ or at least one of $u(i)$ and $u(i+1)$ is zero.

In wireless networks [12], the channel conditions can change frequently, affecting the bit error rate and consequently the channel transmission capacity. The transmitter and receiver can monitor the channel capacity and change the transmission rate accordingly. To coordinate the change in transmission rate, a change-over protocol is used, resulting in a temporary loss in transmission of data. This is modeled by setting the transmission rate at one time period to zero. (For further details of this application setting, see [2,3,5,7,8,9,11,13].)

This problem can be described from a geometric view point. The sequence of channel capacities corresponds to a sequence of "columns" of unit-width such that the bases of all columns are on the x-axis and the height of column i is

---

* This research was supported in part by NSF Grant CCF-0515203.

C. Kaklamanis and M. Skutella (Eds.): WAOA 2007, LNCS 4927, pp. 274–287, 2008.

$h(i)$ (see Fig. 1). The transmission rates correspond to a "used-height" in each column, from the base to a height $u(i)$. The penalty for changing the transmission rate implies that in any solution, consecutive columns with nonzero used-height values have the same used-height and thus form a rectangle (see Fig. 2). The objective is, given the online sequence of columns, to fill the region between the column "skyline" and the x-axis with rectangles in a manner that maximizes the total area covered by the rectangles. The constraints are that each rectangle lies on the x-axis and respects the skyline, i.e., the height of a rectangle is at most the height of the lowest column intersecting it, and that any two distinct rectangles are separated by at least one column with zero used-height. We call this the *online rectangle filling problem*. To be exact, we are interested in the version with a lookahead of one; we assume that this is true for the rest of the paper, unless mentioned otherwise.

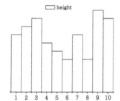
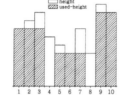

Fig. 1. An example   **Fig. 2.** A feasible solution   **Fig. 3.** Zero columns

The first known algorithms for the online rectangle filling problem had a 4-competitive ratio [3,5]. (See [1,6] for formal definitions of online algorithms, offline optimal algorithms, and competitive analysis.) It was also shown [3,5] that for any online algorithm with a finite lookahead, the competitive ratio is strictly larger than one. (It is easy to see that with no lookahead, the competitive ratio is unbounded.) Besides, the authors gave an offline algorithm that computes an optimal solution for any n-length sequence in $O(n^3)$ time. In a subsequent paper [4], one of the original algorithms (called Wait-Dominate-Hold) was shown to be $(8/3 \approx 2.667)$-competitive and a lower bound of $8/5 = 1.6$ on the competitive ratio of any online algorithm was given. In the same paper, an upper bound of 2 and a lower bound of $1+1/(k+1)$ were shown for an algorithm with a lookahead of $k \geq 2$. In this paper, we improve upon the previous best known results for the problem with a lookahead of one, as described below.

Our Contributions. We present significant improvements on the known solutions [3,4,5] for the online rectangle filling problem. In the next section, we present our main result: an online algorithm (with a lookahead of one) that has a competitive ratio of 1.848. In Section 3, we present a lower bound: no online algorithm (with a lookahead of one) can have a competitive ratio less than 1.6358. In Section 4, we present another improvement over a best known result, an offline algorithm for the problem that takes $O(n^2)$ time.

2   Online Algorithm for Rectangle Filling

Let t be the index of the current column whose used-height $u(t)$ is to be determined. To do so, we use three known values: the used-height $u(t-1)$ of the previous column, the height $h(t)$ of the current column, and the height $h(t+1)$ of the next column. For ease of presentation, we add an extra column at the beginning and the end of the sequence, both with zero height and zero used-height. The algorithm's decisions depend on which of the three situations holds: (1) the previous used-height is larger than the current height, (2) the previous used-height is zero, and (3) the previous used-height is positive and is no bigger than the current height. We describe the decisions of the algorithm in terms of two parameters q_1 and q_2, with $1 < q_1$ and $0 \le q_2 < 1$; we will determine the values of q_1 and q_2 later (based on the analysis in order to obtain the lowest competitive ratio).

Algorithm RecFilling(t)

1. If $u(t-1) > h(t)$, then set $u(t) = 0$.
2. If $u(t-1) = 0$, then
   (a) If $q_1 \cdot h(t) < h(t+1)$, then set $u(t) = 0$.
   (b) If $h(t) > h(t+1)$ and $q_2 \cdot h(t) \le h(t+1)$, then set $u(t) = h(t+1)$.
   (c) If $q_2 \cdot h(t) > h(t+1)$, then set $u(t) = h(t)$.
   (d) If $h(t) < h(t+1)$ and $q_1 \cdot h(t) \ge h(t+1)$, then set $u(t) = h(t)$.
3. If $u(t-1) > 0$ and $u(t-1) \le h(t)$, then
   (a) If $q_1 \cdot u(t-1) < h(t+1)$, then set $u(t) = 0$.
   (b) If $q_1 \cdot u(t-1) \ge h(t+1)$, then set $u(t) = u(t-1)$.

Theorem 1 (Upper Bound). *The algorithm RecFilling with the parameter values q_1 set to 1.848 and q_2 set to 0.667 attains a competitive ratio of 1.848.*

The above theorem will be proved in the rest of this section. For the simplicity of statement we always assume $q_1 > 1/q_2$ and $q_1 \ge 5/3$, as is true for the values $q_1 = 1.848$ and $q_2 = 0.667$ in Theorem 1.

2.1   Idea Behind the Proof

Since the proof for Theorem 1 is long and detailed, we begin with a skeleton description which will also serve as a rationale for our approach. For a given input sequence, let S and S' denote the sequences of used-heights in the solutions computed by algorithm RecFilling and the offline optimal, respectively, and let ALG and OPT denote the throughput values for the solutions of RecFilling and the optimal, respectively. For the ith column, let $u(i)$ denote its used-height in RecFilling and $u'(i)$ denote its used-height in the optimal solution.

We call the columns in S with zero used-height values the *zero columns* (Situations 1, 2(a) or 3(a) of RecFilling). Similarly, the columns in S with nonzero used-heights are *nonzero columns* (Situations 2(b-d) or 3(b)). A column is said to be *fully used* if its used-height equals its height (Situations 2(c-d)). The following observation indicates an important characteristic about nonzero columns.

Observation 1. *The used-height of any nonzero column is at least a $1/q_1$ fraction of its height.*

The proof of the above observation is in our full paper [10]. Observation 1 implies that to prove S is q_1-competitive, we need not be concerned about nonzero columns, inasmuch as they are not dependent on other columns having a large used-height so that the entire sequence S is q_1-competitive. On the other hand, zero columns are dependent, precisely in this manner, on other columns. Intuitively, a zero column is *dependent* on another column if the algorithm creates the zero column so that the other column can (potentially) have a larger used-height. In order to demonstrate S is competitive, the used-heights of both the zero column and the other column have to be considered together. The dependency of a zero column can extend beyond its immediate neighbors when, e.g., its neighbors are themselves zero columns or "barely competitive" ($u(t) = h(t)/q_1$). Thus "dependency chains" can be formed that are propagated through columns. The proof is structured based on the different kinds of dependency chains associated with different kinds of zero columns. We choose not to formalize this intuitive notion of dependency between columns here as we only use it in this outline and not in the actual proof.

Based on the different situations in the algorithm RecFilling, we have three kinds of zero columns:

Type 1 zero columns are those resulting from Situation 1 , i.e., when $u(t-1) > h(t)$. Here, the current used-height $u(t)$ is required to be zero. It occurs when in the previous step $t-1$ the algorithm decides to sacrifice the used-height of column t for a greater used-height in column $t-1$.
Type 2 zero columns are those resulting from Situation 2(a) in RecFilling, i.e., when $u(t-1) = 0$ and $h(t) < h(t+1)/q_1$. The algorithm decides to sacrifice the current column for a potentially greater used-height in the next column.
Type 3 zero columns are those resulting from Situation 3(a) in RecFilling, i.e., when $u(t-1) > 0$ and $u(t-1) < h(t+1)/q_1$. Here again, the algorithm decides to sacrifice the current column for a potentially greater used-height in the next column. This decision is based not on $h(t)$, but on $u(t-1)$.

Fig. 3 gives an example of a solution produced by RecFilling: the 2nd and 6th columns are of Type 1, the 3rd, 4th, and 9th columns are of Type 2, and the 8th column is of Type 3. A simple fact about our algorithm is stated below.

Observation 2. *No zero column of Type 1 or Type 3 can occur immediately after another zero column. A zero column of Type 2 can only occur at the very beginning of S or immediately after another zero column.*

Zero Columns and Dependency Chains. The three types of zero columns and their significances in dependency chains play an important role in the proof. Consider a zero column of Type 1. Its used-height is sacrificed for the previous column, and this is done irrespective of later columns. Thus, intuitively speaking, it is dependent on earlier columns and not on future ones. Further, the dependency chain does not propagate from previous columns through the Type 1 zero column to any future columns. In fact, since the future columns are not affected adversely by the Type 1 zero column, the dependency chain also does not propagate from future columns to the zero column. (This last property is shared by all zero columns.) As we shall see later, it is rather simple to account for Type 1 zero columns.

The used-height of a Type 2 zero column is sacrificed for the next column, and in this decision there is no adverse effect of any previous column. Accordingly, the zero column is dependent on future columns and not on previous ones.

Lastly, the used-height of a Type 3 zero column is sacrificed for the next column, and in this decision it is affected by the previous column. Thus the zero column is dependent on both the future columns and the previous ones. The dependency chains may propagate in either direction (in any particular instance we won't have chains "crossing" each other but in general they may propagate in either direction). Further, a dependency chain from previous columns may propagate through the zero column to future columns. This makes accounting for Type 3 zero columns the most difficult of the three types of zero columns.

Partitioning into Blocks. We make use of the structures in the dependency chains for different types of zero columns in the proof by partitioning S into mutually exclusive blocks of subsequences. Sweeping through the columns of S in order, we create a new block whenever we encounter a zero column of Type 1 or Type 3, in the following manner. If t is a zero column of Type 1, then the current block ends at t and a new block starts at column $t + 1$. If, instead, t is a zero column of Type 3, then the current block ends at $t - 1$ and a new block starts at column $t + 1$. The Type 3 zero column at t itself is not part of any block. Type 2 columns play no role in defining the blocks. For each block in S, we partition a corresponding block in S'. In the rest of the paper, unless otherwise indicated, we refer to the blocks in S.

Observation 3. *A Type 1 column can occur only at the end of a block, and any zero column at the end of a block is of Type 1. Type 2 zero columns can occur within a block, but only as part of a continuous series of zero columns beginning from the first column of the block. Type 3 columns cannot occur within a block. There are no zero columns between any two nonzero columns in a block.*

The partitioning of S into blocks allows us to proceed with a proof consisting of cases that address increasingly complex situations. Essentially, in each case we consider a particular kind of block and show that the total used-height of RecFilling in that block is *competitive* with respect to the used-height of the optimal in the corresponding block, i.e., if we look at the competitive ratio restricted to the block, it is at most 1.848. We consider the following cases:

Case 1. The used-heights of all columns in the block are nonzero.

Case 2. The used-heights of all columns in the block, except for the last column, are nonzero.

Case 3. The block consists of a series of Type 2 zero columns starting from the first column and at least two other columns. Here, the column following the last Type 2 column must be nonzero, but the last column of the block can possibly be a Type 1 zero column. Remember that Type 2 zero columns are dependent only on future columns. If the last column is a Type 1 zero column, this dependency chain stops within the block and so the block can be shown to be competitive.

Case 4. The remaining blocks. These consist of a series of Type 2 zero columns starting from the first column and one other column which is necessarily nonzero. Here, every block, except possibly the very last block in S, is followed by a Type 3 column in S. The dependency chain may propagate either towards previous columns or towards future columns and can cross the block boundaries. Thus we cannot show that the block by itself is competitive. We can however show that the entire sequence S (including the not-in-any-block Type 3 zero columns) is competitive by using an inductive argument on the number of Type 3 zero columns.

The example in Fig. 3 has four blocks: One consists of the 1st and 2nd columns (Case 2); one consists of the 3rd, 4th, 5th, and 6th columns (Case 3); one consists of only the 7th column (Case 1); one consists of the 9th and 10th columns (Case 4). The 8th column is a Type 3 zero column.

The need for multiple cases. In the rest of the section, we present the formal proof using the above case structure. The proof requires quite a few cases and subcases. There are three reasons for this: (1) we need to consider the different kinds of blocks that can occur in S; (2) to obtain tighter bounds we need to consider the different ways the optimal can choose its zero columns; (3) the general bounds for a block are often not tight when the number of columns in a block is small, so we have different cases for long blocks and for short blocks.

Our proof is longer than a bare proof for Theorem 1. This is because we also illustrate the constraints for obtaining the parameter values of q_1 and q_2. We analyze the first three cases, keeping q_1 and q_2 as unspecified parameters. Then we determine the values for these parameters to minimize the worst-case competitive ratio for the three cases. Finally, we verify that the remaining case remains competitive with these parameter values. In the following we use OPT and ALG to denote the the throughput value achieved by the optimal and Rec-Filling for the particular block in the case or subcase in question. Similarly we use $R = \text{OPT}/\text{ALG}$ to denote the competitive ratio of the block in question. For some cases, we distinguish a particular ratio using a subscript, such as R_1 for the ratio in *Case 1*, when it is referred to elsewhere. To save on the symbols needed, we often choose a particular column of a block in a case and WLOG assume its height is 1. Other column heights are then assigned corresponding values.

2.2   Case 1: All Columns Are Nonzero

Since all columns are nonzero, they have the same used-height H. Assume the length of the block is l. According to Observation 1, for each column $1 \leq i \leq l$,

$H \cdot q_1 \geq h(i)$. So we have $\text{ALG} = H \cdot l$ and $\text{OPT} \leq \sum_{i=1}^{l} h(i)$. Thus the competitive ratio for this case is

$$R_1 = \text{OPT}/\text{ALG} \leq q_1 \tag{1}$$

2.3   Case 2: All But the Last Column Are Nonzero

If the block contains at most two columns, then the first one must be fully used since the second one is a Type 1 zero column. According to our algorithm, that case occurs if and only if Situation 2(c) is satisfied, i.e., $h(2) < h(1) \cdot q_2$. For the block, since $\text{OPT} = \max\{h(1), 2 \cdot h(2)\}$, we have

$$\frac{\text{OPT}}{\text{ALG}} \leq \frac{\max\{h(1), 2 \cdot h(2)\}}{h(1)} \leq \max\{1, 2 \cdot q_2\} \tag{2}$$

When the block has $l \geq 3$ columns, let H be the (same) used-height of all nonzero used-height columns of the block. Then for the block, $\text{ALG} = (l-1) \cdot H$. Since the last zero column is of Type 1, its height is less than H. For the block, depending on whether there is a zero used-height column in the corresponding block of the optimal solution S', there are two cases.

1. If the corresponding block of S' has at least one zero used-height column, then according to Observation 1, $h(i) \leq H \cdot q_1$ for $1 \leq i \leq l$. Thus we have

$$\frac{\text{OPT}}{\text{ALG}} \leq \frac{(l-1) \cdot H \cdot q_1}{(l-1) \cdot H} \leq q_1 \tag{3}$$

2. If the block of S' has no zero used-height column, then the used-heights of all columns in that block must be the same, which is at most equal to the height of the last column (let it be H'). Since $H > H'$ and $l \geq 3$, we have

$$\frac{\text{OPT}}{\text{ALG}} = \frac{l \cdot H'}{(l-1) \cdot H} < \frac{l}{l-1} \leq \frac{3}{2} \tag{4}$$

Combining all the above subcases together, we have the competitive ratio R_2 for *Case 2*:

$$R_2 \leq \max\{q_1, 2 \cdot q_2, \frac{3}{2}\} \tag{5}$$

2.4   Case 3: A Series of Type 2 Columns and at Least Two Additional Columns (A Sketch)

Recall that the Type 2 zero columns in the beginning are due to the Situation 2(a) of our algorithm. Fig. 4 shows such an example. Assume the first nonzero column is the kth column. Our approach for this case is as follows: (1) we partition the original block into two sub-blocks (the first sub-block consists of columns from the first one to the $(k-2)$th one inclusively; the second sub-block is from the $(k-1)$th column to the end); (2) analyze the competitive ratio of the two sub-blocks individually; (3) combine the two sub-blocks together to obtain the competitive ratio of the whole block. The complete analysis of this case, including the determination of the particular parameter values $q_1 = 1.848$ and $q_2 = 0.667$, is in [10]. Here we just prove a lemma that plays a central role.

Lemma 1. *Given a sequence of columns S, if each column's height is at least $q > 1.65$ times the height of the previous one, then the value of the optimal solution is at most $\frac{q^2}{q^2-1} \cdot H$ where H is the height of last column in S.*

Proof. Assume S is given from left to right and $H = 1$. We extend S to S^* as follows: (1) starting from the last column to the left, extend every column to a height such that it is equal to $\frac{1}{q}$ of its right column's height; (2) starting from the first (leftmost) column, theoretically add infinite columns to the left such that every column is equal to $\frac{1}{q}$ of its right column's height. So the optimal solution of S is at most equal to that of S^*.

We claim the optimal solution of S^* is to use every other column fully from the rightmost column to the left, as shown in Fig. 5 and whose value is $\frac{q^2}{q^2-1}$. We prove the claim as follows.

Let $SOL(i)$ denote the maximum feasible solution in which from the rightmost column to the left all columns are nonzero except the $(i \cdot k + 1)$th columns for all $k > 0$ (assume the rightmost column is the first one). So there are i consecutive nonzero columns between two adjacent zero columns and the optimal solution we claimed is $SOL(1)$. Fig. 6 shows an example of $SOL(2)$. Since for each group of i consecutive nonzero columns the optimal solution is that each column has a used height equal to the lowest column (i.e. the leftmost one), we can obtain $SOL(i) = \frac{i \cdot q^2}{q^{i+1}-1}$.

We can prove that $SOL(i)$ attains maximum when $i = 1$ if $q > 1.65$ by simply using calculus to show that $SOL(2) > SOL(i)$ for any $i > 2$ and $SOL(1) > SOL(2)$, respectively.

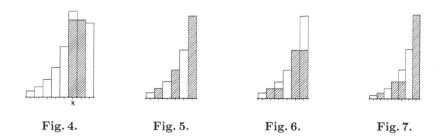

Fig. 4.   **Fig. 5.**   **Fig. 6.**   **Fig. 7.**

We have proved that $SOL(1)$ is the best solution among all solutions where there are a fixed number of consecutive nonzero columns between two adjacent zero columns. In the following we will show by contradiction that $SOL(1)$ is also the best among all solutions including those in which the number of consecutive nonzero columns between two adjacent zero columns is not the same.

Suppose S' is a solution in which the number of consecutive nonzero columns between two adjacent zero columns is not the same and S' is better than $SOL(1)$. Then starting from the first column (the leftmost one), we can find the first place where there are $m > 1$ consecutive nonzero columns. Assume the jth column is the zero column immediately after them. Let S_1 denote the columns from the first column of S' to the $(j-1)$th one and let ALG_1 denote the solution of S' in S_1. Then in S_1, if we use every other one column fully from the rightmost column to the left, then we can obtain another solution ALG_2. By the similar method to the proof $SOL(1) > SOL(i)$, for $i > 1$, it is easy to verify that $\text{ALG}_2 > \text{ALG}_1$ when $q_1 > 1.65$. Then from the $(j+1)$th column, we continue to find the second place where the selection is not one column at a time. By the same method as above, we can get a better solution again. We can continue to do the same thing until the rightmost column. At the end we can obtain a solution which is better than S', implying a contradiction.

So the claim that $SOL(1)$ is optimal is proved, which leads to the lemma.   □

2.5   Case 4: A Series of Type 2 Columns Followed by One Other Column, and Type 3 Columns

We have seen that, when $q_1 = 1.848$ and $q_2 = 0.667$, blocks falling in *Cases 1, 2, and 3* are competitive, i.e. the competitive ratio R for each such block is at most 1.848. What remains is to prove that if S, possibly in addition to the three kinds of blocks above, contains zero columns of Type 3 and blocks falling in *Case 4*, the competitive ratio R is at most 1.848. Unlike the previous cases, here we do not show this is true for any block in *Case 4* per se, rather we use an inductive argument on the number of Type 3 zero columns in S to prove it for the whole sequence.

We noted earlier that a *Case 4* block is always followed by a Type 3 zero column, except possibly when it is the last block in S. Consider two Type 3 zero columns at positions t and r, where $t < r$ and there is no other Type 3 zero column between them. Let S_i be the sequence after column t, and S_{i-1} the (sub)sequence after r. Let B_i be the block just prior to the column r. Note B_i may fall in any of *Cases 1, 3, and 4*, but not in *Case 2*. S_i can possibly consist of blocks other than B_i, occurring prior to B_i, but none of such blocks can fall in *Case 4*. By assuming inductively that S_{i-1} is competitive, we'll show that the sequence S_i' composed of B_i, the Type 3 zero column at r, and S_{i-1} is also competitive. All blocks in S_i, other than in S_i' are already known to be competitive. Thus the entire sequence S_i will be shown to be competitive.

In the following we call B_i the *before-block*, and S_{i-1} the *after-block*. (In spite of the name, remember that S_i is not necessarily a single block.) For the basis step of our inductive argument we need to show that the last block B_1 is competitive. We have seen that this is true for the first three cases. What remains to be shown is that if the last block is in *Case 4*, it is competitive. Such a block is called the *exceptional Case 4* since this is the only block in *Case 4* that is competitive per se—does not require the inductive argument.

In an *exceptional Case 4* block, the last column is fully used. Let its height be 1. The other columns are of Type 2. Applying Lemma 1, we get $\text{OPT} \le 1/(q_1^2-1)+1 \le 1.848$. ALG is 1, and thus the block is competitive. We now build the inductive argument through different cases for the before-block, assuming that after-block has a competitive ratio at most 1.848. WLOG, we'll assume $h(r-1)$ is 1 and describe other heights relative to this.

Before-block is in Case 4. When the before-block belongs to *Case 4*, its last column, at position $(r-1)$, is the only nonzero column. Since RecFilling has a Type 3 column at position r, it follows that $q_2 \leq h(r) \leq q_1$. We proceed using three subcases, based on the value of $h(r)$.

Subcase (a): $\mathbf{0.7652 \leq h(r) \leq 1}$. Here $h(r) \geq q_2$, and so $u(r-1) = h(r)$, as shown in Fig. 8. The figure is for illustration, the actual before and after blocks may have other columns not shown. Applying Lemma 1 to the before-block, OPT $\leq 1 + \frac{1}{q_1^2 - 1}$. Since ALG ≥ 0.7652, we get $R \leq 1.848$ for the before-block. Now since the rth column is of Type 3 column and $u(r-1) = h(r)$, from our algorithm, it follows that $h(r+1) > q_1 \cdot h(r)$. If we attach the rth column to the beginning of the after-block, and run RecFilling on it, it will fall in Situation $2(a)$ of the algorithm. Thus the rth column can be looked upon as a Type 2 zero column attached to the after-block. This new composite block, consisting of the rth column and the after-block, can fall in *Case 3* (when the after-block falls in *Cases 1, 2* or *3*), or can fall in *Case 4* (when the after-block falls in *Cases 1* and *4 (exceptional)*). All these cases are competitive. Thus S_i, consisting of the before-block and the new composite block, is competitive as well.

Subcase (b): $\mathbf{1 < h(r) \leq q_1}$. Here $h(r) \leq q_1 \cdot h(r-1)$ and so $u(r-1) = h(r-1)$, as shown in Fig. 9. For the analysis, we create an pair of blocks which we show are equivalent. We separate the rth column out and set its used-height the same as $u(r-1) = h(r-1)$ and set the used-height of the $(r-1)$th column to zero. So there are two parts, as shown in Fig. 10 and Fig. 11. Assume ALG_1 is the value of our solution in the before part and ALG_2 is for the after part. Obviously $\text{ALG}_2 = h(r-1) = 1$ and $\text{ALG} = \text{ALG}_1 + \text{ALG}_2$.

Let ALG_1' denote the value of our solution if we run the algorithm RecFilling on the before part. Since $h(r-1) \cdot q_1 < h(r+1)$, which means that the $(r-1)$th column can be considered as a new Type 2 zero column, then in the solution obtained by our algorithm the used-height structure will be the same as that in Fig. 10, implying $\text{ALG}_1' = \text{ALG}_1$. Assume the optimal solution for the original structure Fig. 9 is OPT, the optimal solution for the before part (Fig. 10) OPT_1', and the optimal solution for the after part (Fig. 11) OPT_2'. Obviously $\text{OPT}_2' = h(r) \leq q_1$. We have the following lemma.

Lemma 2. *$OPT \leq OPT_1' + OPT_2'$.*

Proof. We prove it by showing that for the optimal solution of the original column sequence (Fig. 9), we can find a solution S_1 for the before part (Fig. 10) and a solution S_2 for the after part (Fig. 11) such that $S_1 + S_2 \geq \text{OPT}$. For the optimal solution OPT, if one of the $(r-1)$th column and $(r+1)$th column is a zero column in OPT, then we just use the same used-height sequence as OPT for the two parts. So $S_1 + S_2 = \text{OPT}$. If both of the two columns are nonzero in OPT, then if the two columns have the same used-height, then we can also use the same used-height sequence as OPT for the two parts. So $S_1 + S_2 = \text{OPT}$. Otherwise, the rth column used-height must be zero in OPT. Let OPT_1 and OPT_2 denote the value of OPT in the before and after part respectively. In the before part, we can set the used-height of the $(r-1)$th column to zero and set the used-height of the after part the same as the $(r-1)$th column used-height in OPT since $OPT_2 = 0$ and $h(r) > h(r-1)$. So $S_1 = OPT_1 - u(r-1)$ and $S_2 = OPT_2 + u(r-1)$, where $u(r-1)$ is the used-height of the $(r-1)$th column in OPT. Then we still have $S_1 + S_2 = OPT$.

Since $OPT_1' \geq S_1$ and $OPT_2' \geq S_2$, our lemma follows. □

Since we have already known that the before part (Fig. 10), which may belong to *Case 3* or *Case 4* (the exception case), satisfies $R \leq 1.848$, which means $\frac{OPT_1'}{ALG_1'} \leq 1.848$. So the competitive ratio for the original composite block (Fig. 9) R is

$$R = \frac{OPT}{ALG} \leq \frac{OPT_1' + OPT_2'}{ALG_1 + ALG_2} = \frac{OPT_1' + OPT_2'}{ALG_1' + ALG_2}$$

$$\leq \max\{\frac{OPT_1'}{ALG_1'}, \frac{OPT_2'}{ALG_2}\} \leq \max\{1.848, q_1\} \leq 1.848$$

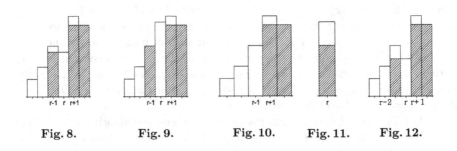

Fig. 8.   Fig. 9.   Fig. 10.   Fig. 11.   Fig. 12.

Subcase (c): $q_2 \leq h(r) < 0.7652$. Here $u(r-1) = h(r)$, as shown in Fig. 12. Since the before-block is a *Case 4* block, there is at least one Type 2 zero column in front of the $(r-1)$th column. We use the similar technique to the last case. However, this time we separate the $(r-2)$th and the $(r-1)$th columns out, as shown in Fig. 13 and Fig. 14. Assume ALG_1 is the value of our solution in the before part (Fig. 13) and ALG_2 is for the after part (Fig. 14). Obviously $ALG_2 = h(r) \geq q_2$ and $ALG = ALG_1 + ALG_2$.

Let ALG_1' denote the value of our solution if we run our algorithm on the before part. If there is another Type 2 zero column in front of the $(r-2)$th column, which is the $(r-3)$th column. Since $q_1 \cdot h(r-3) < h(r-2) < h(r)$, then in the solution obtained by our algorithm, the used-height structure will be the same as that in Fig. 13, implying $ALG_1' = ALG_1$. If there are no other Type 2 columns in front of the $(r-2)$th column, since $q_1 \cdot h(r) < h(r+1)$, we also have $ALG_1' = ALG_1$. Since $h(r-2) \leq 1/q_1$, we extend the height of the $(r-2)$th column in the after part to $1/q_1$ to obtain a new sequence, which makes the optimal solution even greater, implying that our obtained competitive ratio upper bound is always correct. Assume Fig. 14 is the new sequence. Let OPT be the optimal solution for the original sequence (Fig. 12), OPT'_1 the optimal solution for the before part (Fig. 13), and OPT'_2 the optimal solution for the after part (Fig. 14). One can easily find that $OPT'_2 = \frac{2}{q_1}$. By using the similar proof to Lemma 2, we have the similar result.

Lemma 3. $OPT \leq OPT'_1 + OPT'_2$

The complete proof is in [10]. Since we have already known that the structure in Fig. 13 satisfies the inequality $R \leq 1.848$, which means $\frac{OPT'_1}{ALG'_1} \leq 1.848$. So the competitive ratio for Fig. 12,

$$R = \frac{OPT}{ALG} \leq \frac{OPT'_1 + OPT'_2}{ALG_1 + ALG_2} \leq \frac{OPT'_1 + OPT'_2}{ALG'_1 + ALG_2}$$

$$\leq \max\{\frac{OPT'_1}{ALG'_1}, \frac{OPT'_2}{ALG_2}\} \leq \max\{1.848, \frac{2}{q_1 \cdot q_2}\} \leq 1.848$$

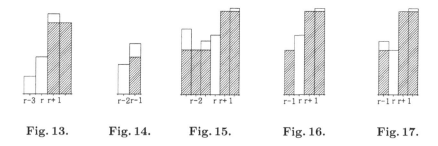

| Fig. 13. | Fig. 14. | Fig. 15. | Fig. 16. | Fig. 17. |

Above all, in the case where the before-block belongs to *Case 4*, $R \leq 1.848$ is satisfied for the new composite block.

Before-block is in Case 1. If the before-block belongs to *Case 1*, as shown in Fig. 15, then $H \cdot q_1 \geq h(r)$ where H is the used-height of each column in the before-block.

If the before-block has only one column, which is the $(r-1)$th column, then depending on the value of H, there are two cases.

1. If $H = h(r-1)$, as shown in Fig. 16, then there must be $h(r-1) \cdot q_1 < h(r+1)$ and $h(r-1) \leq h(r)$. We separate the rth column out with the used-height of $h(r-1)$ and set the used-height of the $(r-1)$th column to zero. By using the similar analysis to second subcase of last case (**before-block is in Case 4**), we can prove the new composite block satisfies $R \leq 1.848$.
2. If $H < h(r-1)$, as shown in Fig. 17, then there must be $h(r) \cdot q_1 < h(r+1)$. So the rth zero column can be considered as a Type 2 zero column of the after-block whose competitive ratio satisfies our previous result. For the before block, $OPT \leq h(r-1)$ and $ALG \geq \frac{h(r-1)}{q_2}$, so $R \leq \frac{1}{q_2} < 1.848$.

If there is more than one column in *Case 1* before-block, such as shown in Fig. 15, then we add the Type 3 zero column (the rth column) to the end of the before-block to create a new block. We prove the new block satisfies $R \leq 1.848$ as follows. Assume the length of the before-block is l, so ALG $= H \cdot l$.

1. If there is no nonzero column in the optimal solution of the new block, then OPT $\leq H \cdot (l+1)$, then $R \leq \frac{\text{OPT}}{\text{ALG}} \leq \frac{l+1}{l} \leq \frac{3}{2}$ since $l \geq 2$.
2. If there is at least one zero column in the optimal solution, then OPT $\leq l \cdot H \cdot q_1$, since the height of each column is at most $H \cdot q_1$. So $R \leq q_1$.

In both cases, $R \leq 1.848$ is satisfied for the before-block. For the after-block, we already know it satisfies the $R \leq 1.848$, so the new composite block also satisfies upper bound.

Above all, if the before-block is in *Case 1*, the upper bound is also satisfied.

Before-block is in Case 3. The analysis of this case is in [10].

Above all, we obtain that $R \leq 1.848$ for the new composite block which is formed by the before-block, the last Type 3 zero column, and the after-block. Note that in the above analysis the only assumption we made is that both the before-block and the after-block satisfy $R \leq 1.848$ except if the before-block is *Case 4*. Since the new composite block also satisfies the ratio upper bound, we can continue to analyze other Type 3 zero columns from the last second Type 3 zero column to the beginning by the same technique as above. Finally, we can obtain the whole solution satisfies $R \leq 1.848$.

It is not difficult to see from the analysis that there are several situations where our algorithm can approximately achieve the obtained competitive ratio, which means that the analysis is tight for the algorithm RecFilling.

3   Lower Bound

Theorem 2 (Lower Bound). *The competitive ratio of every online algorithm with one lookahead for the rectangle filling problem is at least 1.6358.*

We construct an adversarial input pattern as follows. At the beginning two columns are given with their heights $h(1)$ and $h(2) = q \cdot h(1)$ where $q = 2.1638$. Suppose the height of the $(r+1)$th column is known and the algorithm is about to decide the used-height of the rth algorithm. The $(r+2)$th column is given depending on $u(r)$ and following the rules below.

1. If $u(r) = 0$, then we set $h(r+2) = q \cdot h(r+1)$.
2. If $u(r) > 0$, then we set $h(r+2) = h(r+1)$.
3. At the end of the input, assume the last column is r. We add two "extra" columns following the last column with the heights $h(r)$ and $u(r) - \epsilon$ respectively, where ϵ is a very small positive real number. However, if all columns except the first one have the same height, then the two extra columns described above are not necessary.

The proof of Theorem 2 is in [10] due to the space limitation.

4 Offline Algorithm for Rectangle Filling

Theorem 3. *The optimal solution of the offline Rectangle Filling problem can be obtained in $O(n^2)$ time by dynamic programming.*

The algorithm for above theorem is in [10].

References

1. Albers, S.: Competitive Online Algorithms. In: BRICS Lecture Series LS-96-2, BRICS, Department of Computer Science, University of Aarhus (September 1996)
2. Andrews, M., Zhang, L.: Scheduling over a Time-varying User-dependent Channel with Applications to High Speed Wireless Data. In: Proceedings of 43rd Symposium on Foundations of Computer Science, pp. 293–302 (2002)
3. Arora, A., Choi, H.: Channel Aware Scheduling in Wireless Networks. Technical Report 002, The George Washington University (2006)
4. Arora, A., Jin, F., Choi, H.-A.: Scheduling Resource Allocation with Timeslot Penalty for Changeover. Theore. Computer Science 369(1-3), 323–337 (2006)
5. Arora, A., Jin, F., Sahin, G., Mahmoud, H., Choi, H.: Throughput Analysis in Wireless Networks with Multiple Users and Multiple Channels. Acta Informatica 43(3), 147–164 (2006)
6. Borodin, A., EI-Yaniv, R.: Online Computation and Competitive Analysis. Cambridge University Press, Cambridge (1998)
7. Borst, S.: User-level Performance of Channel-aware Scheduling Algorithms in Wireless Data Networks. In: Proceedings of IEEE INFOCOM 2003, pp. 321–331 (2003)
8. Catreux, S., Erceg, V., Gesbert, D., Heath, R.: Adaptive Modulation and MIMO Coding for Broadband Wireless Data Networks. IEEE Communications Magazine 40, 108–115 (2002)
9. Jin, F., Sahin, G., Arora, A., Choi, H.: The Effects of the Sub-Carrier Grouping on Multi-Carrier Channel Aware Scheduling. In: BROADNETS 2004. Proceedings of the First International Conference on Broadband Networks, pp. 632–640 (2004)
10. Wang, H., Chaudhary, A., Chen, D.Z.: Online Rectangle Filling. (manuscript, 2007)
11. Sahin, G., Jin, F., Arora, A., Choi, H.: Predictive scheduling in multi-carrier wireless networks with link adaptation. In: Proceedings of 60th Vehicular Technology Conference (VTC2004-Fall), vol. 7, pp. 5015–5020 (2004)
12. Stallings, W.: Wireless Communication & Networks, 1st edn. Prentice Hall, Englewood Cliffs (2001)
13. Tsibonis, V., Georgiadis, L., Tassiulas, L.: Exploiting Wireless Channel State Information for Throughput Maximization. In: Proceedings of IEEE INFOCOM 2003, pp. 301–310 (2003)

Author Index

Lecture Notes in Computer Science

Sublibrary 1: Theoretical Computer Science and General Issues

For information about Vols. 1– 4588
please contact your bookseller or Springer

Vol. 4707: O. Gervasi, M.L. Gavrilova (Eds.), Computational Science and Its Applications – ICCSA 2007, Part III. XXIV, 1205 pages. 2007.

Vol. 4706: O. Gervasi, M.L. Gavrilova (Eds.), Computational Science and Its Applications – ICCSA 2007, Part II. XXIII, 1129 pages. 2007.

Vol. 4705: O. Gervasi, M.L. Gavrilova (Eds.), Computational Science and Its Applications – ICCSA 2007, Part I. XLIV, 1169 pages. 2007.

Vol. 4703: L. Caires, V.T. Vasconcelos (Eds.), CONCUR 2007 – Concurrency Theory. XIII, 507 pages. 2007.

Vol. 4700: C.B. Jones, Z. Liu, J. Woodcock (Eds.), Formal Methods and Hybrid Real-Time Systems. XVI, 539 pages. 2007.

Vol. 4699: B. Kågström, E. Elmroth, J. Dongarra, J. Waśniewski (Eds.), Applied Parallel Computing. XXIX, 1192 pages. 2007.

Vol. 4698: L. Arge, M. Hoffmann, E. Welzl (Eds.), Algorithms – ESA 2007. XV, 769 pages. 2007.

Vol. 4697: L. Choi, Y. Paek, S. Cho (Eds.), Advances in Computer Systems Architecture. XIII, 400 pages. 2007.

Vol. 4688: K. Li, M. Fei, G.W. Irwin, S. Ma (Eds.), Bio-Inspired Computational Intelligence and Applications. XIX, 805 pages. 2007.

Vol. 4684: L. Kang, Y. Liu, S. Zeng (Eds.), Evolvable Systems: From Biology to Hardware. XIV, 446 pages. 2007.

Vol. 4683: L. Kang, Y. Liu, S. Zeng (Eds.), Advances in Computation and Intelligence. XVII, 663 pages. 2007.

Vol. 4681: D.-S. Huang, L. Heutte, M. Loog (Eds.), Advanced Intelligent Computing Theories and Applications. XXVI, 1379 pages. 2007.

Vol. 4672: K. Li, C. Jesshope, H. Jin, J.-L. Gaudiot (Eds.), Network and Parallel Computing. XVIII, 558 pages. 2007.

Vol. 4671: V.E. Malyshkin (Ed.), Parallel Computing Technologies. XIV, 635 pages. 2007.

Vol. 4669: J.M. de Sá, L.A. Alexandre, W. Duch, D. Mandic (Eds.), Artificial Neural Networks – ICANN 2007, Part II. XXXI, 990 pages. 2007.

Vol. 4668: J.M. de Sá, L.A. Alexandre, W. Duch, D. Mandic (Eds.), Artificial Neural Networks – ICANN 2007, Part I. XXXI, 978 pages. 2007.

Vol. 4666: M.E. Davies, C.J. James, S.A. Abdallah, M.D. Plumbley (Eds.), Independent Component Analysis and Blind Signal Separation. XIX, 847 pages. 2007.

Vol. 4665: J. Hromkovič, R. Královič, M. Nunkesser, P. Widmayer (Eds.), Stochastic Algorithms: Foundations and Applications. X, 167 pages. 2007.

Vol. 4664: J. Durand-Lose, M. Margenstern (Eds.), Machines, Computations, and Universality. X, 325 pages. 2007.

Vol. 4661: U. Montanari, D. Sannella, R. Bruni (Eds.), Trustworthy Global Computing. X, 339 pages. 2007.

Vol. 4649: V. Diekert, M.V. Volkov, A. Voronkov (Eds.), Computer Science – Theory and Applications. XIII, 420 pages. 2007.

Vol. 4647: R. Martin, M.A. Sabin, J.R. Winkler (Eds.), Mathematics of Surfaces XII. IX, 509 pages. 2007.

Vol. 4646: J. Duparc, T.A. Henzinger (Eds.), Computer Science Logic. XIV, 600 pages. 2007.

Vol. 4644: N. Azémard, L. Svensson (Eds.), Integrated Circuit and System Design. XIV, 583 pages. 2007.

Vol. 4641: A.-M. Kermarrec, L. Bougé, T. Priol (Eds.), Euro-Par 2007 Parallel Processing. XXVII, 974 pages. 2007.

Vol. 4639: E. Csuhaj-Varjú, Z. Ésik (Eds.), Fundamentals of Computation Theory. XIV, 508 pages. 2007.

Vol. 4638: T. Stützle, M. Birattari, H. H. Hoos (Eds.), Engineering Stochastic Local Search Algorithms. X, 223 pages. 2007.

Vol. 4630: H.J. van den Herik, P. Ciancarini, H.H.L.M.(J.) Donkers (Eds.), Computers and Games. XII, 283 pages. 2007.

Vol. 4628: L.N. de Castro, F.J. Von Zuben, H. Knidel (Eds.), Artificial Immune Systems. XII, 438 pages. 2007.

Vol. 4627: M. Charikar, K. Jansen, O. Reingold, J.D.P. Rolim (Eds.), Approximation, Randomization, and Combinatorial Optimization. XII, 626 pages. 2007.

Vol. 4624: T. Mossakowski, U. Montanari, M. Haveraaen (Eds.), Algebra and Coalgebra in Computer Science. XI, 463 pages. 2007.

Vol. 4623: M. Collard (Ed.), Ontologies-Based Databases and Information Systems. X, 153 pages. 2007.

Vol. 4621: D. Wagner, R. Wattenhofer (Eds.), Algorithms for Sensor and Ad Hoc Networks. XIII, 415 pages. 2007.

Vol. 4619: F. Dehne, J.-R. Sack, N. Zeh (Eds.), Algorithms and Data Structures. XVI, 662 pages. 2007.

Vol. 4618: S.G. Akl, C.S. Calude, M.J. Dinneen, G. Rozenberg, H.T. Wareham (Eds.), Unconventional Computation. X, 243 pages. 2007.

Vol. 4616: A.W.M. Dress, Y. Xu, B. Zhu (Eds.), Combinatorial Optimization and Applications. XI, 390 pages. 2007.

Vol. 4614: B. Chen, M. Paterson, G. Zhang (Eds.), Combinatorics, Algorithms, Probabilistic and Experimental Methodologies. XII, 530 pages. 2007.

Vol. 4613: F.P. Preparata, Q. Fang (Eds.), Frontiers in Algorithmics. XI, 348 pages. 2007.

Vol. 4600: H. Comon-Lundh, C. Kirchner, H. Kirchner (Eds.), Rewriting, Computation and Proof. XVI, 273 pages. 2007.

Vol. 4599: S. Vassiliadis, M. Bereković, T.D. Hämäläinen (Eds.), Embedded Computer Systems: Architectures, Modeling, and Simulation. XVIII, 466 pages. 2007.

Vol. 4598: G. Lin (Ed.), Computing and Combinatorics. XII, 570 pages. 2007.

Vol. 4596: L. Arge, C. Cachin, T. Jurdziński, A. Tarlecki (Eds.), Automata, Languages and Programming. XVII, 953 pages. 2007.

Vol. 4595: D. Bošnački, S. Edelkamp (Eds.), Model Checking Software. X, 285 pages. 2007.

Vol. 4590: W. Damm, H. Hermanns (Eds.), Computer Aided Verification. XV, 562 pages. 2007.